NEVADA BUREAU OF MINES AND GEOLOGY SPECIAL PUBLICATION 26

Traveling America's Loneliest Road

A Geologic and Natural History Tour through Nevada along U.S. Highway 50

by
Joseph V. Tingley and Kris Ann Pizarro

publication design:
Jack P. Hursh

foreword:
Neal Brecheisen
and Karen Malloy

2000

Mackay School of Mines

UNIVERSITY
OF NEVADA
RENO

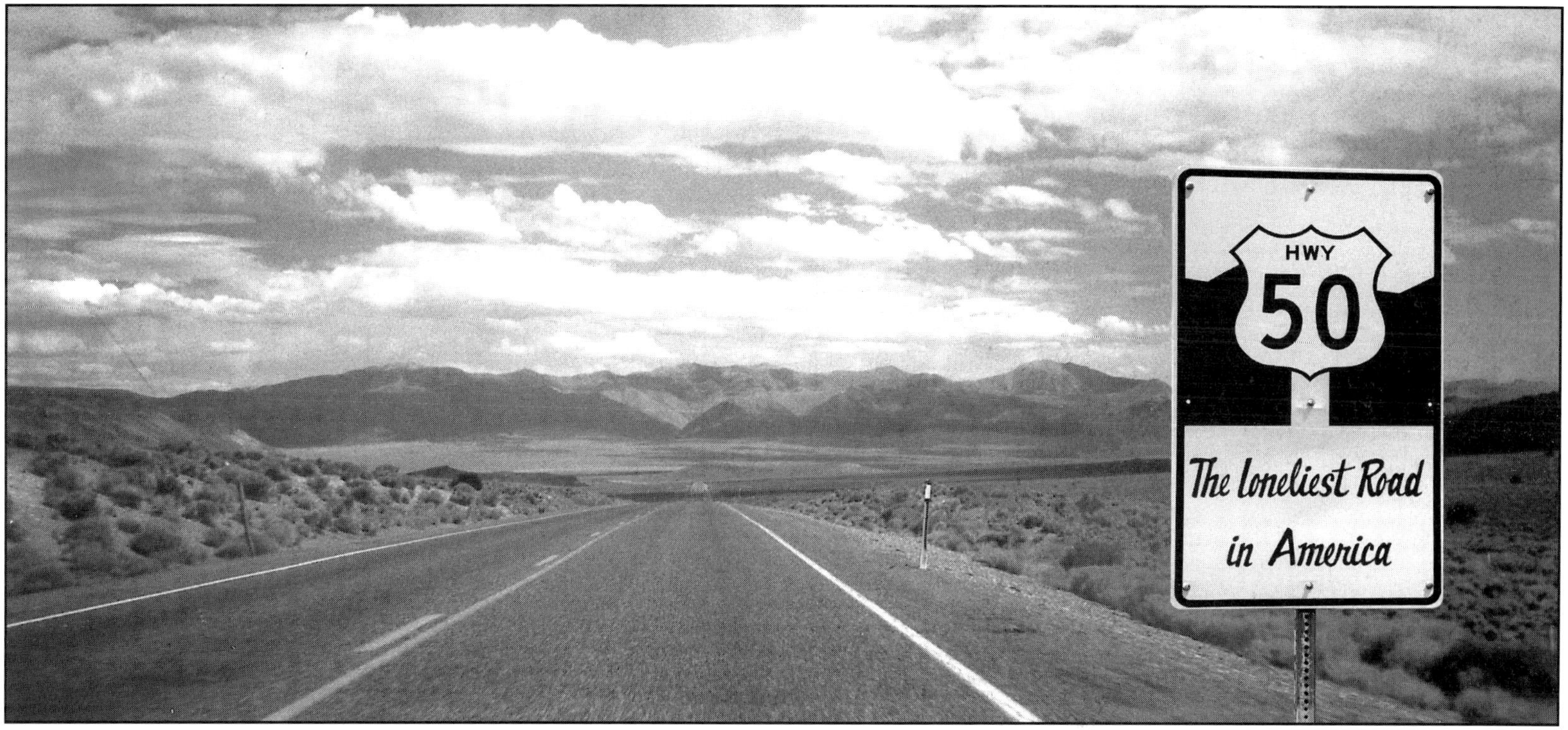

Photo: Kris Pizarro

Under big Nevada skies heading into the Reese River Valley with the Toiyabe Range beyond.

USING THE ROAD LOG

To accommodate the extensive length of the route and the many geologic features that are seen along U.S. Highway 50, this road log is arranged in four sections, each covering an area of special geologic interest. The log goes from west to east, beginning at the California-Nevada state line and ending at the Nevada-Utah border. Mileage is keyed to the highway mile markers. These mileposts begin with zero each time a county line is crossed and they are numbered west to east. Some county mileposts are better than others; we found mileposts in some counties to be right on with our odometer while others seemed to drift a bit. Where we state a feature is at a certain milepost, that is what it says on the post; we even give milepost readings to two decimal points if that is painted on the marker. Be patient and realize that all odometers and road markings are not created equal. We are confident, however, that by observant use of the markers and judicious use of your odometer, you will be able to successfully navigate U.S. 50 using our road log.

Throughout the log, we reference directions by use of numbers as they appear on the face of a clock: 12:00 is straight ahead; 3:00 is to the right, perpendicular the direction of travel; and 9:00 is to the left, perpendicular the direction of travel. Also throughout the log, we point out and often directly quote Nevada Historical Markers. These are placed at points of interest along highways in the state by the Nevada Department of Transportation. The markers are large metal signs, painted "Nevada blue," containing plaques with a short narrative explaining the interesting feature or place. Each marker is numbered and we refer to them in the text by their number (such as NHM #225).

CONTENTS

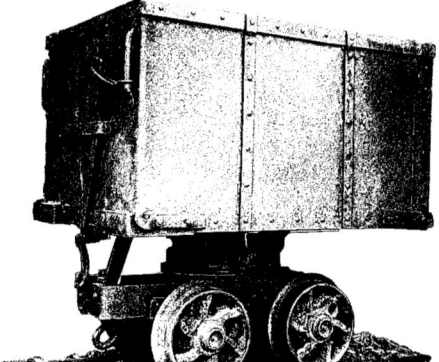

Ore car

FOREWORD

Practically every aspect of human occupation in this land served by U.S. Highway 50 is due in whole or part to minerals. All of the towns, roads, support facilities and even the existence of Nevada's State government itself are the result of minerals. The greatest impacts to this area's landscape and natural history came from the largest silver mining district the world had ever seen up to its time of discovery in 1859. The Comstock Lode, located in Virginia City, Nevada, was a bonanza orebody that drove every aspect of the white man's civilization. For instance, all of the tall pines you see covering the mountains around Lake Tahoe and even the bushy piñon pines seen in the hills around Virginia City are second growth. The original forests were cut down to provide timber for the mines and fuel for hungry boilers and furnaces. Roads, flumes, and tramways were built to transport logs from the high Sierra into Eagle Valley and back up to the rich "diggins" at the foot of Mt. Davidson. Nevada was brought into the Union at a time when President Lincoln needed the wealth of silver to continue the Civil War. There is no doubt that the history of this region was shaped by the demand for minerals.

This U.S. 50 road log gives the traveler a glimpse of the history of the State of Nevada that will hopefully enlighten and spur even more interest. The road log takes you back in time, causing you to imagine the days of rugged men with primitive equipment living a hardscrabble life of bare-boned existence juxtaposed to fabulous wealth. This road log travels through rough and primitive country where people struggled to cross alkali deserts, drank sulfurous waters, and plodded through unforgiving lands. On this journey, you will discover ghost towns and near ghost towns, and will travel empty barren stretches of highway that are filled with interesting history. Take time to stop along the way, smell the sagebrush and piñon, and watch an eagle or a band of wild horses. You will pass through the "Pittsburgh of the west," a mining town so filled with smelters belching black smoke in its heyday that the downwind soils and gardens still contain high levels of lead. You will discover areas of pristine beauty that are practically untouched, and find public-use areas which allow digging for your own minerals such as semiprecious garnets.

If you are lucky enough to sleep under a blanket of stars away from the noise and hustle of civilization, you may begin to appreciate some of the charms of this place. It is big and wide and, for the most part, empty. But if you train yourself to observe what is around you, you will see a new and more interesting world than you may have thought possible. Enjoy your trip.

ACKNOWLEDGMENTS

Many individuals helped us as we assembled this guidebook over the past three years. Special thanks are due to Tom Leshendok, Bureau of Land Management (BLM), Reno, for recommending this project for funding and for patiently seeing it through to its completion. Neal Brecheisen, BLM, Carson City, also instrumental in supporting the project, provided us with information used in the Stateline to Reese River segment of the roadlog. Neal, along with Karen Malloy, also with the BLM, Carson City office, wrote the foreword. Karen Malloy provided descriptions of the cultural history of Grimes Point and Hidden Cave. Bobbie McGonagle, with the Battle Mountain BLM office, supplied information on the Hickison Summit petroglyph site and added to our knowledge of the carbonari of the Eureka area. Jon Price, Director of the Nevada Bureau of Mines and Geology (NBMG), also deserves thanks for allowing us to pursue this project. Jon and Beth Price provided us with helpful notes collected on one of their many journeys along U.S. 50. Some of the material has been compiled from the many geologic roadlogs written for various segments of the route and we include these sources in our bibliography. Some stories have been repeated so many times in so many places that it is difficult to assign credit. One such story, the Chicken Craw gold rush, was traced to Bert Slemmons in a 1966 guidebook, but beyond that the trail is lost.

Our reviewers provided us with many differing insights into the manuscript, and we gratefully acknowledge the time that they gave to their reviews and the resulting clarifications and improvements. Reviews were provided by Mario Desilets, Jack Hursh, and Jon Price, NBMG, and by Eldon L. Allison, Jr., Josh Alpert, Neal Brecheisen, Lee Douthit and Dan Netcher of the BLM.

Jack Hursh designed and typeset this book. He also shared his personal research, tracked down hard-to-find photographs, and furnished us with many of his own. His initiative and unshakable enthusiasm throughout this project deserve special recognition. Dick Meeuwig edited both text and figures and provided many helpful suggestions on content and organization. Susan Tingley reviewed geographic names in the text and figures for correct usage. Maps and illustrations are by Kris Pizarro unless otherwise noted. The topographic data for base maps are from the U.S. Geological Survey. The plant and animal descriptions are derived in part from field guides and related publications listed in the bibliography, notably those by Brown, Dunne and others, Hall, Harrison and Greensmith, MacMahon, Mack, Milne and Milne, Mozingo, Ryser, Savage, Scott, Terres, Upadhaya and others, Whitney, and Whitson and others.

Roy Cazier was our host and guide during several trips to Eureka and White Pine Counties. He generously shared his knowledge of the countryside and its wildlife and gave us free access to his personal collection of photographs.

Thanks to those who took the time to share their knowledge of Nevada's history, plants, and wildlife, particularly Bill Kohlmoos (The Turkey Vulture Society), Ted Floyd (Great Basin Bird Observatory), and Sharon Tilley (Nevada Northern Railroad Museum). Jim Yoakum (Nevada Wildlife Federation), Terry Nelson, Larry Jacox, and Aleta Hursh provided us with a fine selection of photos and artwork.

Finally, we'd like to express special appreciation to the many local residents we met during the course of this project. Their friendliness and generosity were highlights of our travels.

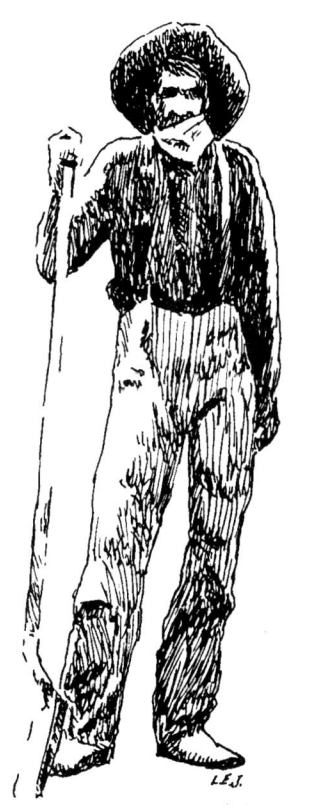

artwork: Larry Jacox

THE ROUTE AND ITS HISTORY

U.S. Highway 50 always has been the best way to travel across Nevada. You can cross Nevada a lot faster by following Interstate 80, but that route, with its four lanes of asphalt and clusters of fast food dispensaries at convenient interchanges, is not an experience unique to Nevada. U.S. 50, on the other hand, is a true Nevada adventure.

Historically, the flow of travel on the U.S. 50 corridor has been from east to west. John Frémont, in 1845–1846, came south from the Humboldt River country and followed a short segment of the present highway route west across Monitor Valley and the Toquima Range before continuing south into Big Smoky Valley. Next came James H. Simpson of the U.S. Army Corps of Topographical Engineers who, in 1859, led an expedition across the Great Basin from Camp Floyd, Utah to the Carson Valley of Nevada. Simpson was charged with finding a wagon route between Great Salt Lake and California that was shorter and faster than the circuitous emigrant route that followed the drainage of the Humboldt River across the northern end of the Great Basin. Westbound, Simpson's expedition entered Nevada about 40 miles north of the entrance point of present-day U.S. 50. Just west of Lone Mountain, Eureka County, Simpson intersected and travelled a route that eventually became the present day U.S. 50, west to Genoa in Carson Valley. On the return trip east, the expedition stayed to the south, closely following what is now the U.S. 50 route all the way to the present Utah border before turning northeastward toward Salt Lake. By Simpson's calculations, the northern route cut 283 miles from the old Humboldt River route. The southern route, now U.S. 50, was slightly longer but still saved some 254 miles. When Simpson submitted the report of his expedition in 1861, he commented that "both of the new routes had been since traveled by emigrants and herds of cattle, and continue to be traveled by them, and upon the more northern route is now running the mail and the pony express."

George Chorpenning's "Jackass Mail" was probably the first commercial venture to use the Simpson route on a regular basis. Chorpenning's company had been hauling mail under contract to the U.S. Post Office since 1851 over the emigrant trail following the Fortymile Desert and Humboldt River route. In 1859, Chorpenning relocated his stations along Simpson's northern track only to lose his contract and business to the newly formed Pony Express in 1860. The Simpson—later U.S. 50—route was a natural for the Pony Express concept of overland mail carried by fast riders on horseback between closely spaced horse exchange stations. Even with numerous mountain ranges to be crossed, the shorter distance and abundance of water and stock feed along the way gave the route an overwhelming advantage over the Humboldt River route.

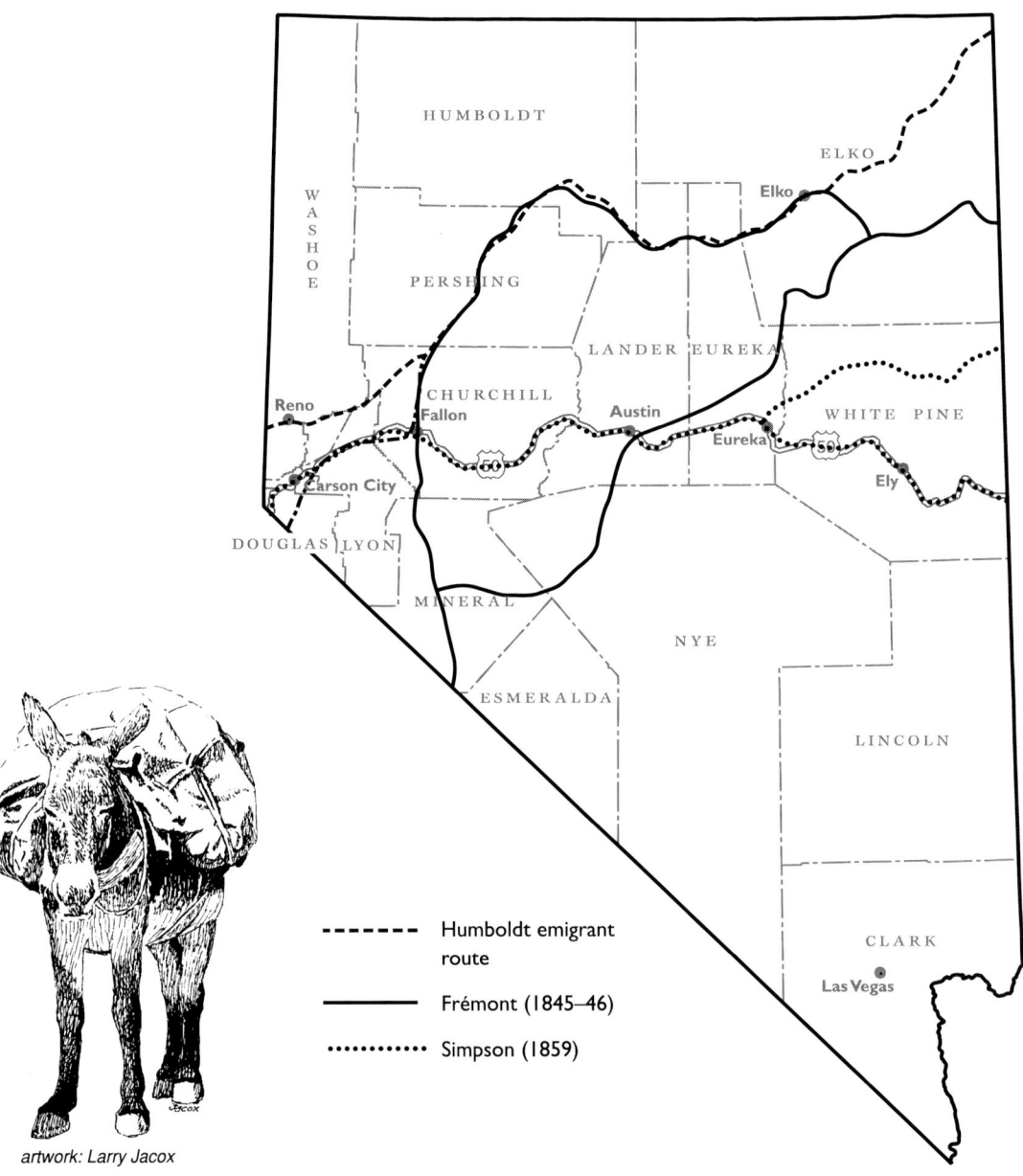

artwork: Larry Jacox

- - - - - Humboldt emigrant route

———— Frémont (1845–46)

··········· Simpson (1859)

Early emigrant and explorer trails across Nevada.

The Pony Express only lasted 18 months in 1860–61, but it established Simpson's route as a major element in travel across Nevada. The transcontinental telegraph system also followed the Simpson-U.S. 50 trail and the completion of its lines in October 1861 sounded the death knell for the Pony Express. The Overland Mail Co. also came along in 1861 on the Simpson route, establishing passenger and mail service using heavy Concord stages capable of transporting six or more passengers with appropriate baggage. The life span of the Overland Mail, 1861 to 1869, coincided with silver discoveries at locations all across central and eastern Nevada. Boom camps such as Austin, Eureka, Hamilton, and Ely quickly turned into full-fledged mining towns and stage routes were adjusted to serve them.

In 1869 the completion of the Central Pacific Railroad across Nevada removed some of the action from the Simpson route. The Central Pacific tracks followed the old emigrant route along the Humboldt River and, later, a system of narrow-gage railroads was built that ran south from the new transcontinental line to serve the major mining camps. Mine and freight traffic moved north over these local rail lines, then east and west on the Central Pacific, leaving the future U.S. 50 route to those few who wanted to quietly travel the mountains and valleys between Austin, Eureka, and Ely.

The evolution of U.S. 50 from an abandoned Pony Express and stage route to a major U.S. highway began in 1912 when the concept of a transcontinental highway system took form. In 1913, an association was formed to promote what would become the nation's first transcontinental route—the Lincoln Highway, a coast-to-coast route from New York to San Francisco. There was, at first, no federal funding for this project and the "highway" was pieced together by linking the best of the existing roads from east to west. Crossing Nevada, the route chosen more or less followed the old Simpson-Pony Express route. Entering the state west of Ibapah, Utah, the new "Highway" followed the Pony Express trail across the Schell Creek Range, then turned south through Steptoe Valley to Ely. At Ely, the highway turned west to follow the present U.S. 50 to a few miles west of Fallon where it split into two main branches. One branch followed the present U.S. 50 route into Carson City, Lake Tahoe, and Placerville, California. The other branch continued west through Fernley and then followed the Truckee River canyon (present-day Interstate 80) to Reno. Traveling the Lincoln Highway became a "thing to do" for those seeking a western adventure, and guidebooks were written containing lists of hazards, precautions, and travel tips for those brave or foolhardy enough to make the trip. Sections of the highway were not always in the best of condition. In 1915, a traveler described the Lincoln Highway as "an imaginary line like the equator!" Even the Lincoln Highway Associations' own guidebooks pointedly warned that "A journey from the Atlantic to the Pacific is still something of a sporting proposition..." The highway was marked by concrete posts bearing an "L" and a profile

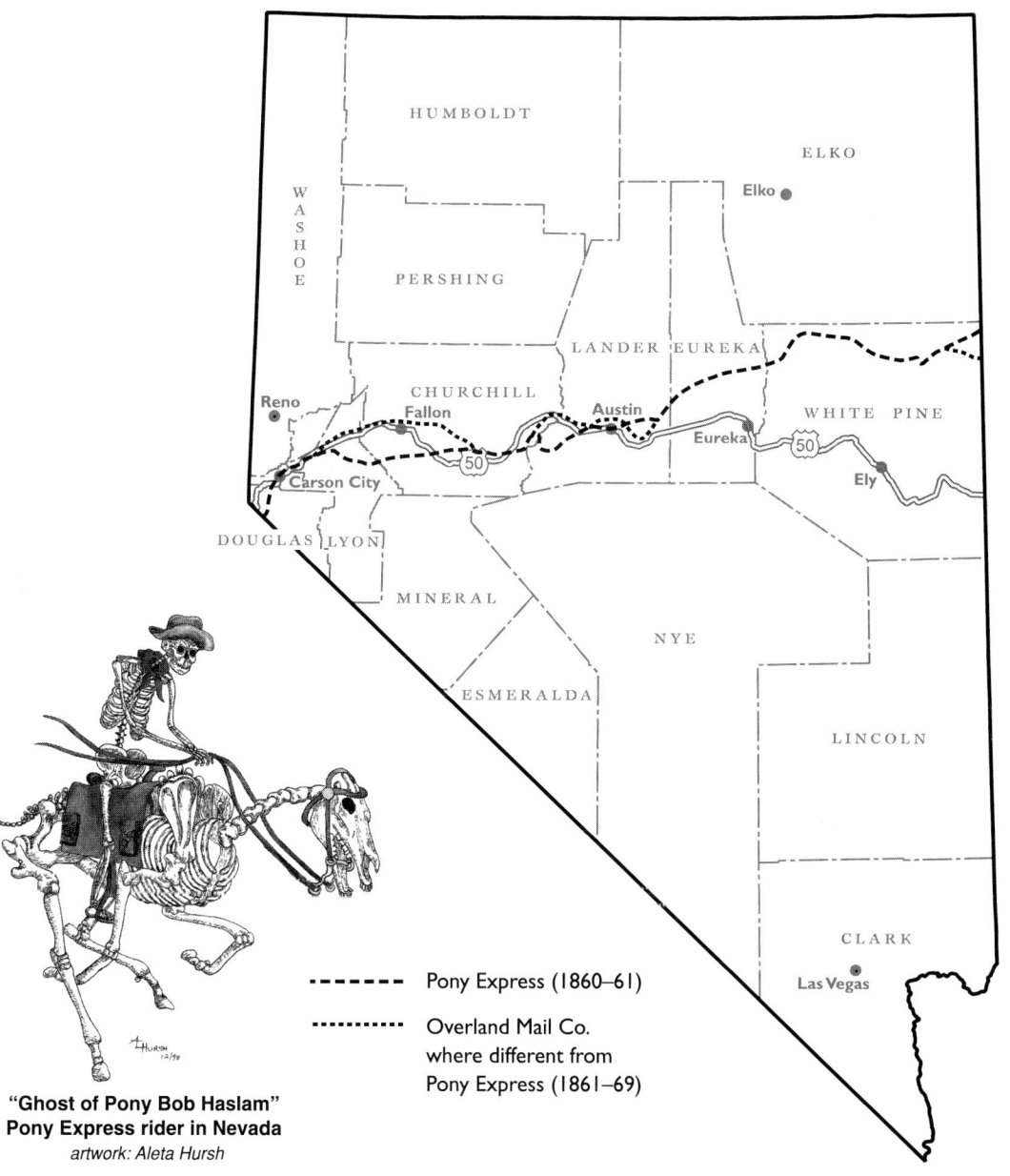

"Ghost of Pony Bob Haslam"
Pony Express rider in Nevada
artwork: Aleta Hursh

------- Pony Express (1860–61)

·········· Overland Mail Co.
where different from
Pony Express (1861–69)

Pony Express and Overland Mail routes across Nevada.

likeness of Lincoln in a circle on the front of the post. Later, some of the bridges on the highway were marked with concrete castings reading "Lincoln Highway." In Nevada, few of these markers have survived; one of the posts is said to be at the Cold Springs maintenance station, between Austin and Fallon, and one of the sets of bridge markers can be seen at a rest stop on Interstate 80 west of Reno.

With a proliferation of other national highways in the 1920s, the Lincoln Highway lost its uniqueness. A numbered system of U.S. highways was established in 1926, and the portion of the Lincoln Highway west of Salt Lake City, Utah, including the Nevada section, was designated U.S. 50. Various changes in alignment have taken place since the 1920s to give us the present route that extends from the Utah-Nevada state line east of Baker, through Ely, Eureka, Austin, Fallon, and Carson City to the Nevada-California state line at the south shore of Lake Tahoe.

So how did the name "the loneliest road" become attached to this historical and intermittently busy route across central Nevada? Some people, those more used to the closely spaced concrete piles and neon-trimmed views of city life, equate solitude and wide-open space with loneliness. Sometime in 1986, an article appeared in a national magazine accusing the stretch of U.S. 50 between Fallon and Ely of "providing the traveler with neither services nor anything to do along the way," and christened the route "The Loneliest Road in America." Locals considered this a compliment rather than an unkind cut, and promptly turned the name into a slogan promoting this special slice of Nevada. A "Highway 50 Survival Kit" was put together that promoted attractions and services along the route. There is even a scorecard included that, when stamped in each town along the way and then sent in to Carson City, qualifies the traveler to receive a handsome "Highway 50 Survivors Certificate" along with a decal and a letter of congratulations. The Nevada State Legislature entered the picture and, in 1987, passed an act officially designating the route "The Loneliest Road in America." Special highway markers were set out and U.S. 50 settled back hoping to attract travelers seeking more adventure than a mind-numbing trip on the Interstate would provide. You will still see an occasional "loneliest road" highway marker along U.S. 50 today, but the novelty has worn off. The highway hasn't changed however, the scenery is still spectacular, the towns are still mostly quiet, and the solitude is far from lonely.

THE GEOLOGY

Nevada geology is about as complex as geology gets anywhere in the world and, as minerals go, Nevada is probably richer than most other places. It took the gold fields of neighboring California and the hope of finding mineral riches there, however, to entice the first miners to cross Nevada. Later, bonanza silver discoveries on the Comstock brought them back to Nevada where many remained, first to work the bonanza silver mines, later to work mines of lead, zinc, copper, tungsten, and most recently to mine the huge deposits of gold that have carried Nevada to its present place as the largest gold-producing state in the United States.

U.S. 50 spends a little over 400 miles crossing Nevada. While doing so, the route samples almost every aspect of the state's geology and passes through or near many historical mining towns. The route passes by several large gold and copper mining sites. Although the mines are not open to the general public, some viewpoints are provided where the traveler may see a modern mine at work.

To better understand Nevada geology as it unfolds along our route, some basic principles of geology are outlined in the following section where we talk about minerals and rocks, geologic structures, and get involved—just a little—in some mega-thinking like plate tectonics and calderas. Once we have the setting laid out, we then

---

IT'S ALIVE!

The lesson is that the whole thing—the whole Basin and Range, or most of it—is alive. The earth is moving. The faults are moving. There are hot springs all over the province. There are young volcanic rocks. Fault scarps everywhere. The world is splitting open and coming apart. You see a sudden break in the sage like this and it says to you that a fault is there and a fault block is coming up. This is a gorgeous, fresh, young, active fault scarp. It's growing. The range is lifting up. This Nevada topography is what you see during mountain building. There are no foothills. It is all too young. It is live country. This is the tectonic, active, spreading, mountain-building world. To a non-geologist, it's just ranges, ranges, ranges.

Kenneth Deffeyes *in Basin and Range,* John McPhee (1980)

---

talk—also just a little—about ore deposits and how they formed. While reading these sections, it will be helpful to refer to the glossary at the back of the book for definitions of specific terms. For a more detailed look at the geology of Nevada and the Great Basin, check into *Geology of Nevada* (Nevada Bureau of Mines and Geology Special Publication 4) by John Stewart, and *Geology of the Great Basin* by Bill Fiero. Feel free to jump forward to page 18 if you are anxious to start your trip.

Minerals and Rocks

To earth scientists, the word mineral has a very specific definition; a mineral is "any naturally formed, solid, chemical substance having a specific composition and a characteristic crystal structure." The most common mineral, one that we all probably know and recognize, is quartz. At the less common but still well known end of the scale are diamond and gold, both minerals but much more valuable than quartz. Iron pyrite, "fool's gold," is a mineral, as are chalcopyrite, galena, and magnetite, which are mined for their copper, lead, and iron contents, respectively. So much for high-powered definitions; the important point to remember is that minerals, brought together by various chemical and physical processes, form rocks. Rocks are the basic building blocks of the solid part of our earth, the stuff that helps hold everything else together and gives us something to walk on when they are underfoot and to marvel at when they are uplifted into mountains.

Rocks are divided into three major types—igneous, sedimentary, and metamorphic—each defined by the process by which it is formed. **Igneous rocks**, named from the Latin *igneus* meaning "fire," are formed by the cooling of magma (molten rock) from deep below the Earth's surface. Most rising magma never reaches the surface, but cools slowly at depth, forming masses of buried igneous rocks called plutons (after Pluto, the Greek god of the underworld), or intrusive rocks because of their intruding nature. Plutonic rocks are further divided into batholiths, stocks, dikes, and other categories based on size and shape. Batholiths are really big! All of the plutonic rocks of the Sierra Nevada are collectively known as the "Sierra Nevada batholith." Stocks are smaller, usually underlying a mountain peak or two rather than an entire range. Dikes, smaller yet, are pipes or irregular sheets of igneous rocks that crosscut and fill cracks in older plutonic or other rocks.

Magma that reaches the surface through cracks in the Earth's crust produces volcanic rock. The various products of volcanic eruptions range from quiet lava of the sort we see flowing in molten rock rivers from Hawaiian volcanoes into the sea to violently formed pyroclastic rocks tossed out from eruptions like that of Mount St. Helens. Along our route across Nevada, we see lots of ash-flow tuff, a special type of pyroclastic rock, so we need to look at the definition of pyroclastic in a little more detail. *Pyros* in Greek means "fire," and clastic derives from *klastos*, meaning "broken." Pyroclastic rocks, then, form from hot, broken fragments exploded from volcanoes and dropped in a pile around the base of the vent (like cinder in cinder cones). Ash-flow tuffs form sheets of rock that can cover many square miles. The name is descriptive: ash refers to small particle size; tuff is the rock composed of the small volcanic particles; and the term "ash flow" gives us a clue as to how the rock arrived at its final location—the incandescent (but not molten) ash particles mixed with hot volcanic gases forming a dense cloud that flowed like water down the side of the vent and came to rest in surrounding lower areas. As the ash flow came to rest, the super-hot particles fused together to form dense rock layers.

Both plutonic and volcanic igneous rocks are given specific rock names based on mineral composition. The many different rock types are composed mostly of a small number of minerals; most are silicates (compounds of silica, oxygen, and a handful of other elements such as sodium, potassium, calcium, and iron). Common minerals include quartz, feldspar, mica, hornblende, iron oxide minerals, and many minerals present in only trace amounts. Here is where things start to get complicated. We need to know exact mineral composition, grain size, and other things not easily determined by the eye to put an igneous rock in its proper category. We can get by with knowing only that lighter-colored rocks usually have lots of quartz in them, dark rocks are quartz-poor, plutonic rocks are usually coarse-grained, and volcanic rocks are usually fine-grained or composed of crystal fragments in a fine-grained matrix. Granite is an example of a plutonic rock that contains relatively large amounts of quartz; rhyolite is the volcanic rock of the same general composition.

Most of the entire first section of our route crosses plutonic igneous rock—granite and granodiorite of the Sierra Nevada batholith. The best exposures of these rocks are seen from the highway along the eastern shore of Lake Tahoe, and in road cuts as the road crosses the Carson Range between Lake Tahoe and Carson City. Along sections of the highway further to the east, especially through the Clan Alpine and Desatoya Mountains, volcanic igneous rocks such as rhyolitic ash-flow tuffs are common.

Sedimentary rocks are formed in two ways: clastic sedimentary rock (remember, broken pieces) such as sandstone and siltstone form from particles of preexisting rocks broken down and transported by water, wind, ice, or just plain gravity, and deposited in new locations; and nonclastic sedimentary rock forms from material precipitating, by either chemical or biological means, from the water of seas or lakes. Chemical precipitation usually happens in warm, shallow water where evaporation causes the water to become oversaturated with certain minerals like calcium carbonate and salt. The minerals drop out of solution as fine particles that build up into sediments. Some limestone and salt beds form this way. Sediments formed by biological means had a little help from the local population. Sediments formed from masses of decayed plant life produced thick coal beds (hardly any of these in Nevada). Active sea life, both plant and animal, that used calcium for protective shells built up thick deposits of reef limestone, coquina (shell limestone), and chalk. Freshly laid down as sand or mud, all of these deposits are called sediments. With time, as the sediment layers become thicker and older material is deeply buried, pressure builds up and the sediments are squeezed and cemented into sedimentary rock.

Just like igneous rocks, sedimentary rocks are categorized based on mineral content, method of formation, and particle size. Coarse-grained sediments, such as gravel, form rock known as conglomerate; sand is transformed into sandstone; and finer sediments become siltstone, mudstone, or shale, depending on how much clay they contain. The sediments formed from chemical and biological precipitates usually start out as limy mud that, when hardened to stone, becomes limestone or dolomite. Most sedimentary rocks show layering (called bedding) that is a relict of their method of formation. Since sedimentary rocks are really just solidified trash heaps of older material, they commonly contain evidence of the life and environment that existed at the time of their formation-fossil remains of plants and animals and traces of their existence, such as tracks and burrows. Siltstone and mudstone, which may have started out as mud in shallow deltas or mudflats, sometimes preserve things like mud cracks, ripple marks, and even raindrop impressions. All of these features make sedimentary rock outcrops fun places to stop and wander about with eyes to the ground for awhile. Who knows what you will find! Remember, however, that collecting is always best done in moderation, and never in State or National Parks and Monuments, or other protected sites.

You will see sediments, sands and fine silt in the basin of Pleistocene Lake Lahontan, mainly through the lowlands of the second segment of the U.S. 50 log. Sedimentary rocks will be most prominent later in road cuts and outcrops in the many ranges between Austin and the Utah state line.

Metamorphic rocks are the third major rock type. As their name indicates (from the Greek *meta* meaning "change," and *morphe*, meaning "form"), the form of these rocks has been changed by heat and pressure. Under sufficient pressure, limestone recrystallizes into marble. Mudstone, siltstone, and shale transform to slate and schist. Sandstone becomes a hard, dense rock known as quartzite. Metamorphosed igneous rocks sometimes just get the prefix "meta" tacked in front of their original name, like meta-andesite or metadiorite. When the old rock gets really heated and squeezed, crystals move around in the rock, and the stretched, striated result is called gneiss.

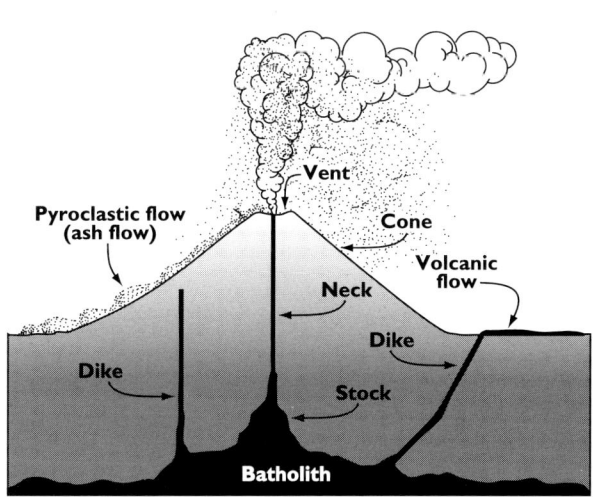

Igneous rock forms.

There is a progressive change, with time and changing conditions, of rock from one type into another. Since we think that everything was molten at one time in Earth history, the earliest rocks were, therefore, igneous. With weathering, transport, and redeposition, came sedimentary rocks. With heat and pressure generated by depth of burial, sedimentary rocks (and also igneous rocks that are buried and heated by various means) are transformed into metamorphic rocks. If this process continues to the extreme, the rocks melt and the process starts over again with fresh (recycled) magma. This is called the **rock cycle**:

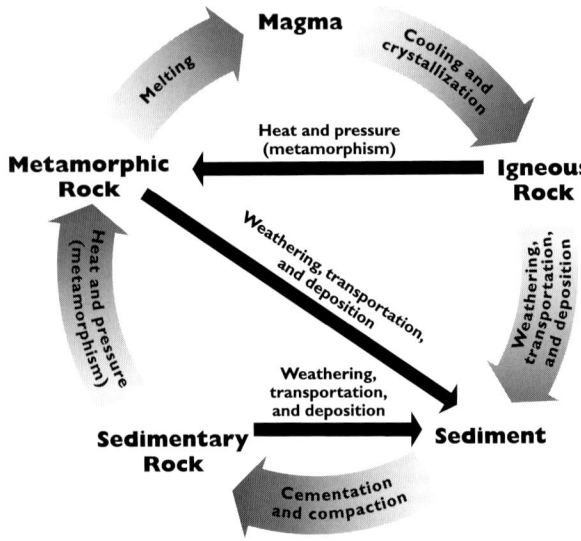

On our journey across Nevada on U.S. 50, we will not see much metamorphic rock until the last segment of the trip, where we will find that much of the higher parts of the Snake Range (in Great Basin National Park) is composed of quartzite.

Geologic Structures

Rock formations are not static; they move, get shoved and pushed about, bend, fold, and eventually break apart. The evidence of all this activity that is seen in the rocks today is called structure. Folding is fairly self explanatory; folds can be tight, with all of the bending within the span of a hand or arm-reach, or they can be broad, with flexure extending over hundreds or thousands of feet. We see folded rocks many places along the way across Nevada.

Nevada also has faults, lots of them (only in the rocks, however!). When rock masses break and one or both of the two resulting pieces move relative to each other, the surface of breakage is called a fault plane. Faults are classified by which way the broken pieces moved, and by the angle of the break. The block of rock above the fault plane is called the hanging-wall block, the block below the fault plane is the footwall block (these are old mining terms, a miner hung his candlestick or lantern from the hanging wall and stood on the footwall).

Downward movement of the hanging wall relative to the footwall produces a normal fault. Upward movement of hanging wall relative to the footwall produces a reverse fault. Movement of the blocks right or left along the fault plane defines lateral faults, either right- or left-lateral. Faults can be high-angle or low-angle; low-angle reverse faults have a special name—thrust faults—and they have great significance in Nevada. Thrust faults have played an important role in Nevada's gold belt, and we will discuss them in some detail later in the road log segment between Austin and Ely. Normal faults of regional extent, called detachment faults, occur near the end of our trip, in the area of Great Basin National Park, and we comment on these special faults in that segment of the log.

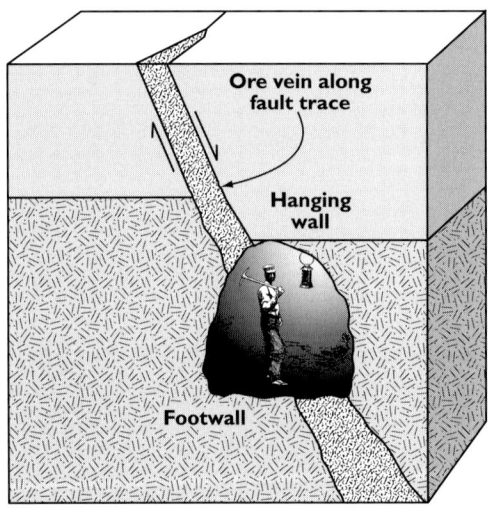

Relationship of hanging wall, footwall, vein, and fault.

Plate Tectonics

In 1782, Benjamin Franklin hypothesized: "The crust of the Earth must be a shell floating on a fluid interior. Thus the surface of the globe would be capable of being broken and disordered by the violent movements of the fluids on which it rested." The modern concept of plate tectonics, more or less the same as presented by Ben Franklin over 200 years ago, is that all of the Earth's surface is adrift. This surface, the Earth's lithosphere, consists of large rigid plates that move in response to the flow of the heat-softened asthenosphere

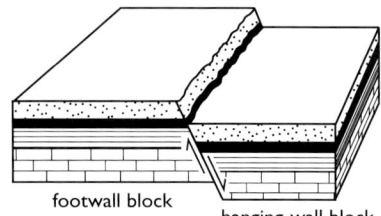

footwall block   hanging-wall block

Normal fault

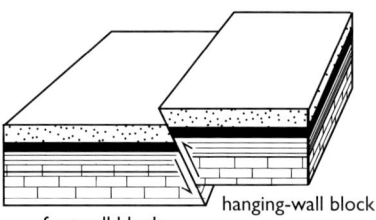

footwall block   hanging-wall block

Reverse fault

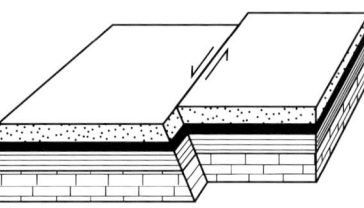

Left-lateral fault

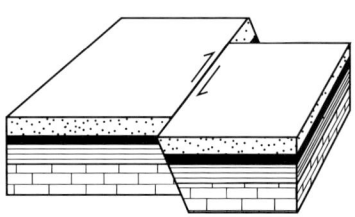

Right-lateral fault

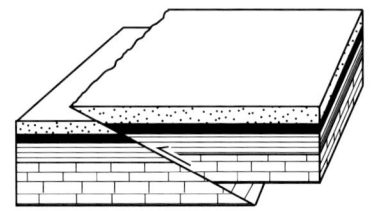

Thrust fault

Types of faults.

8

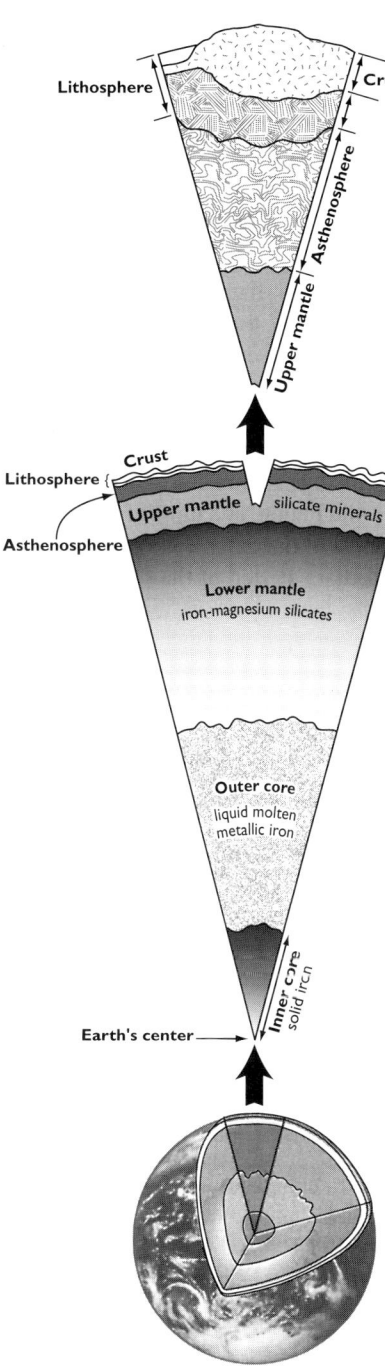

Structure of the Earth's interior.
(modified from R.W. Christopherson 1998)

beneath them. Over millions of years, plate movements have created the Earth's ocean basins, continents, and mountains, all of which essentially ride as passengers on the shifting plates.

The plates can be very large. For example, nearly all of North America, including Nevada, is part of the North American plate. To the west of us, underlying the entire Pacific Ocean, is the huge Pacific plate. When masses as large as these collide, great amounts of inertia and force are involved. It's no surprise, therefore, that most of the world's large-scale geological activity, such as earthquakes and volcanic eruptions, occurs at or near plate boundaries.

At the present time, our two local plates are sliding together along a boundary geologists call the San Andreas Fault; the Pacific plate is moving to the northwest in relation to North America. In the (geologically) recent past (until about 20 million years ago), the Pacific plate was moving eastward and colliding directly with the North American plate. The North American plate came out on top of this collision, so to speak, and the Pacific plate was forced beneath the North American, becoming a "subducted" plate.

Forceful plate collision and subduction are responsible for many of the major geologic features now exposed along our route. As the Pacific plate moved under the North American plate down into the Earth's mantle, it became hot, eventually melting to form magma (remember, the rock cycle—with heat and pressure rock melts and forms new magma). This magma rose into the near-surface crust; some was spewed out on surface to form volcanic rock, some solidified below surface to become plutonic rock. In the western segment of our route, faulting has exposed plutonic rock formed deeper in the crust—this is the Sierra Nevada batholith complex. Further to the east, we move into volcanic rock and the calderas from which they came, both surface expressions of the subduction that triggered the volcanic activity.

Tectonic setting along the present U.S. 50 corridor showing (a) collision and subduction of Pacific and North American plates during the Mesozoic Era, and (b) right-lateral faulting and formation of Basin and Range beginning about 20 million years ago. Heavy arrows show northwest movement of the block of oceanic crust west of the San Andreas Fault, and the resulting extensional forces in the Basin and Range to the east. ▶

Where does the Basin and Range structure that dominates the geologic picture across most of our route fit into plate tectonics? The present northwestward sliding movement of the Pacific plate tends to create tensional (pull-apart) forces in the North American block. Plate interactions have resulted in northwest-trending, right-lateral faults (parallel to the San Andreas Fault) and generally north-trending normal faults that define "fault-block" mountain ranges with down-dropped valleys between them—our Basin and Range province. Nevada's major earthquakes (about one magnitude 7 every 30 years) result from plate tectonics.

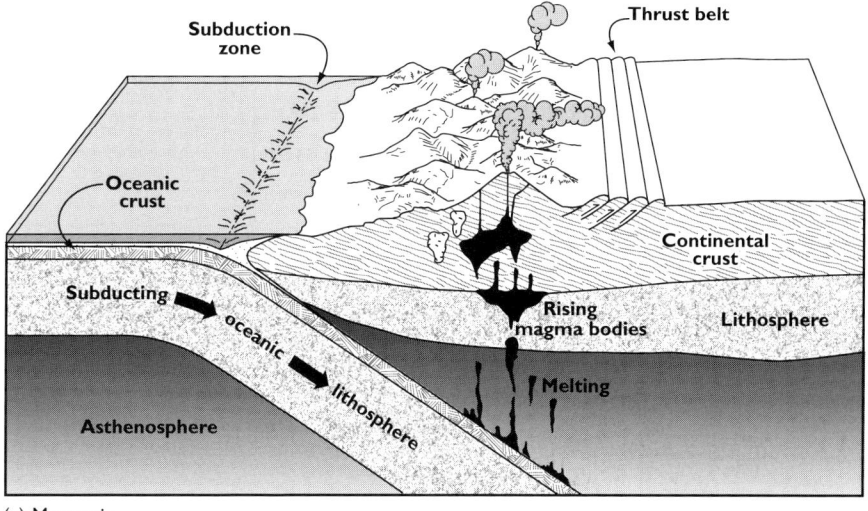

(a) Mesozoic

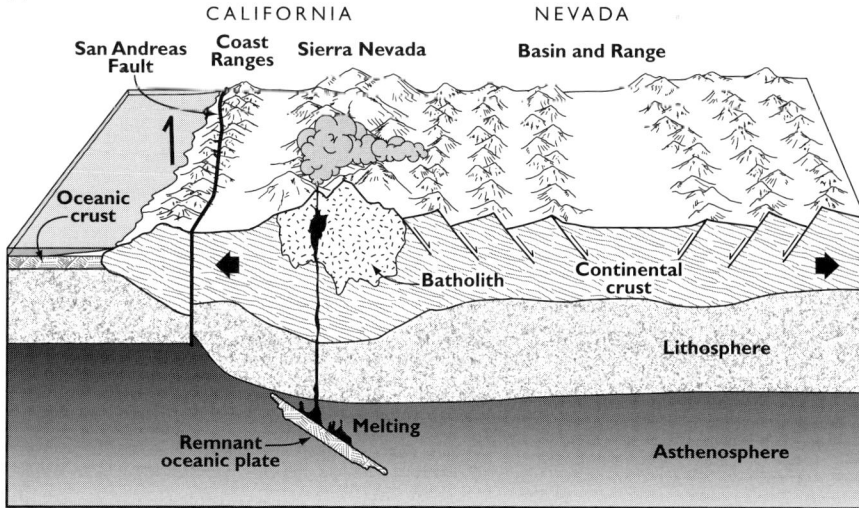

(b) Late Cenozoic (about 20 million years ago to the present)

Geologic Time

Geology is mainly a world of relative time. Geologists tend to describe time as being before or after certain events, the presence or absence of life forms comes to mind as perhaps the most notable of these. The names of the geologic "eons" were originally based on the absence, presence, and complexity of life forms recorded in the rocks: the

Time Units of the Geologic Time Scale				Development of Plants and Animals
Eon	Era	Period	Epoch	
PHANEROZOIC	CENOZOIC	Quaternary	Holocene — 0.01 Ma	Humans develop
			Pleistocene — 1.8 Ma	
		Tertiary	Pliocene — 5.3 Ma	Age of Mammals
			Miocene — 23.8 Ma	
			Oligocene — 33.7Ma	
			Eocene — 54.8Ma	
			Paleocene — 65.0 Ma	
	MESOZOIC	Cretaceous — 144 Ma		Extinction of dinosaurs and many other species
				First flowering plants
		Jurassic — 206 Ma	Age of Reptiles	First birds
		Triassic — 248 Ma		Dinosaurs dominant
	PALEOZOIC	Permian — 290 Ma		Extinction of trilobites and many other marine animals
		Carboniferous: Pennsylvanian — 323 Ma	Age of Amphibians	First reptiles
				Large coal swamps
		Carboniferous: Mississippian — 354 Ma		Amphibians abundant
		Devonian — 417 Ma		First insect fossils / Fishes dominant / First land plants
		Silurian — 443 Ma	Age of Fishes	
		Ordovician — 490 Ma		First fishes
		Cambrian — 540 Ma	Age of Invertebrates	Trilobites dominant / First organisms with shells
PROTEROZOIC		2500 Ma	Collectively called Precambrian, comprises about 87% of the geologic time scale	First multicelled organisms
AZOIC (Archean)		4600 Ma		First one-celled organisms / Age of oldest rocks / Origin of the earth

Numbers followed by "Ma" indicate millions of years ago.

Geologic time scale. (data from Geological Society of America)

oldest eon, Azoic, was, until recently, thought to be without life; the Proterozoic Eon was the time of "early life"; the Phanerozoic Eon, which extends to the present, is the time of "evident life." Eons are divided into eras. The Phanerozoic Eon is divided into three eras; Paleozoic—old life forms, Mesozoic—middle life forms, and Cenozoic—recent life forms. Eras are divided into shorter "periods" that usually bear the name of some European locality where they were first studied and described. "Cambrian," as an example, is named for Cambria, the Roman name for Wales, and "Devonian" is named for Devonshire, England. The era in which we live, Cenozoic, is divided into two periods; Tertiary and Quaternary, names held over from an older four-part subdivision of geologic time. Humans, being the "splitters" that we are, have further divided both the Tertiary and Quaternary periods into smaller slots called epochs which are shown on the geologic time scale.

The rocks we see along U.S. 50 are mostly confined to the Paleozoic Era and younger. Most of the sedimentary rocks in the eastern segments of our road guide, for example, are Paleozoic age. There are some Mesozoic-age sedimentary rocks exposed in the west segment of the log, around Carson City and Fallon, and most of the granitic rock of the Sierra Nevada is also Mesozoic age. Most of the extensive volcanic rocks in the central part of Nevada formed during the Tertiary Period of the Cenozoic Era, and the extensive inland lakes that once occupied the basins U.S. 50 now crosses east and west of Fallon were full during the Pleistocene Epoch of the Quaternary Period.

Ore Deposits

"Ore" is just a special type of rock that happens to be rich in something that we consider valuable, like gold, silver, copper, lead, or tungsten. Nevada is famous for its metallic ores, and U.S. 50 passes by some of the largest, most famous metal mining camps in the state. Most of the ore types we see in Nevada are directly related to intrusive rocks; the ores formed from hot, metal-rich waters exhaled from cooling, solidifying magma or circulated through the crust by heat from the magma. As the ore fluids found their way into the rocks bordering the intrusions, the fluids cooled and lost their ability to hold the metals in solution. Metallic minerals then precipitated, forming various types of ore deposits. These deposits are zoned outward from their source; certain metallic minerals form in the deep, hotter zones. Other minerals make it to greater distances before becoming trapped in deposits, and still others form deposits so far from their source we have a hard time deciding just exactly which magma body is responsible for them. Sometimes, the leftover fluids find their way all the way to the surface where they spew out, often violently, as hot springs. Even here, the fluids deposit minerals such as sinter (a form of quartz) and travertine (calcite) in hot spring terraces. We don't find many metallic minerals in these surface deposits, but sometimes there are traces of mercury, and there can be lots of sulfur.

These orebodies formed from hot fluids we call "hydrothermal" deposits (hydro, water, and thermal, hot). In Nevada, hydrothermal deposits are all over the place. These include the common gold-silver vein deposits that were mined at most of the bonanza mining camps that made the state famous as well as replacement deposits of lead and zinc, porphyry copper deposits, tungsten deposits, and of most recent interest, the Carlin-type gold deposits. As we have described, these deposits form at increasing distance outward from their source magma. Tungsten, iron, and sometimes copper can occur at the contact of the source intrusive and intruded rocks. These deposits, formed in the deepest, hottest part of the hydrothermal system, are called contact metamorphic deposits. Porphyry copper deposits form a little further from source, replacement deposits are still further, and veins and Carlin-type gold deposits are found in the most distant reaches of the hydrothermal system.

The most famous mining camp we pass is Virginia City, site of the Comstock Lode. Virginia City's mines produced silver—lots of silver, and also a significant amount of gold from hydrothermal veins. The term "lode" is an old Cornish-miner term for a fissure in country rock filled with ore minerals. The word was derived from the verb "to lead": whatever the miner could follow expecting to find ore was his lode. Lodes are usually thought of as tabular, having depth and length as their greatest dimensions with width as the smallest—they are long and narrow. The Comstock Lode was a set of parallel quartz veins that the miners followed for over 13,000 feet along the east side of the Virginia Range. Mining went as deep as 2,900 feet, and the Lode was up to several hundred feet wide in places, although this width was made up of several veins with barren material in between. The veins were fillings of mainly quartz with a small amount of silver- and gold-bearing minerals that occupied fissures in the country rock. The silver was contained mostly in the mineral argentite (silver sulfide), although native silver along with gold was present at the surface.

Two other ore deposit types important in the history of mining along U.S. 50 are lead-silver deposits at Eureka, and porphyry copper deposits near Ely. Both of these are also hydrothermal deposits, but they are quite different from each other. Lead and silver at Eureka were found in massive replacement bodies in limestone. There, the ore solutions dissolved the limestone and deposited sulfide minerals such as galena, tetrahedrite, sphalerite, and pyrite into the space. The result was large, thick ore pods that followed beds in the limestone. Using the miner's definition, these too could be called "lodes," but they were tabular and flat rather than tabular and steep.

The porphyry copper ores at Ely, as a type, are in between contact metamorphic ores and replacement ores. The term "porphyry" comes from the type of rock

that usually hosts these deposits, a plutonic igneous rock with characteristically large mineral crystals floating in a fine-grained matrix. Copper minerals, mainly the copper sulfides chalcocite and chalcopyrite, along with lots of pyrite, are dispersed in thin veins and disseminated as mineral grains and clots in the host rock. These orebodies are large, irregular masses with dimensions in the hundreds to thousands of feet. Their large size lends them to mining in open pits. At Ely, you can drive to the pit overlook at the Robinson Mine. This will give you a good idea of the scale of a copper deposit of this type.

Micron gold, disseminated gold, invisible gold, and Carlin-type gold are all names you will hear applied to Nevada's latest mining specialty. At the present time, this is the most important ore deposit type in Nevada. Exactly where it fits in the family tree of ore deposits is still under debate. These gold deposits are similar in size to the big porphyry copper deposits, but contain gold not copper, and usually occur in sedimentary not intrusive rocks. The gold is so fine grained that it can't be detected by panning, it can't even be seen without the help of the highest-power microscopes. Except quartz, there aren't many other minerals associated with these deposits. This probably explains why we didn't recognize this type of deposit until 1961, when Newmont Mining Co. discovered the original Carlin deposit in northern Eureka County. U.S. 50 passes through Carlin-type gold country from Austin all the way to Ely. We'll talk more about some of the mines as we pass by them.

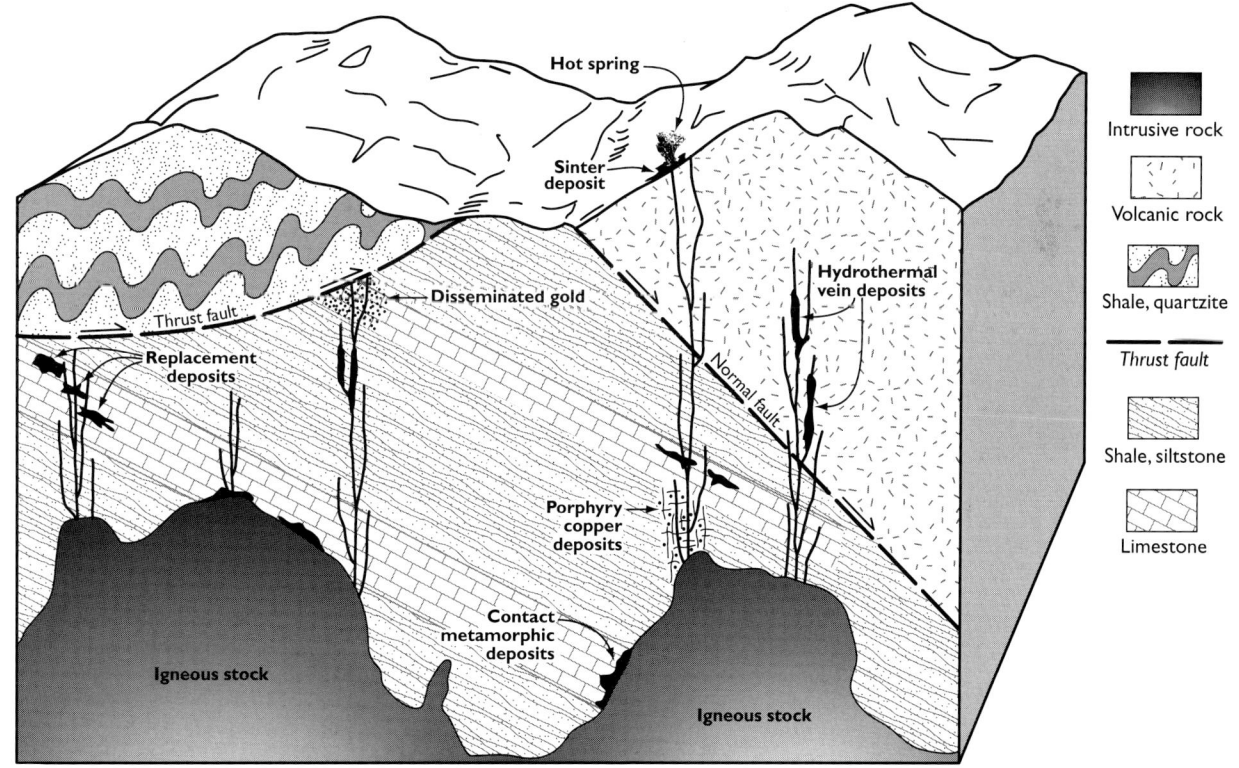

Hydrothermal ore deposits. ▶

COLOR PHOTO CAPTIONS

PLATE 1

Generalized geology along the U.S. Highway 50 corridor in Nevada.

PLATE 2

Vegetation types along the U.S. Highway 50 corridor in Nevada.

PLATE 3

3a **Paintbrush** (*Castilleja spp.*) Because of hybridization and wide intraspecific variation, many species of paintbrush are hard to tell apart. Flower colors range from brilliant red to orange to yellow. Paintbrush can be found from sagebrush flats to above the timberline. *Photo: Jack Hursh*

3b The **western tanager** (*Piranga ludoviciana*) inhabits open coniferous forests throughout the West. The female is yellow-green above and yellow below. *Photo: Mitzi Hultin*

3c The **wild rose** (*Rosa woodsii*) is found near streams and springs and in other relatively moist locations. Under favorable conditions it may form dense thickets. Rose hips are an important winter food source for birds and mammals. *Photo: Roy W. Cazier*

3d **Mourning cloak** (*Nymphalis antiopa*) One of the first butterflies to appear in spring, the mourning cloak inhabits woodlands, meadows, parks, and gardens. Caterpillars feed on willow, elm, and *Populus* species. *Photo: Jack Hursh*

3e **Spooner Lake**, Spooner Lake State Park. *Photo: Kris Pizarro*

3f View of **Lake Tahoe** and the **Sierra Nevada** from Logan Shoals vista point. *Photo: Kris Pizarro*

3g **Rabbitbrush** (*Chrysothamnus nauseosus*) **Height 1–7′, usually less than 3′; many slender, flexible branches covered with dense, feltlike hairs; leaves very narrow and covered with whitish woolly hairs; aromatic; dense clusters of bright yellow flowers appear August–October.**

Rabbitbrush is a common shrub found in dry soils in association with sagebrush and in grasslands, piñon-juniper woodlands, abandoned farmsteads, and along roadsides. It is an invasive shrub, and dense stands indicate overgrazing or other types of disturbance. Cattle don't care for the plant, though sheep, antelope, and deer occasionally eat it. It is utilized more heavily by birds and small mammals.

The genus name *Chrysothamnus* is derived from two Greek words, *chrysos* meaning gold and *thamnus* meaning shrub. Rabbitbrush shows itself at its best in the fall when its yellow blooms light up the landscape. *Photo: Kris Pizarro*

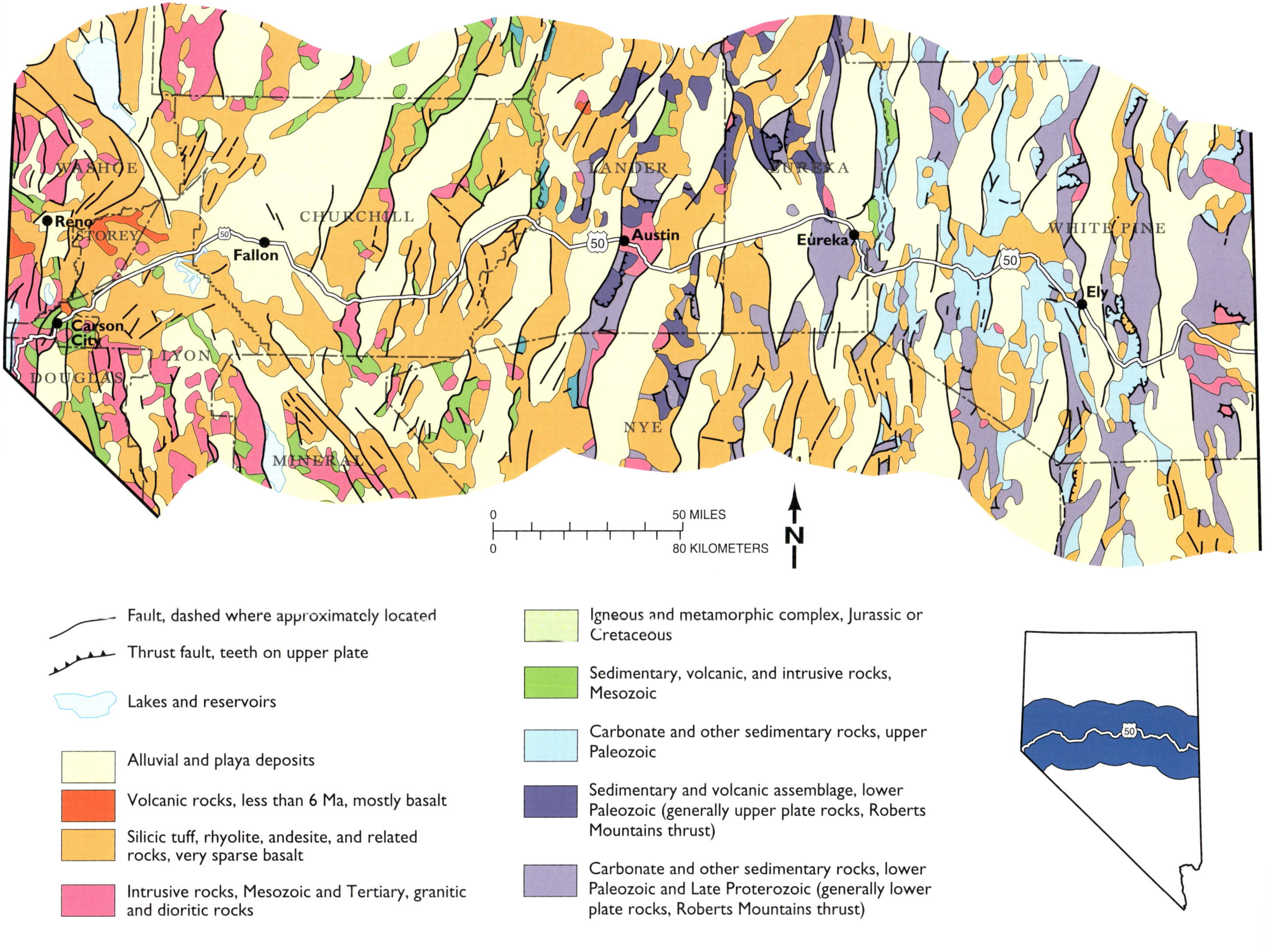

Map showing geology along U.S. Highway 50 across Nevada, with cities Reno, Carson City, Fallon, Austin, Eureka, and Ely labeled. Counties shown include WASHOE, STOREY, CHURCHILL, LANDER, EUREKA, WHITE PINE, DOUGLAS, LYON, MINERAL, and NYE.

0 — 50 MILES

0 — 80 KILOMETERS

N

Fault, dashed where approximately located

Thrust fault, teeth on upper plate

Lakes and reservoirs

Alluvial and playa deposits

Volcanic rocks, less than 6 Ma, mostly basalt

Silicic tuff, rhyolite, andesite, and related rocks, very sparse basalt

Intrusive rocks, Mesozoic and Tertiary, granitic and dioritic rocks

Igneous and metamorphic complex, Jurassic or Cretaceous

Sedimentary, volcanic, and intrusive rocks, Mesozoic

Carbonate and other sedimentary rocks, upper Paleozoic

Sedimentary and volcanic assemblage, lower Paleozoic (generally upper plate rocks, Roberts Mountains thrust)

Carbonate and other sedimentary rocks, lower Paleozoic and Late Proterozoic (generally lower plate rocks, Roberts Mountains thrust)

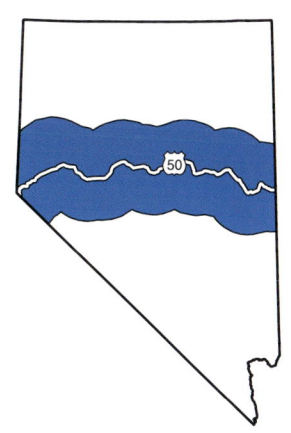

PLATE 1

13

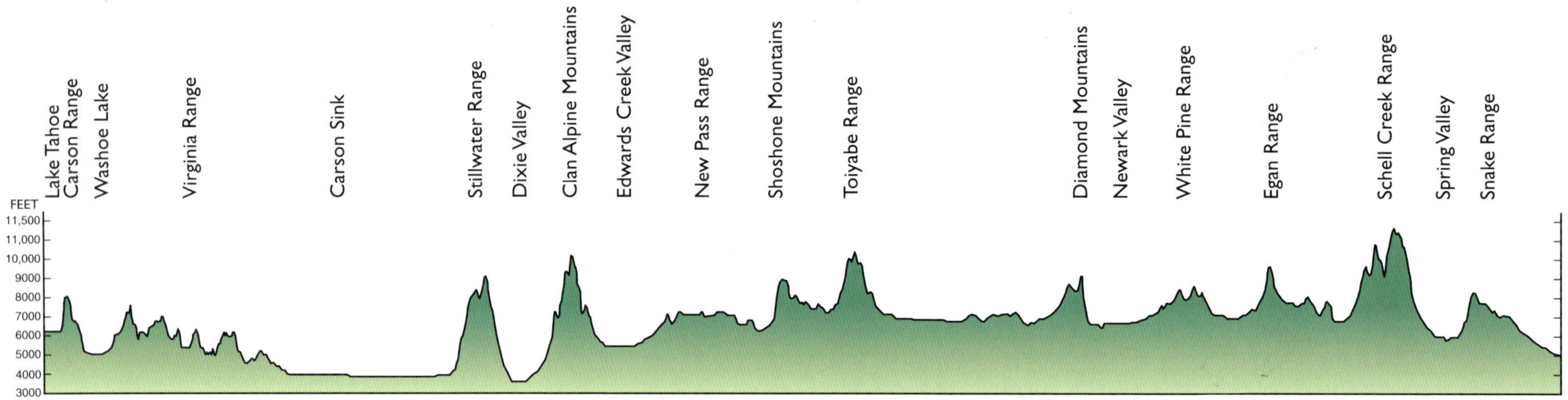
Lake Tahoe
Carson Range
Washoe Lake
Virginia Range
Carson Sink
Stillwater Range
Dixie Valley
Clan Alpine Mountains
Edwards Creek Valley
New Pass Range
Shoshone Mountains
Toiyabe Range
Diamond Mountains
Newark Valley
White Pine Range
Egan Range
Schell Creek Range
Spring Valley
Snake Range
FEET
11,500
11,000
10,000
9000
8000
7000
6000
5000
4000
3000

Schematic transect along U.S. 50

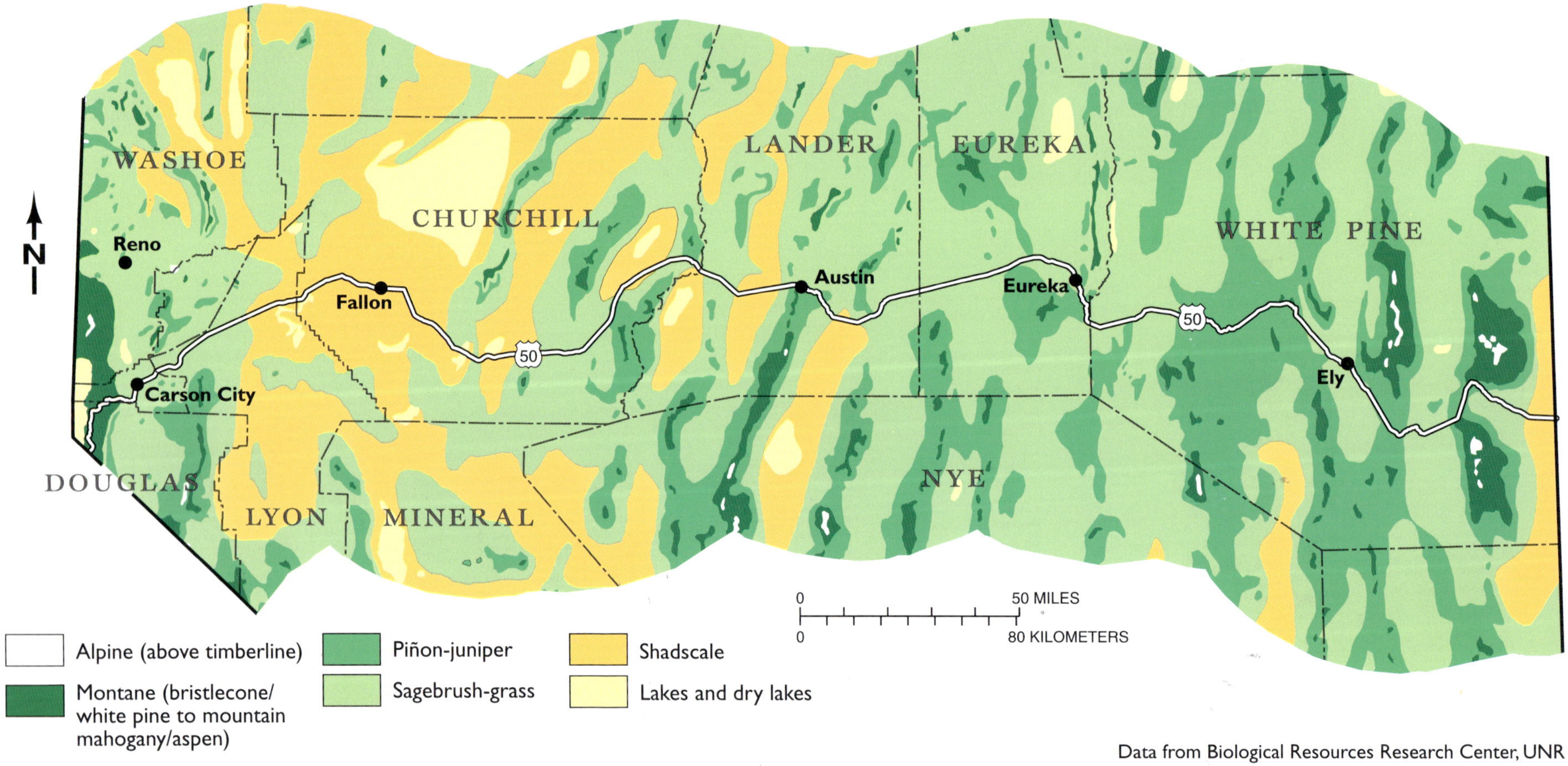
WASHOE
CHURCHILL
LANDER
EUREKA
WHITE PINE
DOUGLAS
LYON
MINERAL
NYE
N
Reno
Fallon
Austin
Eureka
Ely
Carson City
50
0
50 MILES
0
80 KILOMETERS
Alpine (above timberline)
Montane (bristlecone/ white pine to mountain mahogany/aspen)
Piñon-juniper
Sagebrush-grass
Shadscale
Lakes and dry lakes

Data from Biological Resources Research Center, UNR

PLATE 2

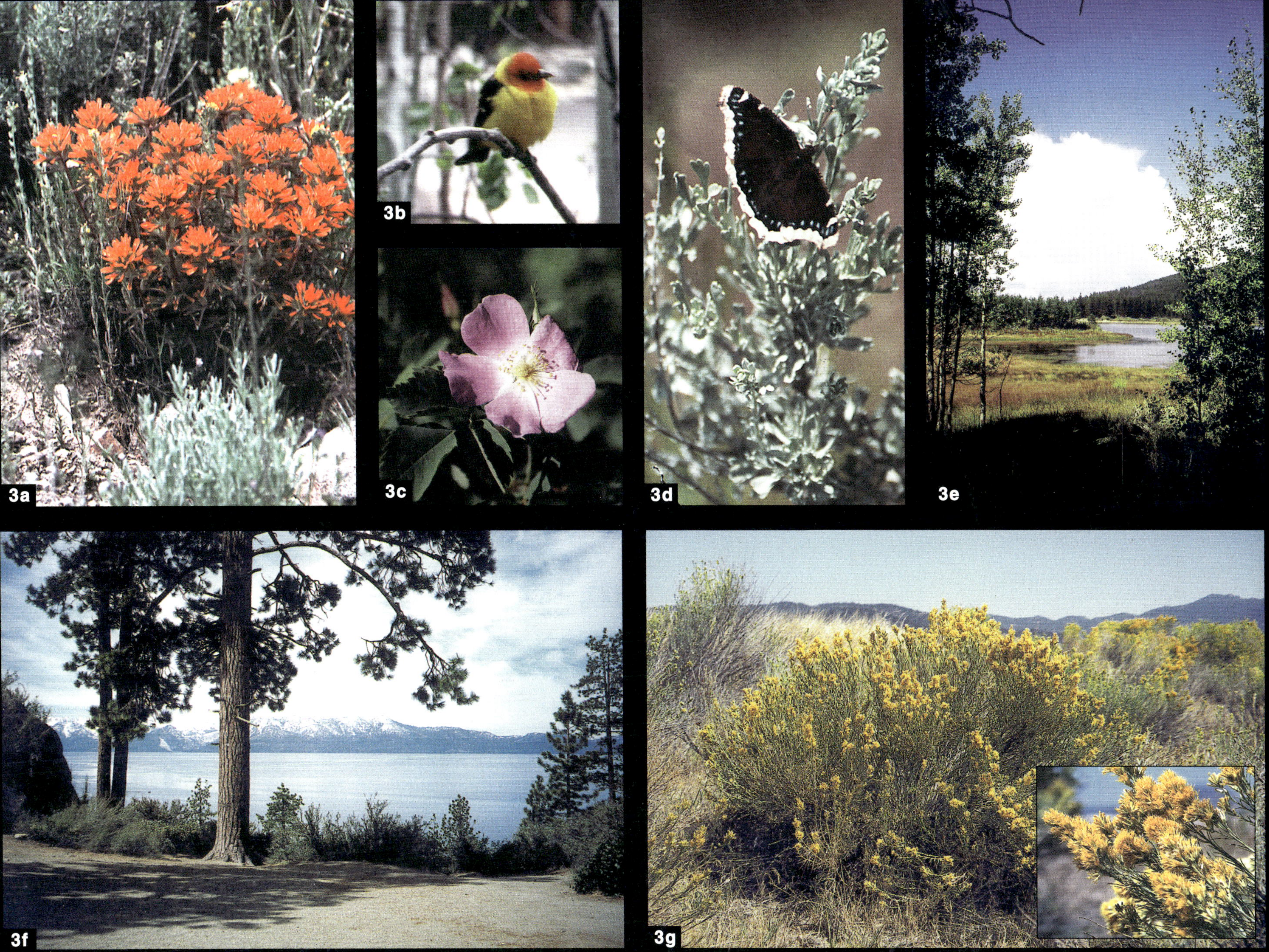

3a

3b

3c

3d

3e

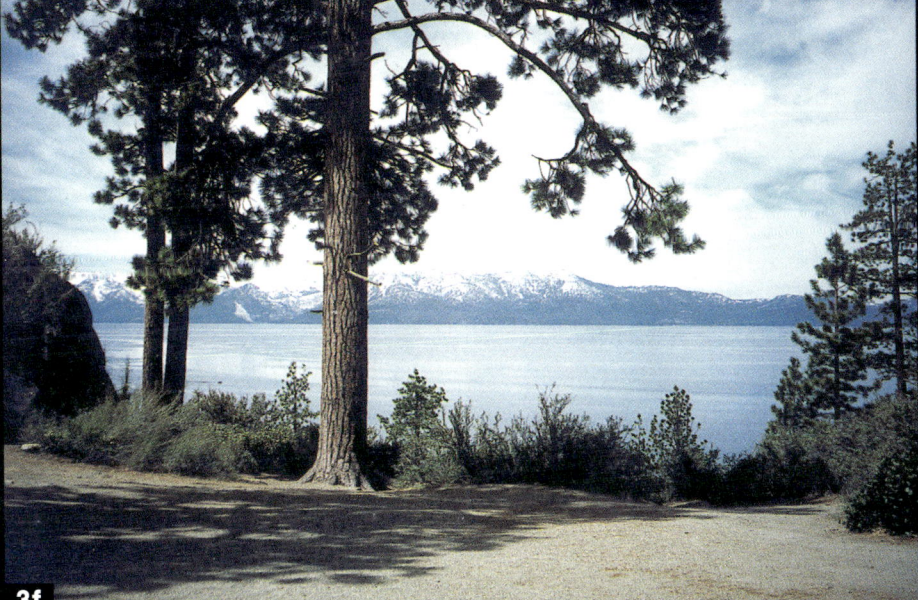

3f

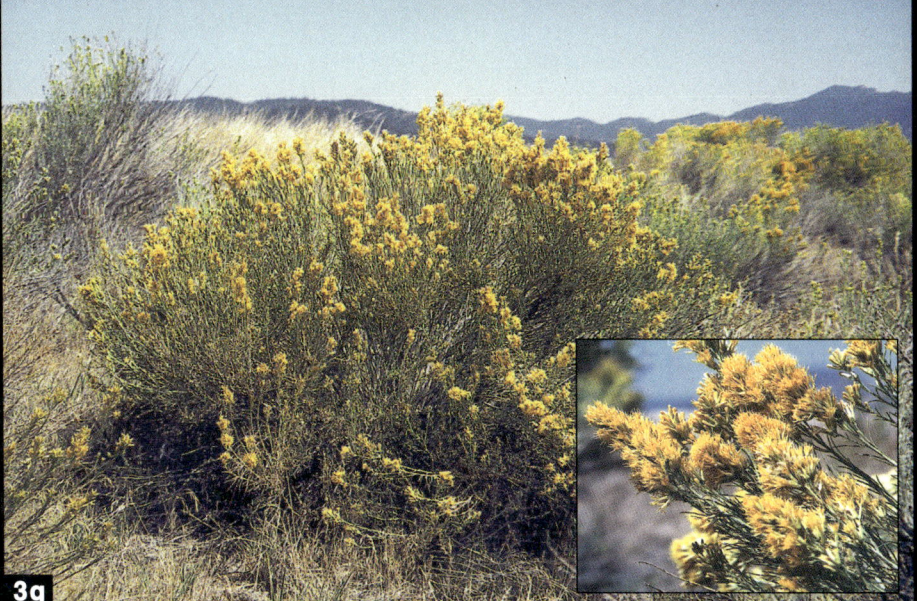

3g

PLATE 3

15

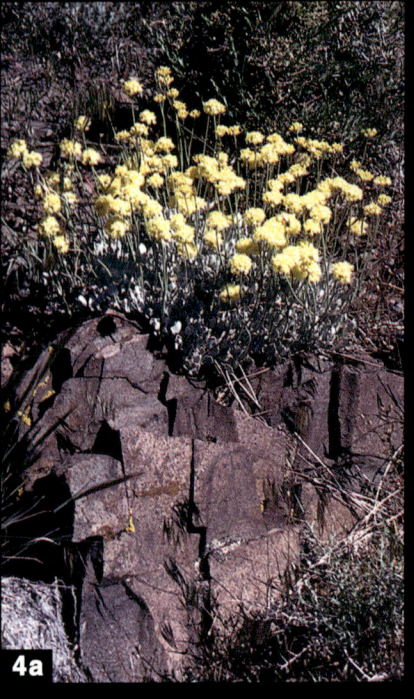

4a

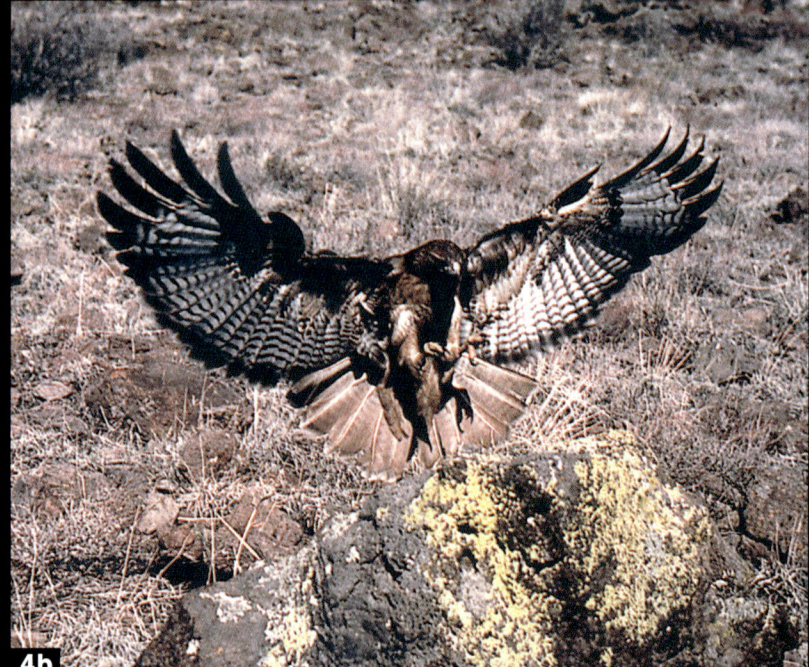

4b

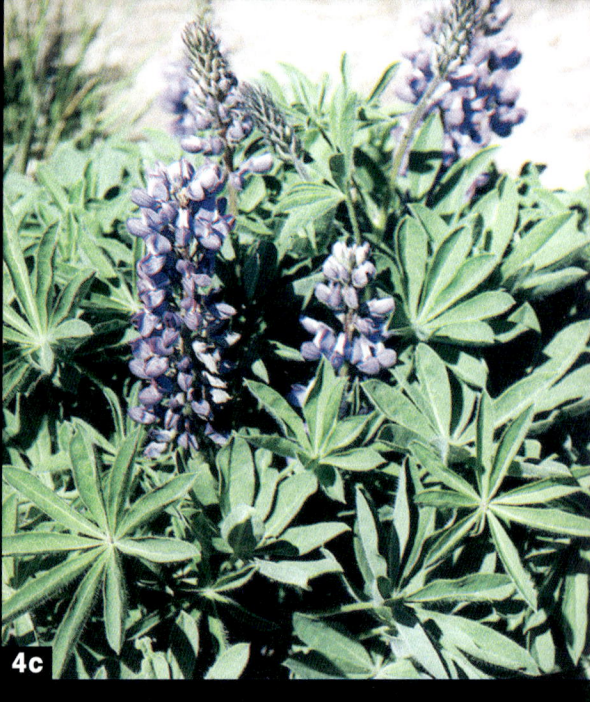

4c

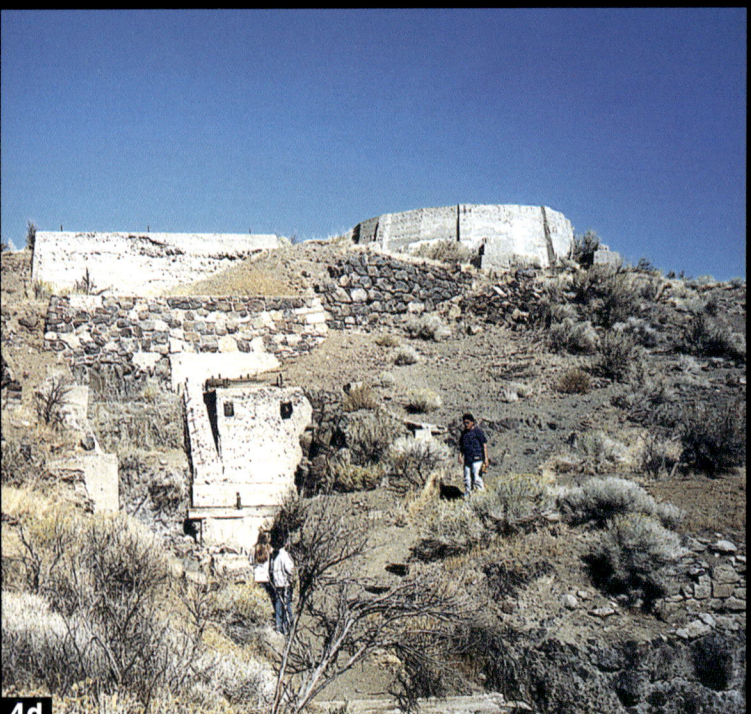

4d

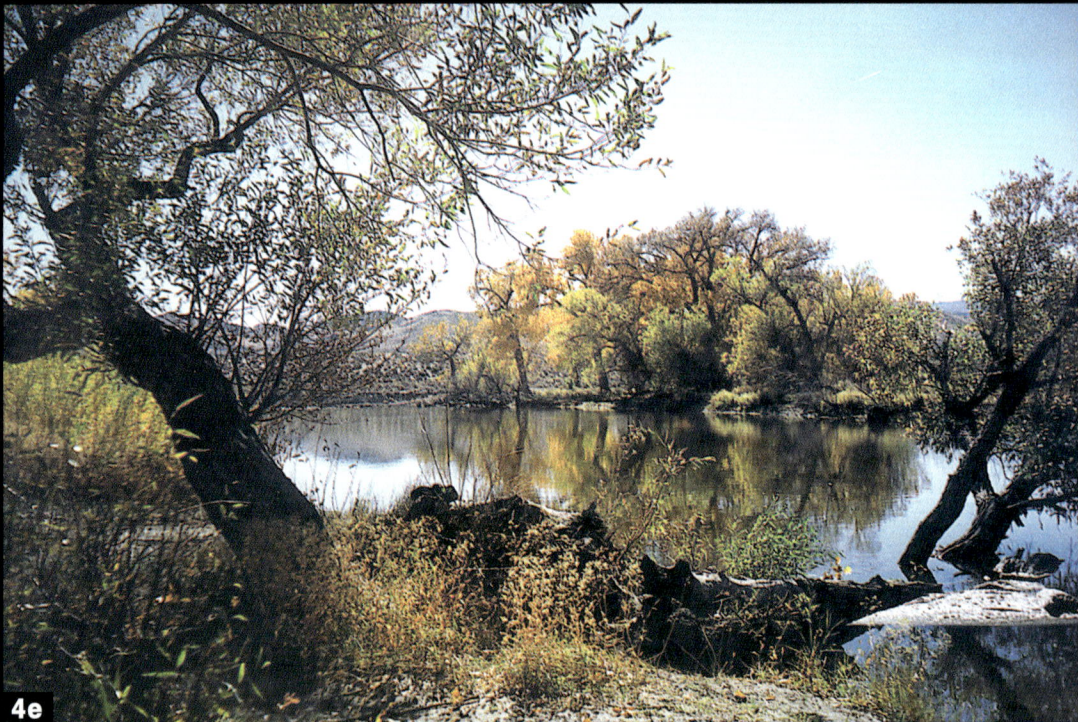

4e

6

PLATE 4

COLOR PHOTO CAPTIONS

PLATE 4

4a **Sulfur flower**, or **desert buckwheat** (*Eriogonum umbellatum*) Eriogonum species are widespread in Nevada, from elevations of 4,000 to 9,000 feet. Flowers are an important source of nectar for bees, and seeds are consumed by a variety of birds and rodents. *Photo: Jack Hursh*

4b **Red-tailed hawk (*Buteo jamaicensis calurus*) Length 19–25", wingspan 46–58"; weight 2.5–3.5 pounds; large, brown, sturdily built hawk; individual coloration varied, but most show broad band of dark streaking across white belly; identifiable in flight by broad rounded wings (and fanned reddish tail; female larger than male; call is a harsh, descending kree-e-e-e.**

This subspecies of red-tailed hawk is a common permanent resident of the Great Basin. The population peaks during the winter months with the arrival of migrants from the north. The redtail is a master of soaring, and in flight it is most often seen with wing and tail feathers fully spread as it circles slowly in thermal updrafts or rides deflection currents along ridge lines.

The redtail is often seen perched on utility poles or fence posts across Nevada. It has exceptional binocular vision and can spot prey at great distances. It will perch for long periods of time, then suddenly glide off after a rabbit or other rodent, bird, lizard, or snake.

Redtails probably mate for life and will return to the same nesting area year after year. They build a stick nest in a tall tree, rocky ledge or transmission tower 15–70 feet above the ground. The male hunts food for the female while she is nesting. Two to three young hatch after about 30 days and make their first flight about 45 days after that. *Photo: Rick Kline, Cornell Laboratory of Ornithology*

4c **Lupine** (*Lupinus spp.*) Lupines are found from the sagebrush steppe to above the timberline. It is often difficult to tell one of the many lupine species from another. Lupines are members of the pea family and fix nitrogen in the soil. *Photo: Ricardo Pizarro*

4d Foundations of the **Rock Point Mill**, Dayton State Park. *Photo: Kris Pizarro*

4e Autumn on the **Carson River** west of Fort Churchill. *Photo: Kris Pizarro*

SECTION I: FROM THE SIERRA FRONT INTO THE GREAT BASIN

The first section of the road log begins at Stateline, a gaming and resort center on the southeast shore of Lake Tahoe, where U.S. 50 enters Nevada from the west, and extends to the Lyon-Churchill county line a few miles east of the desert community of Silver Springs. This first portion of the route crosses the eastern edge of the Sierra Nevada province and then enters the Basin and Range province and the Great Basin.

The Sierra Nevada province is characterized by massive outcrops of erosion-resistant granitic rocks that form the Sierra Nevada and the adjacent ranges along Nevada's western border. The Sierra Nevada is a huge block of the Earth's crust that has broken free on the east along the Sierra Nevada fault system and tilted westward. It is overlapped on the west by sedimentary rocks of California's Great Valley and on the north by sheets of volcanic rock extending south from the Cascade Range in northern California. The core of the Sierra Nevada is composed of large masses of granitic rocks (batholiths) of Mesozoic age that have intruded Paleozoic and Mesozoic metamorphic rocks. Along the Sierra Nevada crest and on the steep eastern front of the range, the rocks have been greatly uplifted and most exposures are of the granitic rocks. Remnants of older metamorphic rocks are seen in places as pendants perched on top of some of the granite batholiths, or as thin lenses caught between two granitic masses.

To the east of the main Sierra Nevada, fault blocks of Sierran rocks have been down-dropped on branching segments of the Sierra Nevada fault system. This marks the transition into the Basin and Range province, a region characterized by generally north-trending mountain ranges separated by wide valleys. The formation of these "basins and ranges" is discussed in the earlier section on plate tectonics. Along our U.S. 50 route across Nevada, the Basin and Range province coincides with the Great Basin.

The Great Basin, a term coined by John Frémont in 1844, is a vast closed basin covering nearly all of Nevada, half of Utah, and parts of California, Oregon, and Idaho. It is defined by its hydrological character—rivers and streams in the Great Basin do not flow out of the basin but end in interior lowlands known as playas and sinks, forming largely ephemeral desert lakes. The Sierra Nevada

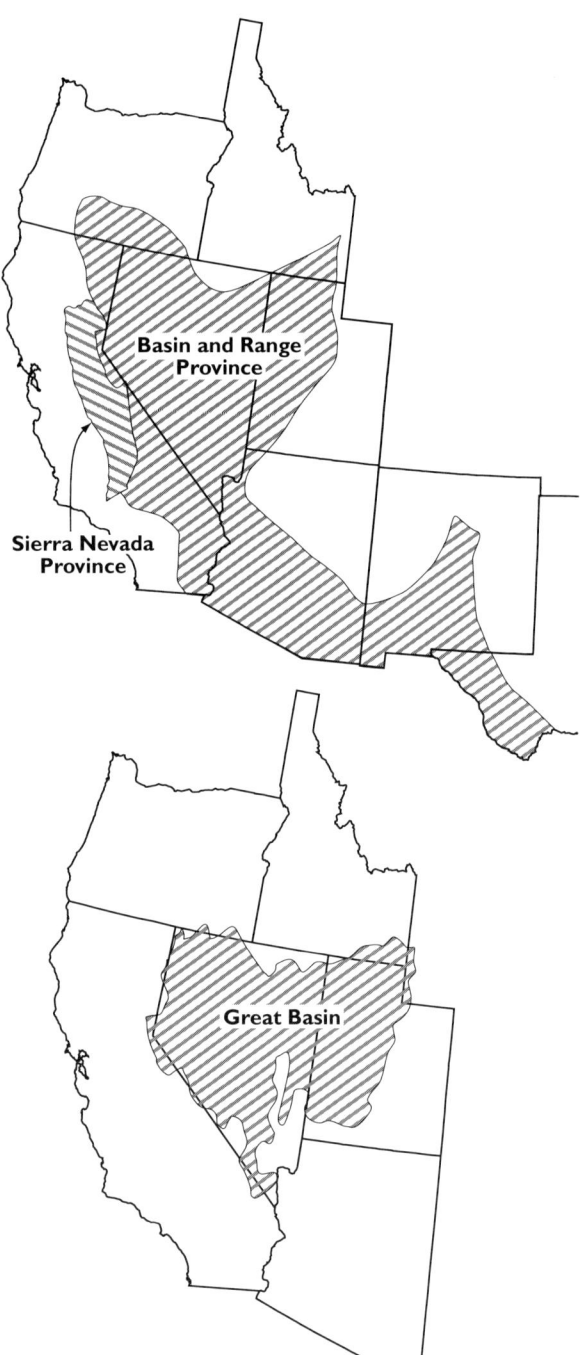

Sierra Nevada and Basin and Range provinces, and the Great Basin.

marks the western edge of the basin, the Wasatch Front forms the eastern edge, and the north edge of the basin is defined by the volcanic fields of northern Nevada, southern Oregon, and southern Idaho. To the south, the basin edge is formed by the northwestern limit of the Colorado River drainage. The Great Basin is not exclusively a basin. It has high mountain ranges within it, such as the Toquima, Toiyabe, and Schell Creek Ranges with crests near 12,000 feet, and the Snake Range with Wheeler Peak reaching 13,060 feet. The intervening valley floors are generally between 4,000 and 6,500 feet in elevation.

For most of the Great Basin, the Sierra Nevada is a tremendous physical barrier to the passage of moisture-laden storms that move eastward from the Pacific onto the continent. In winter, as the air masses rise over the western slope of the Sierra, they expand and cool below their dewpoint, dropping much of their significant load of moisture in the form of snow. On the east side of the Sierra Nevada, toward Nevada, the east-moving winter air masses flow down the mountain front and are warmed above their dew point, and most of the moisture left in the air tends to remain there rather than falling on Nevada's valleys. This creates what is known as a "rain shadow." Although Nevada lies in the rain shadow of the Sierra Nevada, the effect is repeated to a lesser extent all the way to the east across the Great Basin. Each high mountain range across the state collects moisture on its windward side and has a rain shadow on its leeward side.

The climatic conditions present in the Great Basin define the Great Basin desert, a temperate desert with snowy winters and hot, dry summers. The valleys are dominated by sagebrush and shadscale; the biologic communities on the mountain ranges differ with elevation, and the ranges are like islands isolated by seas of desert vegetation in the lower, drier valleys. (plate 2)

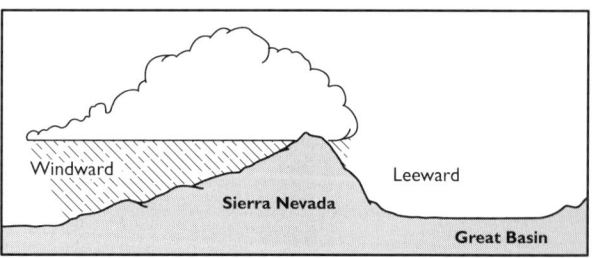

The rain shadow of the Sierra Nevada.

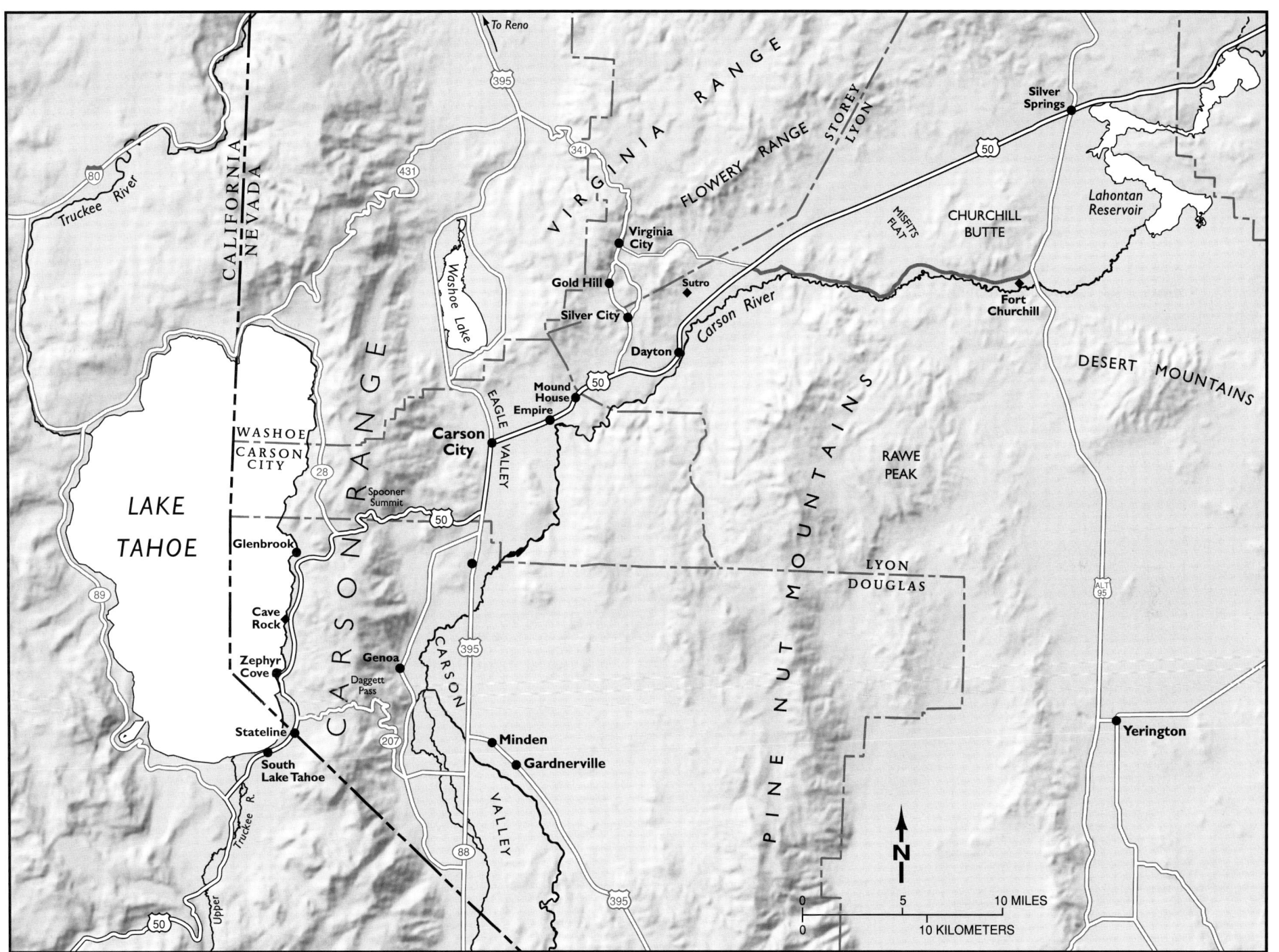

Route map, Douglas-Carson City-Lyon Counties.

19

ROADLOG DESCRIPTION

	0.0	DO 00

Begin at Stateline on U.S. 50 heading north and east. The California-Nevada state line is easily spotted by the contrast of motels and gift shops on the California side with sleek high-rise hotels and casinos on the Nevada side. This is Milepost 0, Douglas County, Nevada. Set your odometer to zero when you are between Harvey's Hotel on the left and Harrah's Hotel on the right.

| 0.4 | 0.4 |

Friday's Station on right, white house centered in large area of lawn at 2:30. Originally a Pony Express station in 1860–61.

| 0.2 | 0.6 |

Junction with Nevada State Route 207. Go straight, stay on U.S. 50. For the next ten miles or so, the highway follows the eastern shore of Lake Tahoe.

SIDE TRIP 1, KINGSBURY GRADE TO CARSON VALLEY

Travel State Route 207 east over Kingsbury Grade (Daggett Pass) to Carson Valley. Spend some time at Walleys Hot Springs, then visit historical Genoa, Nevada's first settlement (see Trip C in Geologic and Natural History Tours in the Reno Area, NBMG Special Publication 19). You can retrace your route back to U.S. 50 and continue north, or you can travel north from Genoa through Jacks Valley to the junction with U.S. 395, then continue north to rejoin U.S. 50 as it descends from Spooner Summit and turns north into Carson City.

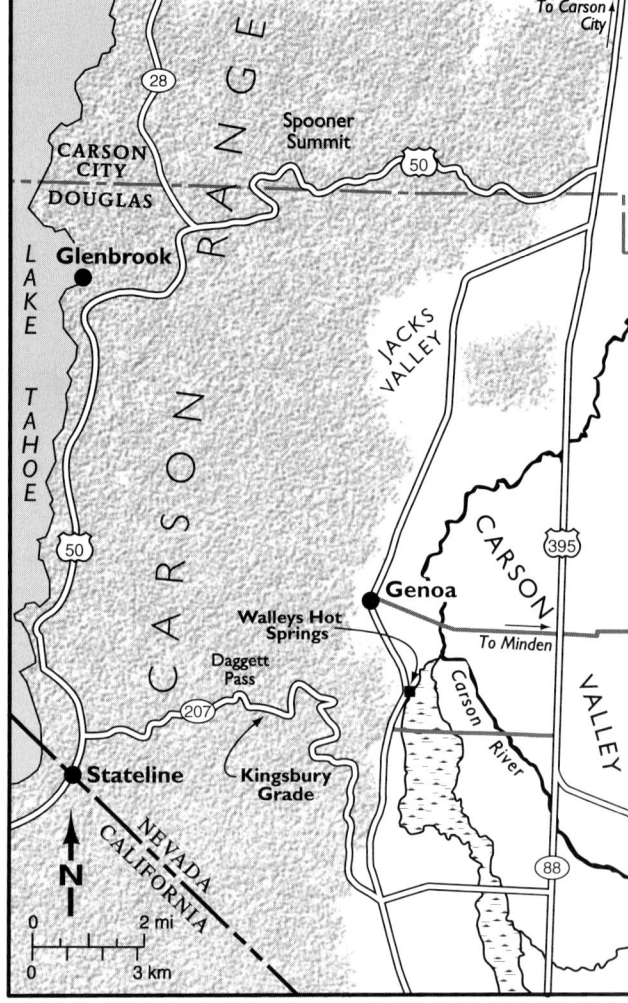

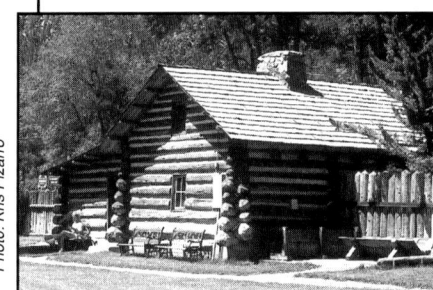

A restoration of the first permanent building built at Mormon Station (present-day Genoa). The building is within Mormon Station State Historic Park. ◄

Photo: Kris Pizarro

Street scene in present-day Genoa. ▼

Kingsbury Grade to Carson Valley ▶

View to the south along the west side of Carson Valley. Jobs Peak is the barren peak to the right of center on the skyline. Genoa is located at the range front at the right margin of the photo. The white patch above the line of trees at the range front near right center is on the trace of the Genoa fault. ▼

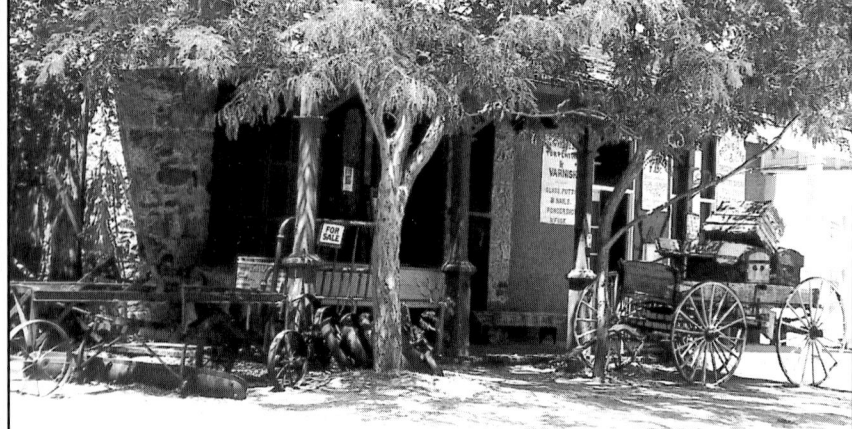

Photo: Kris Pizarro

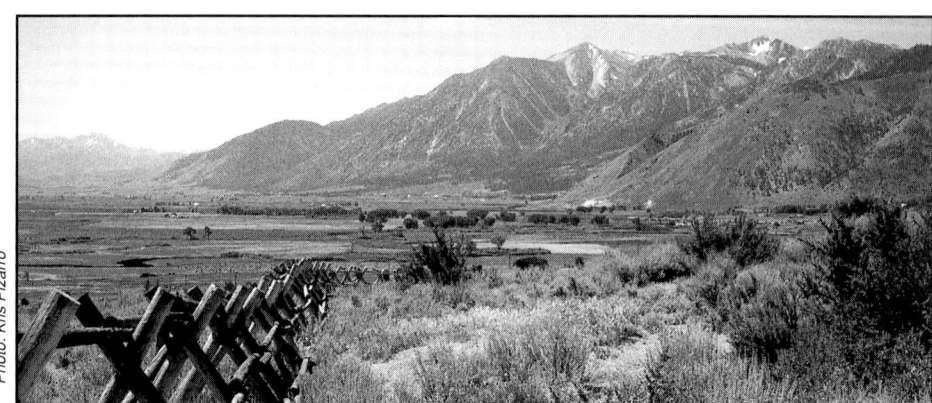

Photo: Kris Pizarro

interval	cumulative	milepost
0.3	0.9	
1.0	1.9	
2.3	4.2	

First good view of Lake Tahoe to the left, some nice aspen trees and a meadow—not much of this type of view left along this part of the lakeshore.

Round Hill-Nevada Beach. This is one of the few places along this section of the lake with easy access to a public beach. For a short side trip, turn left (toward the lake) at the stoplight at Round Hill and drive the short distance to Nevada Beach, a U.S. Forest Service public campground. This is a great place for overnight camping or for just stopping for awhile to walk along a sandy beach and enjoy the view of the lake.

Zephyr Cove is to the left. The beach is granitic sand and gravel washed down from the Carson Range. Home of the M.S. Dixie paddle wheeler, it also has a public beach and riding stables.

LAKE TAHOE

Lake Tahoe occupies a valley between the main crest of the Sierra Nevada and the Carson Range. The lake is about 22 miles long and 12 miles wide, and the average depth of the water is 1,000 feet (the greatest measured depth is 1,645 feet). Only about one-third of the lake is in Nevada. The main feeder is the Upper Truckee River, which flows into the south end of the lake. Water exits at the northwest corner of the lake, forming the main Truckee River, which eventually flows into Nevada. The lake level is controlled by a dam at the outlet, and the maximum elevation of the water level is 6,229.1 feet (a court-mandated level). In times of drought, however, the lake level can fall below the level of the outlet, and the Truckee River below the dam ceases to flow. Groundwater, reservoirs, and tributary streams add to the flow of the river further downstream.

Lake Tahoe is commonly included in the Sierra Nevada province but its formation is due to block-faulting and volcanism, features more commonly associated with the Basin and Range province. The Lake Tahoe basin is a down-dropped block, box-like in shape, bordered by steeply dipping faults. Volcanic flows and sediments at the north end of the basin have effectively dammed the exit channel, and caused the fault-block valley to become a lake. Lake formation probably began about 5 million years ago during the Pliocene Epoch.

The first white men to make note of Lake Tahoe were John Frémont and his topographer, Charles Preuss, who, on February 14, 1844, spotted the lake while making observations from a peak near Carson Pass to the south. In his report of 1845–46, Frémont called the lake "Mountain Lake," but on his map of 1848, he gave it the name "Lake Bonpland" in honor of the French naturalist and companion of Baron Alexander von Humboldt,

an early traveler in the Great Basin. In 1851, the lake was renamed Lake Bigler in honor of California Governor John Bigler. None of these names stuck and the lake came to be called Tahoe, more or less conforming to the original Washoe Indian name for the body of water, which some authorities say means "water in a high place" while others say it means "big lake or water."

Lake Tahoe did not receive much attention from the early cross-country emigrants; it probably got in the way of their objective of quickly reaching the California gold fields. The Donner route over the Sierra Nevada lies north of the Lake Tahoe basin, while the Pony Express-Placerville route now followed by the western extension of U.S. 50 skirts the south end of the lake before crossing the Sierra Nevada at Echo Summit. Following the discovery of the Comstock Lode and the development of deep underground mines there, Tahoe's timber resources became valuable, and a lumbering industry grew along the eastern shore. At the end of the 19th century, when both the mines and the timber became depleted, the scenery and solitude of the lake attracted wealthy visitors from San Francisco and other areas who acquired estates along the lakeshore. There was no road network around the lake at that time, and transportation, supply, and mail service were by lake steamer. These steamers were elegant ships that are only a memory in the present-day era of small power boats and "personal water craft." Some of the mansions built at that time are now in public ownership, and the grounds are state parks. Public use of the lake really expanded during the years following World War II: boating and fishing on the lake became popular, ski resorts were built in the mountains surrounding the lake, and gaming on the Nevada side flourished.

For more information on Lake Tahoe, see Trip C, *Geologic and Natural History Tours in the Reno Area*, NBMG Special Publication 19.

Steamer *Tahoe*, Lake Tahoe, 1906.

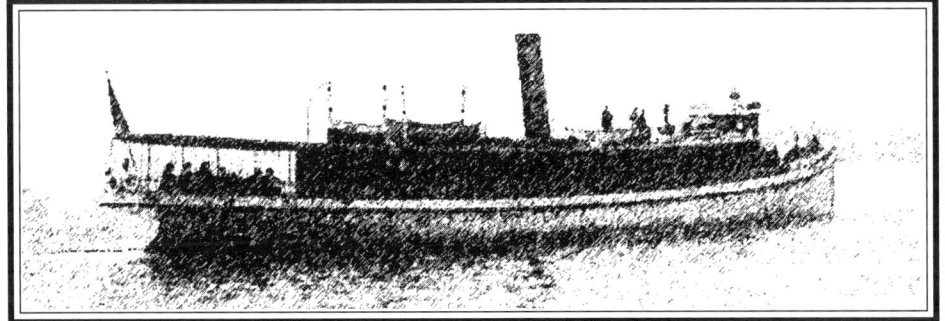

— Nevada Historical Society

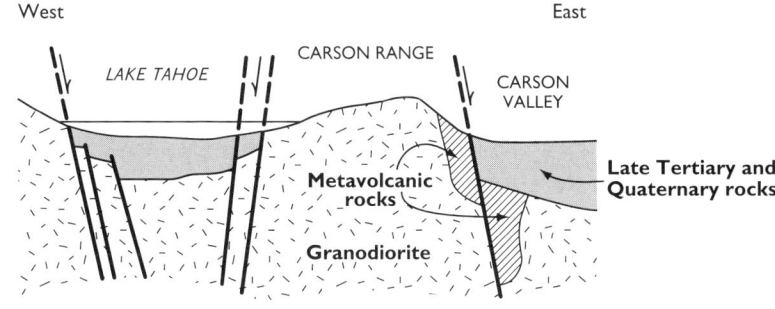

Diagrammatic cross section, Lake Tahoe to Carson Valley, showing block faulting.

interval	cumulative	milepost	
1.8	6.0		Note granite boulders in the road cut that have been rounded by weathering.
0.5	6.5		Cave Rock, ahead, is an eroded Miocene volcanic neck. During the Pleistocene, the lake level was 140 feet higher and caves were cut high on the south side of the rock by wave action; hence the name, Cave Rock. If you look to the right, high in the steep ravine, you can see the contact of the rugged volcanic rock with the older (Cretaceous) granitic rock (the contact area is partly obscured by talus, so don't look too hard and lose track of the road ahead). Romanticized in Indian lore as a sacred place, Cave Rock was a landmark on the Lake Bigler Toll Road, which skirted the rock on its lake side. The rock was first tunneled for the construction of a highway in 1931 and the second tunnel was put through in 1958.
0.3	6.8		Cave Rock boat landing is on the left. A fee site administered by the Nevada Division of State Parks, this is a popular boat landing with a very small beach, rest rooms, and some picnic tables. The water-level view of the lake is spectacular, and the view of Cave Rock is better than from the highway above. A partial profile of the "Lady of the Lake," a natural feature in the cliff face below Cave Rock can be seen from the breakwater, and remnants of the wall of quarried granite blocks that supported the old toll road can still be seen on the west face of the rock.
0.2	7.0		Entrance to highway tunnel through Cave Rock
1.0	8.0		Logan Shoals Vista Point on the left. There is parking on the left (lake) side of the westbound traffic lane. The turnout is wide, but traffic is usually heavy so be careful if you decide to stop here. A short trail leads to a rocky point overlooking the lake. This is about the best view of the lake you will get along our short stretch of highway. To the north, you can see Logan Shoals, a shallows area marked by partially submerged granite outcrops; to the south you can look back on Cave Rock; and to the west there is a good view of the high Sierra crest. There are also some informational signs describing the history of the area. There are no picnic tables or other facilities, but you can perch on the rocks and enjoy the lake. (plate 3f)

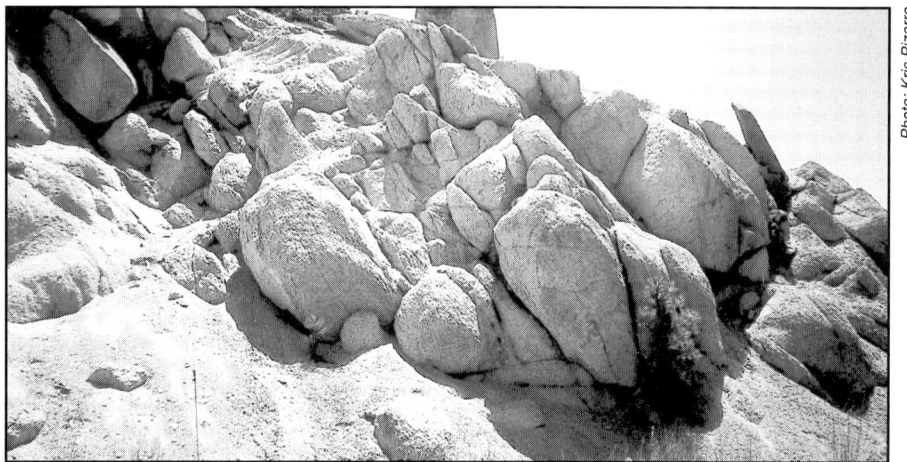

Photo: Kris Pizarro

Spheroidal weathering of jointed granitic rock. Weathering attacks corners and edges ▲ of the jointed blocks more than the flat surfaces, and the rocks take on a spherical shape. Note the rivulets of grus (decomposed granite) cascading from the weathered outcrop (along the bottom of the photo).

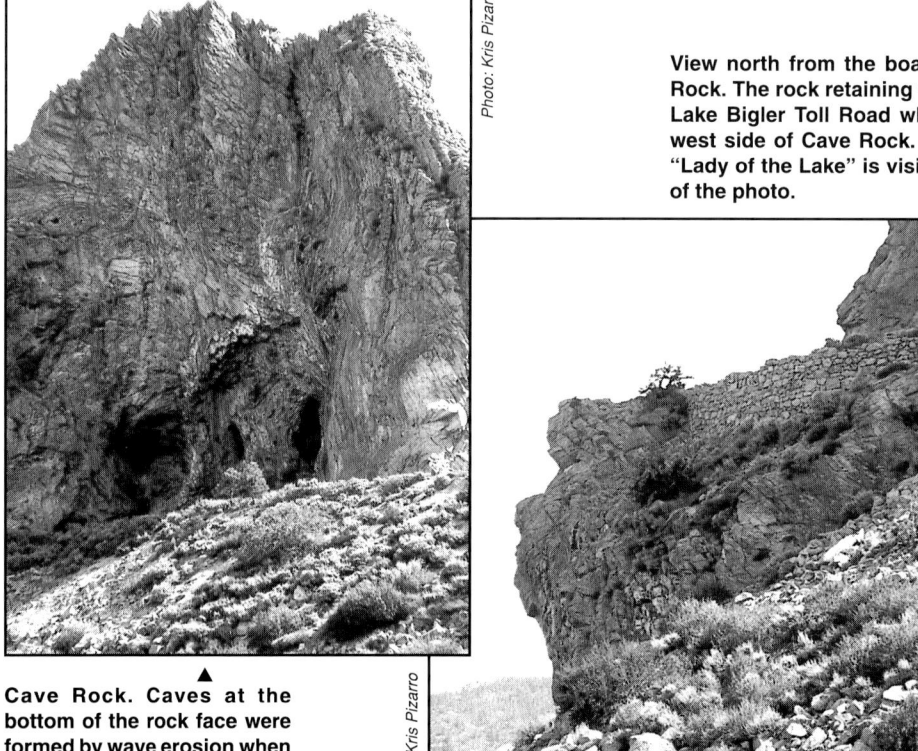

Photo: Kris Pizarro

View north from the boat landing at Cave Rock. The rock retaining wall marks the old Lake Bigler Toll Road where it skirted the west side of Cave Rock. The profile of the "Lady of the Lake" is visible at the left side of the photo. ▼

▲
Cave Rock. Caves at the bottom of the rock face were formed by wave erosion when Lake Tahoe was at a higher level than it is today.

Photo: Kris Pizarro

interval	cumulative	milepost	
1.0	9.0		Glenbrook Post Office, on right
0.6	9.6		Glenbrook turnoff is on the left. Glenbrook is below us and to the left as we come around the corner.
1.2	10.8		Multicolored rocks visible in road cuts in this section are hydrothermally altered granitic rock. The original rock has been altered by heat and hot solutions probably generated by volcanic activity. Silicate minerals such as feldspar have been broken down into soft, white clay. Iron-bearing minerals, such as pyrite, have also been altered, leaving behind rusty-appearing iron-oxide staining.
			To the north of the highway (on your left), on the slope above the ravine, the rugged, linear outcrop that forms the crest of the north-south-trending ridge marks the trace of a narrow latite dike (a light-colored intrusive rock). See if you can spot other similar dikes along the slope to the east of the main ridge—hint, the dikes are not altered and form rugged outcrops in contrast to the subdued outcrop of the altered rocks cut by the dikes.
			There is a wide turnout here if you want to stop and look at some of these rocks.
1.5	12.3		Spooner Junction, State Route 28 to the left. Bear right, stay on U.S. 50.
			This area bears the name of Michele E. Spooner, a French Canadian entrepreneur who, with several partners, operated a shingle mill and sawmill here in 1868. In 1873 Spooner's company, Summit Fluming Co., and all other mills in this area were taken over by the Carson & Tahoe Lumber & Fluming Co. This company, headquartered at Glenbrook, went on to become the largest of the huge combines supplying wood and lumber to the Comstock (NHM #225).
			For an optional short side trip, take State Route 28 north for about one mile and turn right into Spooner Lake State Park. (plate 3e)

GLENBROOK, COMSTOCK-ERA LUMBERING CENTER

Glenbrook was the first permanent settlement at Lake Tahoe, and was the focal point for the logging industry during the Comstock mining boom. Timber was needed in vast quantities in the mines to shore up shafts and tunnels dug to get to the Lode, and to provide ground support as the rich ores were removed. Many of the surrounding slopes were clear-cut to satisfy this appetite (envision all of the hillsides denuded of timber as far as the eye can see). The timber you now see is all second growth.

Lumbering operations in the Glenbrook area began with the construction of a sawmill in 1861. From timbered areas around the lake, logs were floated or barged to Glenbrook to be cut into mine timbers and other lumber for Comstock use. The Carson & Tahoe Lumber & Fluming Co. railroad hauled lumber from Glenbrook up Spooner Summit to the crest of the Carson Range. At that point, the lumber was sent by flume down the east slope of the range to Carson City for shipment to Virginia City. Logging in the area reached a record high in 1875 but essentially died out by the late 1880s because of the decline of mining on the Comstock and the decline of the timber resource. Today Glenbrook is a resort and residential community.

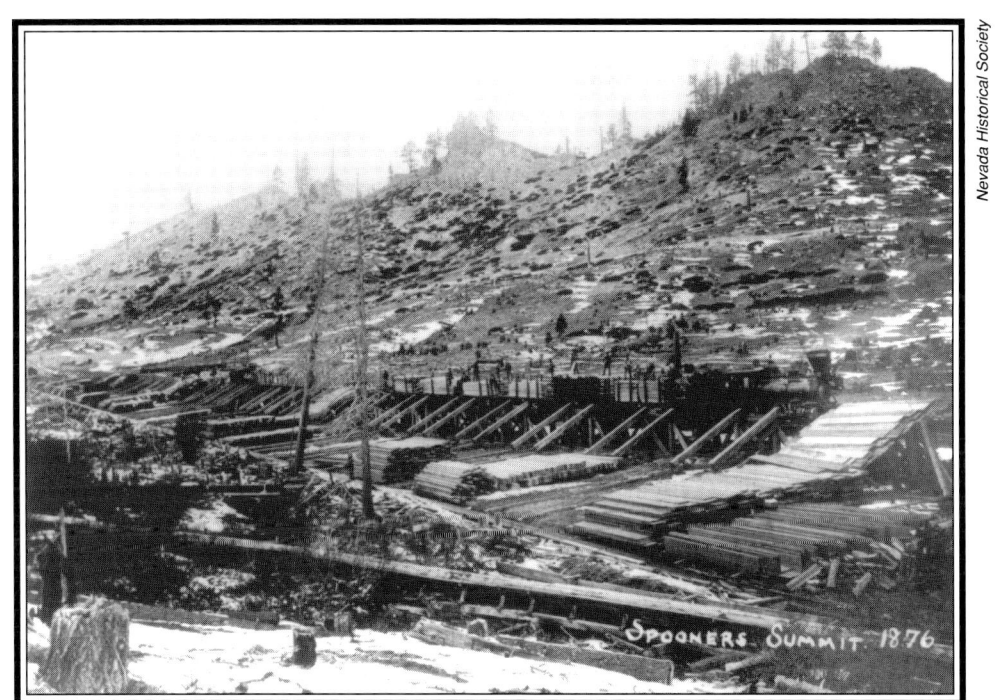

Lumber transfer operation at Spooner Summit, 1876

SPOONER LAKE STATE PARK

The State Park offers a restful place to take a break among the pines. There is an interpretative area offering information on cultural and natural history, and the area provides opportunities for walking and nature study as well as fishing in Spooner Lake. Come back to this area during the winter and you can snowshoe or cross-country ski around the lake or head into the backcountry on trails that begin here.

interval	cumulative	milepost	
1.8	14.1		Turnout. You can see aplite dikes (more narrow, resistant rock outcrops) in deeply-weathered granite.
0.9	15.0	DO 15.0	Spooner Summit (elevation 7,146 feet). The terminus of the lumber railroad from Glenbrook was near this point. The lumber was carried by flume from here to Carson City.
0.2	15.2		Tahoe Rim Trailhead is on the left.
			Spooner picnic area of the Humboldt-Toiyabe National Forests on right. Picnic tables, bathroom, aspens, mountain air—everything a traveler needs!
1.3	16.5	CC 00	Entering Carson City. Originally, this was Ormsby County but in 1969, the name of the "county" was changed to Carson City. The result is always an awkward verbal fumble—is it Carson City or Carson City County, or Carson City rural area and, if so, should we call the population center where all of the buildings are Carson City urban area, or perhaps Carson City City? Whatever, the important point is to note that the county mile markers start again at zero at this point at the county, or is it city, line.
2.1	2.1		Turnout on the right. Good view of the Carson Valley and the rapidly growing towns of Gardnerville and Minden. Unaltered granite of the Sierra Nevada batholith is exposed in the road cut.
0.8	2.9		Telephone turnout.
0.5	3.4		Note fault zone in road cut.
2.9	6.3		Shear zone (crushed rock along zone of parallel faults) in cut.
0.7	7.0		Decomposed granite ("DG") has been mined from the pits to the left.
4.7	7.6	CC 7.6	Junction with U.S. 395. Turn left onto Carson Street towards Carson City (urban area!). This is where the side trip over Kingsbury Grade to Genoa rejoins U.S. 50.

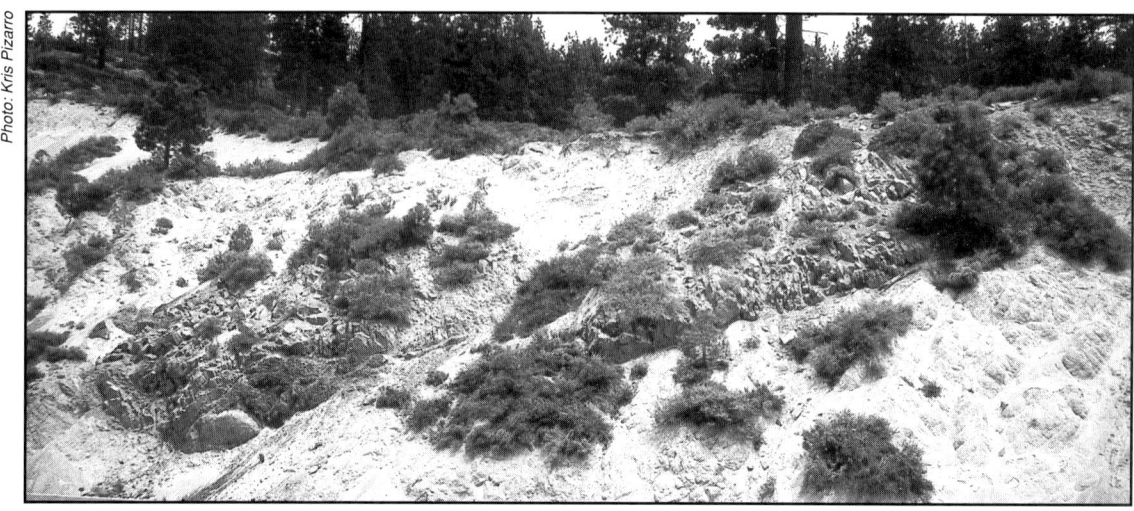

The rough-appearing, more resistant rock is an aplite dike that has intruded granitic rock. Faulting has cut and offset the dike, which dips at a low angle into the hillside.

Bitterbrush (*Purshia tridentata*) Many-branched shrub <3–12' tall; three-lobed, bright or dark green leaves; many small (½" diameter) pale yellow flowers May–June depending on elevation.

Bitterbrush is commonly found in association with sagebrush. It prefers dry, well-drained soils and is widely distributed throughout the sagebrush steppe, piñon-juniper woodlands, and open conifer forests at elevations from 4,000 to 11,000 feet. Bitterbrush is a principal browse for deer, pronghorn antelope, elk, and bighorn sheep.

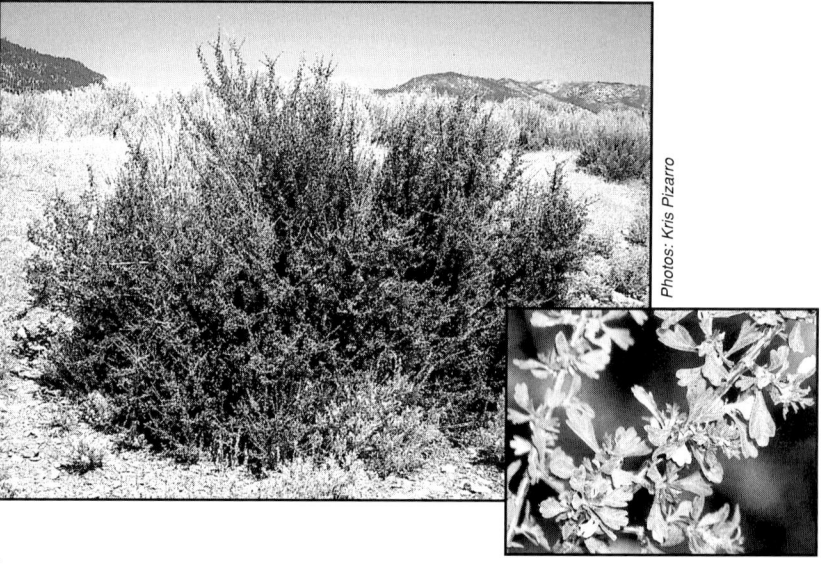

TAHOE RIM TRAIL

The Tahoe Rim trail is a hiking, equestrian, and in some sections, biking trail that circles the Tahoe Basin. It extends for 150 miles and follows ridges and mountaintops through two states. The trail is moderately difficult, passing through elevations from 6,300 feet to 9,400 feet with a 10 percent average grade. Camping is generally allowed, subject to U.S. Forest Service and Lake Tahoe Nevada State Park restrictions. The trail passes through wilderness areas, and a special Forest Service permit is required for those sections of the trail. If you have the time and energy, the trail provides access to breathtaking views of Lake Tahoe, the high Sierra, and the valleys and ranges of the westernmost Great Basin.

| 2.0 | 9.6 | |

Nevada State Railroad Museum on left. Original and beautifully restored cars and engines of the historical Virginia and Truckee Railroad (V&T) are on display, including the steam engine, Glenbrook, which hauled timber from the Glenbrook timber mills to the flumes at Spooner Summit.

The Carson City Chamber of Commerce and Visitor Center is located in a separate building at this site.

CARSON CITY

Carson City, Nevada's capital, was founded in 1858 by Abe Curry, who named it for Kit Carson, John Frémont's celebrated scout. The fortuitous discovery of the Comstock Lode in 1859 gave the city life as a freight and transportation center for the mines, and Carson City was selected as the territorial capital in 1861. The Territorial Legislature also established Carson City as the seat of Ormsby County and leased Curry's Warm Springs Hotel (east of town) as the site of the Territorial Prison. The site (considerably renovated and expanded) is still in use as one of the state's maximum security prisons. When Nevada became a state in 1864, Carson City was selected as the state's capital. In addition to the State Capitol and other government buildings, there are many historical buildings and homes in the city. The Carson City Chamber of Commerce and Visitor Center (ahead on the left at the State Railroad Museum) is a worthwhile stop where information on historical buildings, museums, walking tours—everything to see and do in Carson City—may be obtained.

Carson City is set between the Carson Range, on the left (west), the Virginia Range, straight ahead beyond the city (north), and the Pine Nut Mountains on the right (east). There is a slight complication to Carson City's location; Carson City is not located in Carson Valley! Carson Valley is the next valley to the south and is the location of the towns of Minden and Gardnerville. Carson City is in Eagle Valley, commemorating the site where one Frank Hall shot and killed an eagle in 1851.

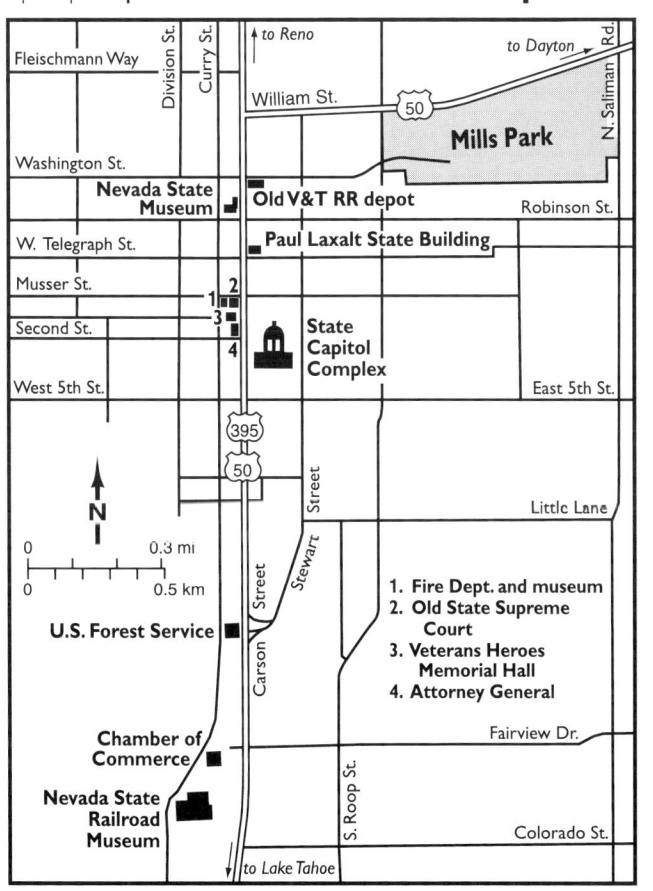

Carson City street map.

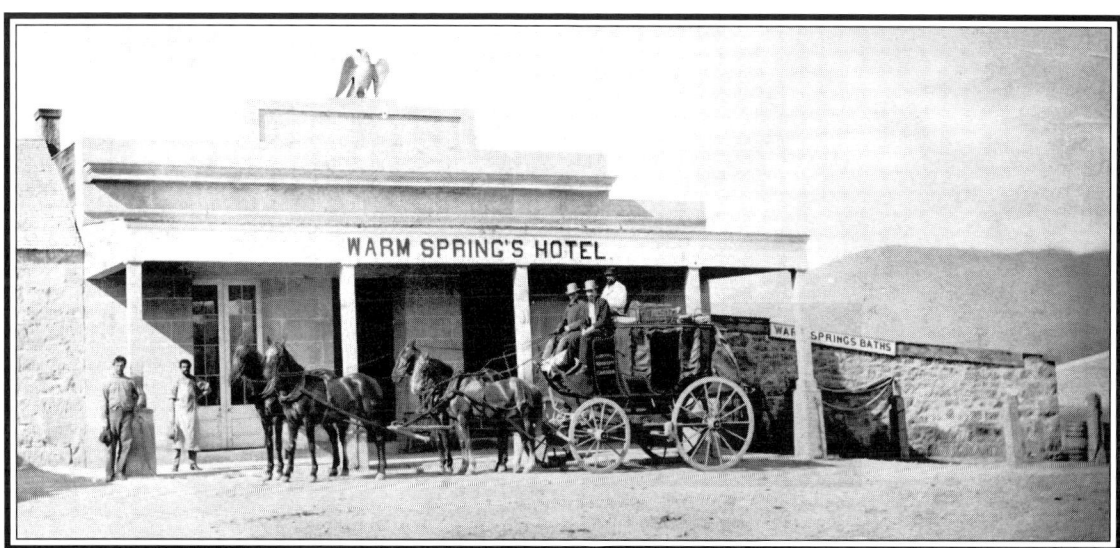

Curry's Warm Springs Hotel and bathhouse, Carson City (about 1870). The first Territorial Legislature (1861) met here in a building adjacent to the baths, possibly in the building shown in this photo.

Nevada Historical Society

VIRGINIA AND TRUCKEE RAILROAD

The V&T, perhaps Nevada's best known mining camp railroad, was built between Virginia City and Carson City in 1869. Ground was broken in February, and construction of "the crookedest railway in the United States" was completed into Carson City by November 1869. According to Smith (1943) "that peculiar adjective crookedest was appropriate to the road's torturous course and to the manner in which it was financed." It seems that Virginia City mine owners, the mill owners, and the all-powerful Bank of California (actually all more or less the same group) devised a straightforward but somewhat questionable scheme to gain complete control of the Comstock

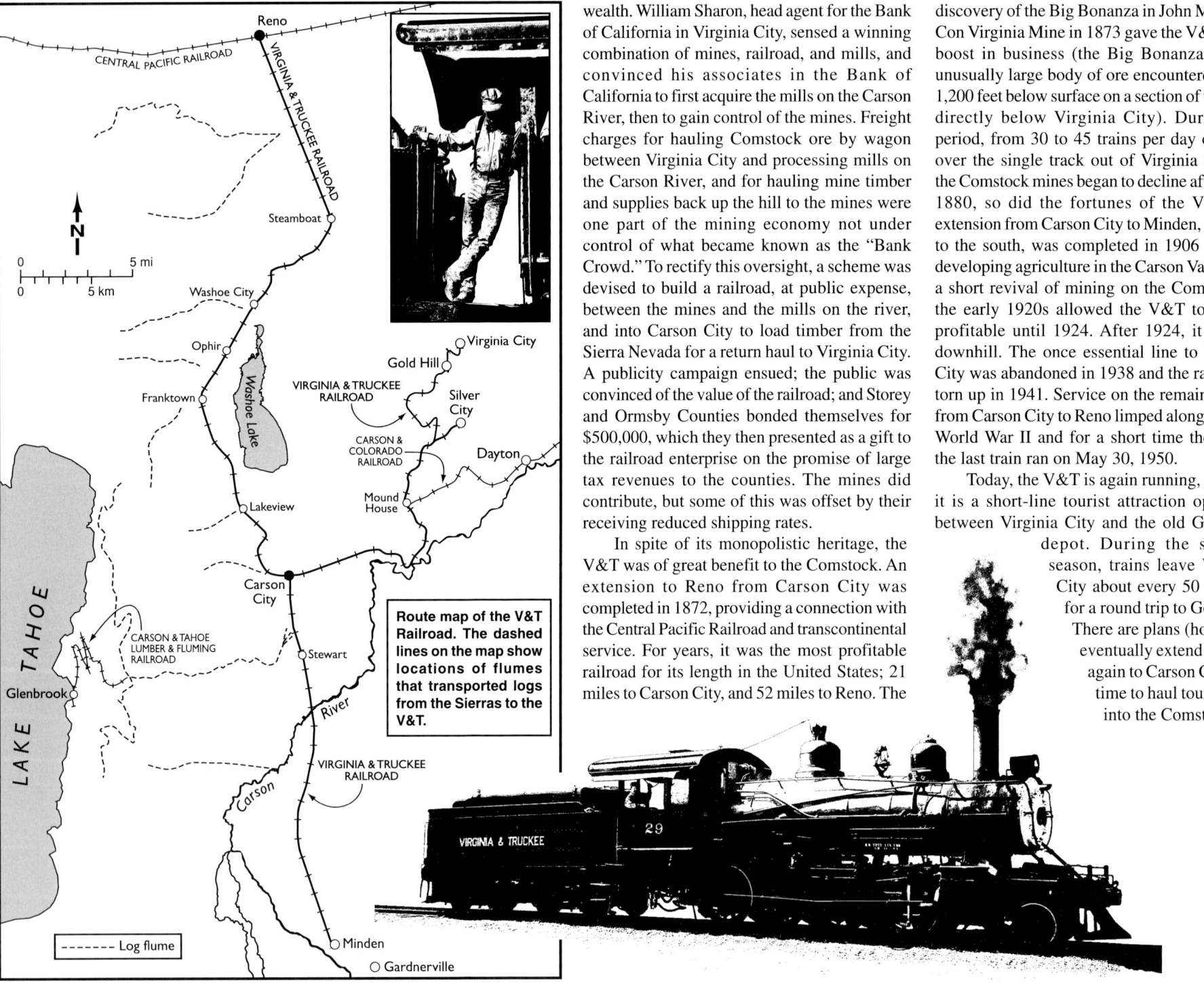

Route map of the V&T Railroad. The dashed lines on the map show locations of flumes that transported logs from the Sierras to the V&T.

wealth. William Sharon, head agent for the Bank of California in Virginia City, sensed a winning combination of mines, railroad, and mills, and convinced his associates in the Bank of California to first acquire the mills on the Carson River, then to gain control of the mines. Freight charges for hauling Comstock ore by wagon between Virginia City and processing mills on the Carson River, and for hauling mine timber and supplies back up the hill to the mines were one part of the mining economy not under control of what became known as the "Bank Crowd." To rectify this oversight, a scheme was devised to build a railroad, at public expense, between the mines and the mills on the river, and into Carson City to load timber from the Sierra Nevada for a return haul to Virginia City. A publicity campaign ensued; the public was convinced of the value of the railroad; and Storey and Ormsby Counties bonded themselves for $500,000, which they then presented as a gift to the railroad enterprise on the promise of large tax revenues to the counties. The mines did contribute, but some of this was offset by their receiving reduced shipping rates.

In spite of its monopolistic heritage, the V&T was of great benefit to the Comstock. An extension to Reno from Carson City was completed in 1872, providing a connection with the Central Pacific Railroad and transcontinental service. For years, it was the most profitable railroad for its length in the United States; 21 miles to Carson City, and 52 miles to Reno. The discovery of the Big Bonanza in John Mackay's Con Virginia Mine in 1873 gave the V&T a real boost in business (the Big Bonanza was an unusually large body of ore encountered about 1,200 feet below surface on a section of the Lode directly below Virginia City). During this period, from 30 to 45 trains per day operated over the single track out of Virginia City. As the Comstock mines began to decline after about 1880, so did the fortunes of the V&T. An extension from Carson City to Minden, 14 miles to the south, was completed in 1906 to serve developing agriculture in the Carson Valley, and a short revival of mining on the Comstock in the early 1920s allowed the V&T to remain profitable until 1924. After 1924, it was all downhill. The once essential line to Virginia City was abandoned in 1938 and the rails were torn up in 1941. Service on the remaining line from Carson City to Reno limped along through World War II and for a short time thereafter; the last train ran on May 30, 1950.

Today, the V&T is again running, but now it is a short-line tourist attraction operating between Virginia City and the old Gold Hill depot. During the summer season, trains leave Virginia City about every 50 minutes for a round trip to Gold Hill. There are plans (hopes!) to eventually extend the rails again to Carson City, this time to haul tourist gold into the Comstock.

interval	cumulative	
0.8	10.4	Intersection with 5th street.

Within the next block you will see the three main houses of government for the State of Nevada. Starting on the right with the Legislature building, which houses both the Senate and the Assembly (Senate on the north end, Assembly on the south end). Next, still on the right and a little further back, is the new Supreme Court Building. The shining dome is atop the Capitol, which houses the offices of the Governor and the Secretary of State. Directly across Carson Street is the old Supreme Court Building, which now houses the office of the Attorney General. Notice that the Capitol and some of the older buildings along this street, including the Mint Building ahead, are constructed of a buff-colored sandstone. This rock was quarried at the State Prison Quarry (Curry's old hotel site) just east of town.

0.3	10.7	

On the right is the old Federal Building, now the Paul Laxalt State Building. This vintage 1890 red brick building housed Carson City's main post office for years. It later housed the Nevada State Library and now, after undergoing renovation, is the home of the Nevada Commission on Tourism and the Nevada Magazine.

0.1	10.8	

The Carson City Mint building is on the left. Home of the famous CC mint mark familiar to coin collectors. It was here that thousands of dollars were minted in both silver and gold. Now the State Museum, the building houses an excellent replica of the underground workings of the Comstock Lode. A stop here is well worthwhile.

0.1	10.9	

On the southeast corner of Carson Street and Washington Street (just past the Texaco Station), stands the old depot for the V&T. From here, freight and passengers went east to Virginia City and the Comstock Lode or north to Reno to connect with the Central Pacific Railroad. The V&T shops and engine house were about a block east of here. The engine house, an impressive structure built of sandstone from the State Prison Quarry, was for years a familiar Carson City landmark. Unfortunately, in 1991 its owners decided liabilities outweighed historical value, and the structure was razed, leaving an empty lot. The now-homeless cut stone was said to have been hauled off to California to become part of a new "historical" winery.

0.1	11.0	

Turn right at the next light and proceed east on U.S. 50 (William St.).

CARSON RIVER MILLS

Mills to process ore from the Comstock first were built in Gold Canyon and Sixmile Canyon, near the mines. Later, as production increased and larger facilities were needed, mills were built along the Carson River to take advantage of an abundant source of water power. Near here, on the east bank of the river near the town of Empire, the first small mill was built in 1860 and later enlarged to become the Mexican Mill. Other large mills were then constructed further downstream, spurring the growth of Empire. Ore was hauled to the mills at first by wagon, later by the V&T. A 7-mile portion of the Carson River, from east of Empire extending into Lyon County toward Dayton, was virtually a continuous strip of mills and settlements, and traces of the old mills can still be seen along the river today.

At a typical 1870s Carson River mill, the ore was delivered by the V&T Railroad and dumped into large bins above the mill building. The ore was fed into batteries of stamps (large iron pestles or piston-shaped weights on rods) which crushed the ore into sand-sized particles. The Carson River provided the power to turn huge water wheels linked to a cam shaft, which caused the individual stamps in each battery to lift, then fall, crushing the rock with each blow. The pulverized ore, or pulp, was then mixed with salt, mercury, and water in steam-heated vats, called pans (this was known as the "Washoe Pan Process"). After mixing for up to 12 hours, gold and silver particles freed from the ore were collected by the mercury. The resulting substance, a gold-silver-mercury alloy known as amalgam, was collected from the bottom of the pans for retorting. Retorting consisted of placing the amalgam in a sealed retort, or still, where it was heated to drive off the mercury, leaving behind the gold and silver as a metallic sponge. The mercury was collected and reused; the gold-silver sponge was melted down and poured into iron molds forming doré bars (a gold-silver mixture) weighing about 100 pounds apiece. These bars were usually shipped directly to a mint for refining into pure gold and silver.

Unfortunately, the milling process did not recover all of the silver and gold in the ore. It is estimated that only about 75 percent of the value of the ore was recovered and the remainder, as much as 3 million ounces of gold and 64 million ounces of silver, was lost in tailings discharged into the river. Along with these tailings, 15 million pounds of mercury were also lost. All of this material, rock tailings, gold, silver, and mercury, ended up scattered along miles of the Carson River channel and floodplain below the old mill sites.

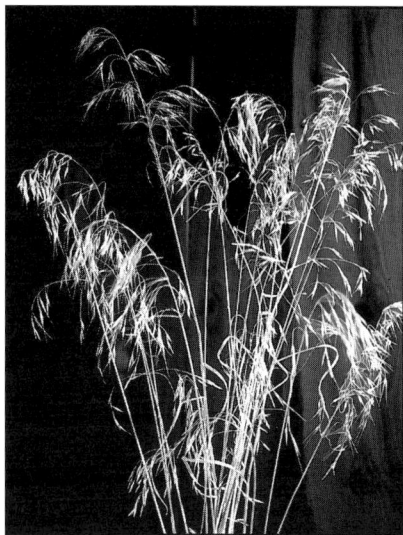

Photo: Kris Pizarro

Cheatgrass (*Bromus tectorum*) Annual; 4–30" tall; self pollinating; produces abundant seeds; very common in sagebrush steppe lands and in overgrazed and disturbed areas.

Cheatgrass was first brought into the United States in the late 1800s, probably in impure seed or as packing material. In just over 100 years, it has dispersed into all of the contiguous 48 states and most of Canada. Cheatgrass competes successfully with many native grasses and shrubs. In a typical sagebrush-grass ecosystem, disturbance such as heavy grazing allows cheatgrass to invade and spread. It germinates in the fall and establishes its root system through the winter, giving it a head start on spring-germinating plants. By mid-June it has set seed and completed its life cycle. Adaptation to fire gives cheatgrass its greatest advantage. In summer extensive stands of dry cheatgrass increase occurrence of fires. In time frequently recurring fires reduce the ability of perennial grasses and shrubs to re-establish themselves and facilitate the increasing dominance of cheatgrass.

interval	cumulative	milepost	
0.4	11.4		Mills Park. Named for Darius Ogden Mills, one of the directors of the V&T Railroad.

Along this section of highway, the crest of the Virginia Range, to the left (north), is capped by red and black basaltic cinder deposits. The cinders occupy the centers of two small volcanic vents and spill over the north and south slopes of the range. These deposits are currently being mined to supply landscaping material for the Reno and Carson City areas.

interval	cumulative	milepost	
2.7	14.1		Intersection with Arrowhead Drive on left, Deer Run Road on right. The BLM Carson City District Office is located one long block south at the intersection of Deer Run Road and Morgan Mill Road. Morgan Mill was one of several water-powered mills located on the Carson River to process worked Comstock ores.
2.5	16.6	LY 00	Lyon County line, note highway mile markers begin at zero at county line.
0.7	0.7		Mound House, on left. Historical Mound House was located 1/2 mile north of this point. Originally constructed in 1871 as a station and siding on the V&T, it served for some time simply as a wood and water stop. In 1880, when the V&T constructed the Carson & Colorado narrow gauge railroad from here to other mining camps of western Nevada and the Owens Valley region of California, Mound House became a booming shipping point.

The Southern Pacific Railroad purchased the Carson & Colorado from the V&T in 1900, just prior to the Tonopah silver strike. In 1905, Southern Pacific built a short line from its new station at Hazen, on the main line, to intersect the C&C at Fort Churchill, taking most of the booming Tonopah-Goldfield business away from the V&T. The narrow gauge line was abandoned from Mound House to Ft. Churchill in 1934. During 1900–1920, extensive gypsum mining and milling operations to produce plaster were carried out immediately northwest of Mound House (NHM #61).

Gypsum is still mined from a large open pit located in the south flank of the Virginia Range about 1½ miles northwest of the highway. The pit is at about 8:00 in the hills beyond the tall shot-tower (lead shot is manufactured here—molten lead is dropped inside the tower where it forms into spheres and solidifies as it falls, making lead shot). The gypsum is found in a small, fault-bounded block of Mesozoic-age metasedimentary rocks surrounded by granitic rocks.

interval	cumulative	milepost	
1.3	2.0	LY 2.0	Junction with State Route 341, Silver City, Gold Hill, and Virginia City are to the left.

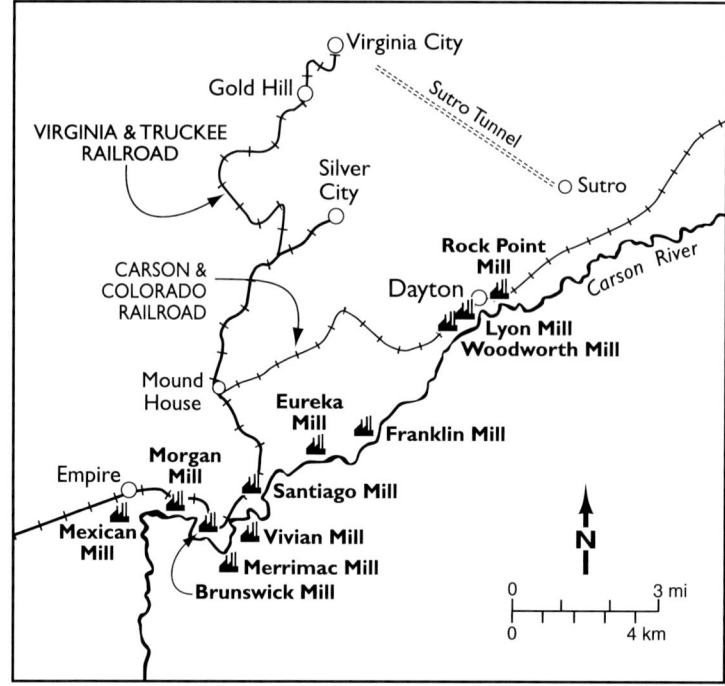

Comstock-era silver mills along the Carson River. Spur lines from the V&T connected with many of the mills near Empire and Mound House.

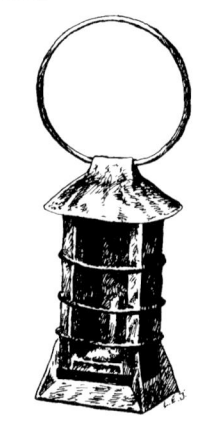

Comstock mine lantern.
artwork: Larry Jacox

SIDE TRIP 2, VIRGINIA CITY-SIXMILE CANYON

Many consider Virginia City to be the cradle of civilization in Nevada and, for sometimes rough and rowdy mining camps, Virginia City and its two Comstock neighbors really were true outposts of civilization. The short 8-mile trip up the canyon to visit the Comstock is well worthwhile. Follow State Route 341 for a couple of miles, to where it forks. Take either fork. State Route 342, to the left, winds through Silver City and Gold Hill and is probably the more scenic of the two choices; State Route 341, to the right, avoids the towns but passes by some old mines on the Occidental Lode, paradoxically east of the main Comstock Lode. Both routes merge at the south end of C Street in Virginia City.

After you get to Virginia City, spend some time walking the board sidewalks and checking out the remains of the "Queen of the Comstock." The town has its tourism facade but, underneath this, the buildings are authentic and you can get a good feel for the way it was in its heyday in the 1870s. Check out a bookstore and pick up one of the many Comstock guidebooks available, or stop at the Visitors Bureau and get information on walking tours to the historical buildings and sites about town.

Return to U.S. 50 either by retracing your path on State Routes 341 or 342, or be adventurous and follow the road down Sixmile Canyon. This road rejoins U.S. 50 near Sutro, east of Dayton. It is a narrow, slow-travel road with lots of curves and a few unpaved sections, but you can see evidence of some of the early silver mills and get a close-up look at Sugarloaf, the prominent, conical-shaped peak formed of andesite that marks Sixmile Canyon.

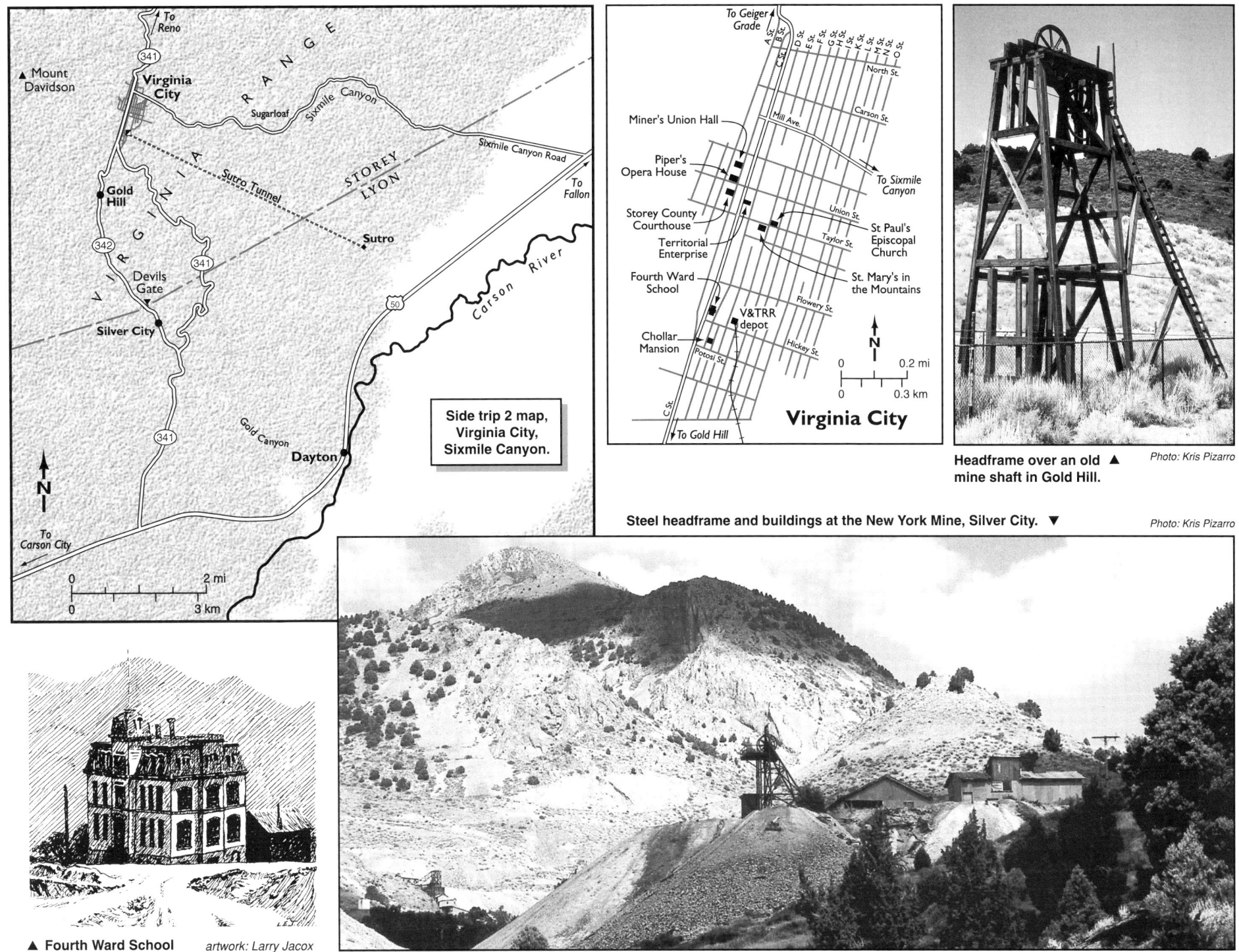

Side trip 2 map, Virginia City, Sixmile Canyon.

Virginia City

Headframe over an old ▲ mine shaft in Gold Hill.

Photo: Kris Pizarro

Steel headframe and buildings at the New York Mine, Silver City. ▼

Photo: Kris Pizarro

▲ **Fourth Ward School**   *artwork: Larry Jacox*

VIRGINIA CITY AND THE COMSTOCK LODE

Virginia City's story began in 1849 with the discovery of placer gold at the mouth of Gold Canyon. Placer miners worked the stream on and off for the next few years, following the gold upstream to its source—many small gold-bearing veins in the Silver City area. The stream was mostly barren above what is now called Devils Gate. The outcrop of the Comstock Lode at what is now Gold Hill was also quickly found, but it too was mostly barren of gold and was ignored for several years. In March or April 1859, prospectors digging alongside the Lode uncovered the top of what later was known as the "Old Red Ledge." The ore, formed in a hanging-wall split of the main Comstock Lode, was crushed and weathered, and consisted of quartz, gold, and a lot of heavy, blue-black material that turned out to be rich silver sulfide. In June 1859, a similar discovery was made a little over 1 mile to the northeast on vein croppings at what became known as the Ophir discovery site. Once the incredibly rich silver ore of the Comstock was recognized, the "Rush to Washoe" began. Virginia City became Nevada's first bonanza boom town, and the first silver-mining camp in the United States. During its main production period, from 1860 to 1880, the Comstock produced more precious metals than all of the rest of the United States. Before it finally came to a (temporary?) rest in 1986, almost $500 million in silver and gold was dug from a roughly 3-mile-long stretch of ground along the base of Mount Davidson. Many excellent books are available on the history of Virginia City and the Comstock. Dan DeQuille's *Big Bonanza*, and Elliot Lord's *Comstock Mines and Miners* are good contemporary descriptions of the boom period. Grant Smith's *The History of the Comstock Lode* is a good overall history of the Lode and its engineering works, and gives comments on the politics that operated behind the mining scene. This book has been recently updated with new sections covering the modern era of mining activity.

The valuable ore deposits of the Comstock Lode are precious-metal-bearing quartz veins. The veins occur in and along a major north-northeast-trending fault (the Comstock fault) that bounds the southeast face of Mount Davidson, the mountain that looms above Virginia City. C Street, Virginia City's main street, runs roughly parallel to the course of the vein system, and outcrops of Comstock quartz can still be seen, and even pounded on and sampled, on the hillside above town. The Lode is essentially a stockwork zone (a zone of narrow, branching and interconnecting veins) of brecciated quartz formed along the Comstock fault and in nearly vertical hanging-wall fractures connected with the main fault. The bonanza ores consisted of quartz and a little calcite along with sphalerite, galena, chalcopyrite, pyrite, and lesser amounts of argentite and gold.

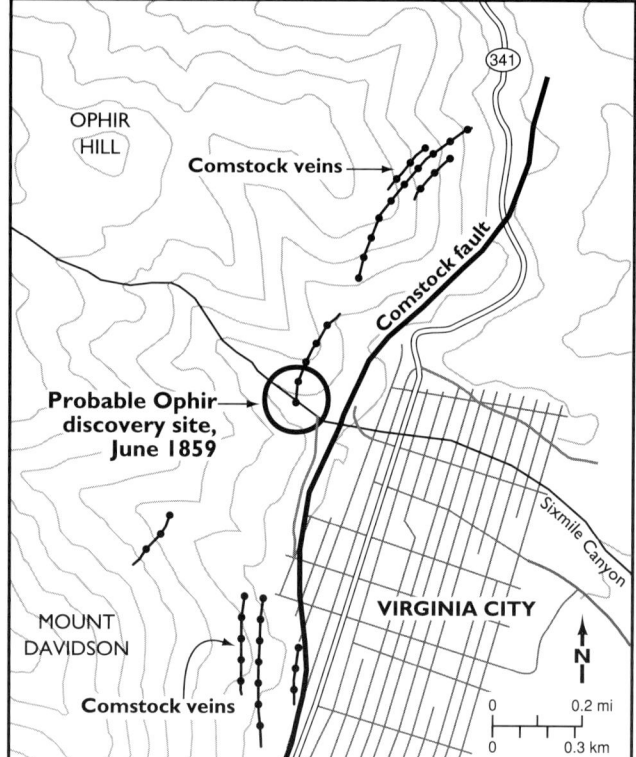

The Comstock fault and outcrops of the Comstock Lode. The Ophir discovery site, circled, is near the head of Sixmile Canyon where a segment of the Comstock vein is exposed in the canyon wall.

Artist's conception of ▲ a Comstock-era mill in Sixmile Canyon. Sugarloaf is in the background.

artwork: Larry Jacox

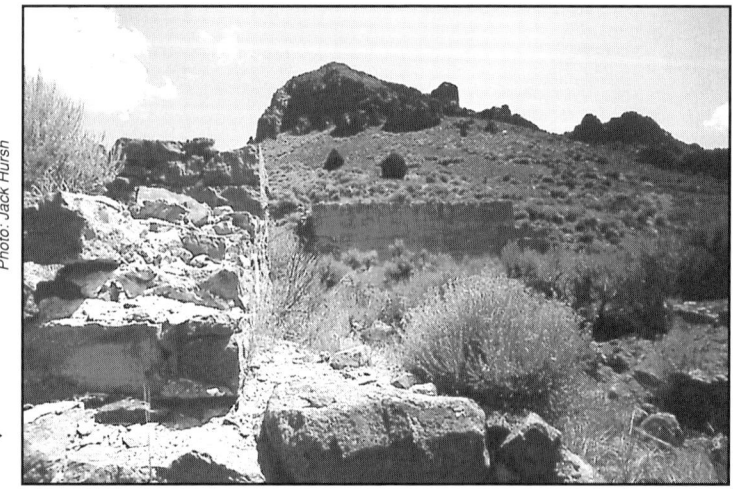

Photo: Jack Hursh

Only rock and concrete mill foundations remain at the base of Sugarloaf today. ▶

0.5	2.5	

Gold Canyon is to the left about 2 miles north of the highway at 9:00–10:00. Gold Canyon begins at Gold Hill, on the south end of the Comstock Lode, skirts along the southwest flank of the Flowery Range (in the background), and joins the Carson River at Dayton. Placer gold discovered at Dayton in 1849, at the mouth of this canyon, led to the discovery of the Comstock ten years later.

1.8	4.3	

Ditch on hillside (left) was to transport water from the Carson River to the gold workings in Dayton. This project was possibly the first use of Chinese laborers in Nevada.

0.7	5.0	LY 5.0

Entering Dayton Township. Mormon emigrants made the first discovery of gold here in 1849.

Rock Point mill, Dayton, 1862
▼

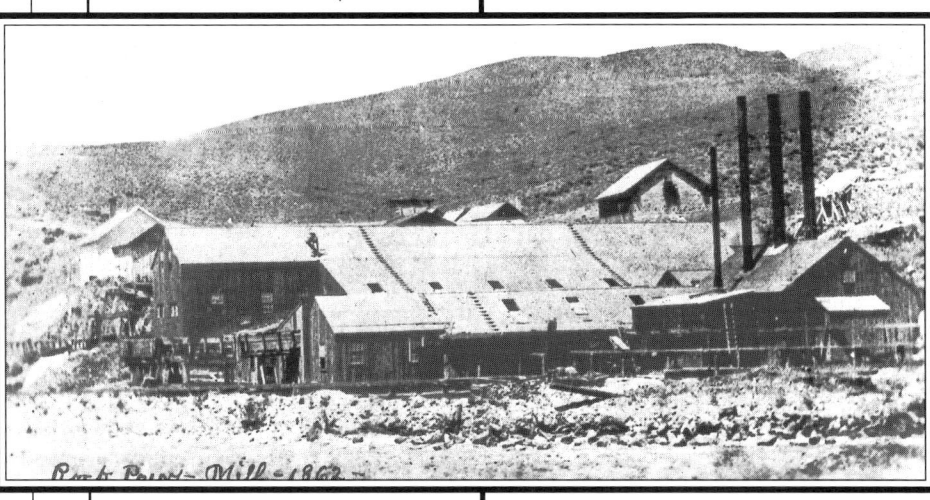

Nevada Historical Society

The Lode was mined in deep underground mines, and the ore was hoisted out through shafts sunk farther and farther to the east of the outcrop to intersect deeper and deeper parts of the lode. The first line of shafts were close to the outcrop, uphill from C Street; the second line of shafts were sunk just downhill from the center of town, along a line to the east of the site of the old V&T depot; and the deep third-line shafts were along the present east edge of town. Only the dumps of the shafts remain; with one exception, none of the surface mine workings have survived to the present (part of the headframe of the Combination Shaft is still in place).

DAYTON

Dayton might be the oldest settlement in Nevada, depending on how you define settlement. According to research by Guy Rocha, Nevada State Archivist, an emigrant diary reported a trading post at Gold Canyon (as Dayton was known following the gold discovery there in 1849) in the summer of 1850 and went on to record that the operator of the trading post, James "Old Virginny" Finney, spent the winter of 1850–1851 in a small dugout at the mouth of Gold Canyon. Mormon Station (at present-day Genoa), considered to be the first permanent structure in Nevada, was not built until 1851.

By the spring of 1851 other miners had joined Finney in the pursuit of placer gold, and by 1852 placer mining was more or less booming in Gold Canyon. A large part of this mining population was Chinese, and the community was generally known as "Chinatown." On November 3, 1861, the residents voted to rename the community "Dayton" in honor of surveyor John Day. Later that month, the Nevada territorial legislature made Dayton the Lyon County Seat.

As the Comstock boomed, so did Dayton; it became a busy commercial center serving the mines. In 1861, the Rock Point Mill was built to treat ore from the Comstock mines. Other mills were erected on the river and Dayton grew, reaching a peak population of 2,500 in 1865. The population decreased slightly after the mid-1860s, but prosperity returned during construction of the Sutro Tunnel between 1872 and 1878. During the height of milling early in the 1870s Dayton and vicinity boasted a dozen mills with 180 stamps. As activity on the Comstock declined, so did Dayton. Fires swept the town on more than one occasion. By 1882, the population had dropped to 500. In 1909, fire destroyed the courthouse. Two years later the county seat was relocated to Yerington.

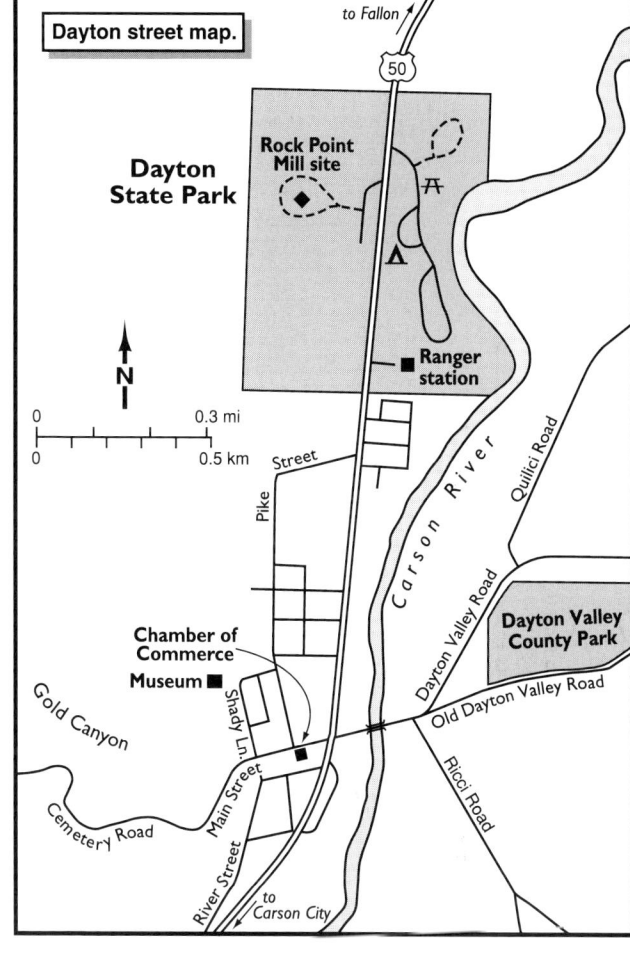

Dayton street map.

to Fallon

50

Dayton State Park

Rock Point Mill site

Ranger station

N

0   0.3 mi
0   0.5 km

Carson River

Pike Street

Quilici Road

Dayton Valley County Park

Dayton Valley Road

Old Dayton Valley Road

Chamber of Commerce Museum

Shady Ln.

Ricci Road

Gold Canyon

Cemetery Road

Main Street

River Street

to Carson City

interval	cumulative	milepost	
0.6	5.6		Main street Dayton (the traffic light). Gold Canyon can be seen as you look to your left, up Main Street and through town. Note the Carson River on the right; this must be very close to the 1849 gold discovery site—the point where water from Gold Canyon flowed into the Carson River. To visit Dayton, turn left at the light and go through downtown. Just past the Lyon County offices is NHM #200. Up the hill another ½ mile you will find the historical Dayton Cemetery.
0.7	6.3		Old Rock Point mill remains are on the left. (plate 4d) Built in 1861 to treat ore from Gold Hill, the mill was described as one of the largest of its day. The height of activity at the site was between 1861 and 1865. Rebuilt several times following fires, the mill was used intermittently into the early 1920s, when it was finally torn down and moved to Silver City. The millsite is now within the Dayton State Park.
0.2	6.5		Entrance, Dayton State Park on right. The park has both overnight and day-use picnic facilities.
1.3	7.8		On the left at about 9:00 you can make out a line of tan to gray mine dumps at the base of the hills. These are rock waste from the Sutro Tunnel which was driven from the valley into the underground workings of the Comstock mines to drain them of hot water. The tunnel opening is amidst the trees at the center of the rock piles.

Placer mining returned to Dayton about 1920, when the Gold Canyon Dredging Co. built a floating dredge, and by 1923 recovered over $300,000 in gold from 3 million yards of gravel from Gold Canyon. Placer mining continued on an intermittent basis until about 1943, but there has been little activity since that time. Large rock piles on the north edge of town are evidence of the placer mining.

Today, Dayton is a small, but again growing community. Many residents commute to Carson City and even to Reno from here, choosing to enjoy the somewhat rural setting on the river away from the bright lights of the gaming centers. Dayton did, however, gain a little bright-lights notoriety in 1960 when Marilyn Monroe and Clark Gable arrived with director John Huston to film shots for the movie "The Misfits." Some of the movies shots were filmed at the Old Corner Bar, still in business on the town's Main Street.

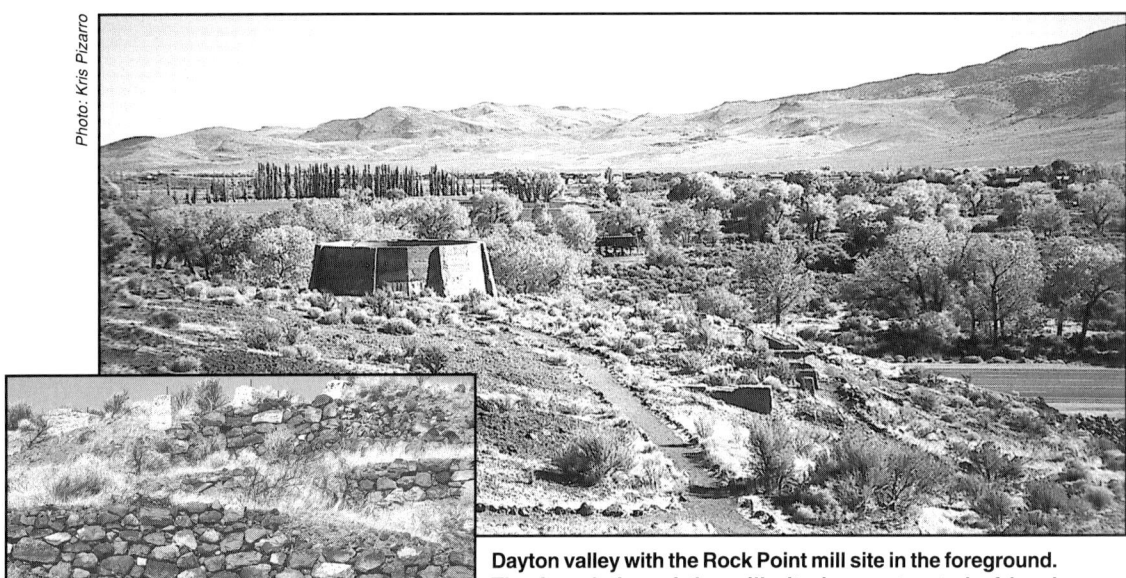

Dayton valley with the Rock Point mill site in the foreground. The foundation of the mill site is constructed of local andesitic rock.

Fremont cottonwood (*Populus fremontii*) Deciduous; grows to 100' tall with broad, rounded crown; trunk can reach up to 4' in diameter; leaves alternate, generally triangular in shape with coarse to finely rounded teeth.

The Fremont cottonwood is a pioneer species that occurs along stream sides, irrigation ditches, springs, and other sites where subsurface water is available. Large numbers of 1-mm-long seeds attached to a cottony tuft are produced in the spring and disseminated in the wind and on water. Seeds usually germinate within 24 to 48 hours on fresh alluvium laid down by receding spring flood waters. Root growth is very fast in order to ensure an adequate supply of water to a seedling as upper layers of the soil dry out. The trees may also regenerate by sprouting from stumps or by producing suckers (shoots on roots). Fremont cottonwoods adjacent to watercourses, lakes, and ponds stabilize banks, help control erosion, and provide shade and debris cover necessary for fish populations.

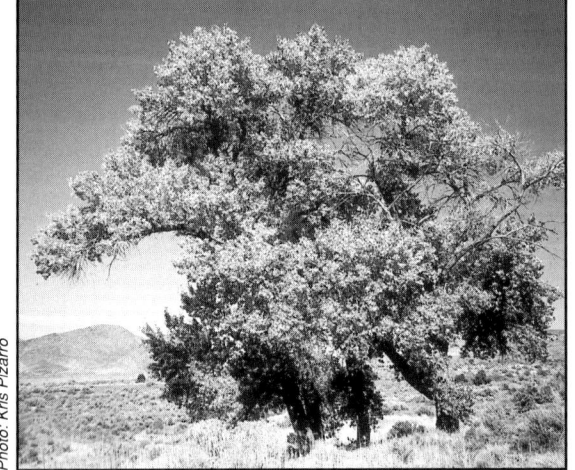

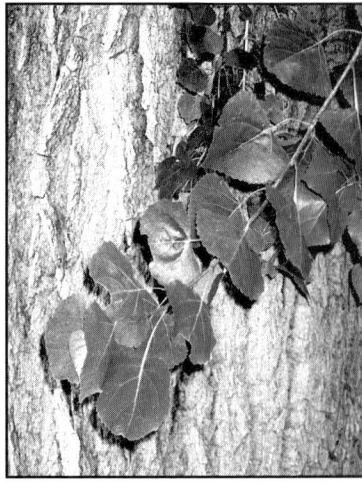

ADOLPH SUTRO AND THE SUTRO TUNNEL

Adolph Sutro was a German emigrant who arrived at Virginia City via San Francisco as a tobacco merchant. He had no formal engineering training, but a natural ability in mechanics had been honed by work in the cloth industry in Germany. Possibly, he was also somewhat familiar with tunnels driven by the Romans in Spain, and by later miners in the Harz Mountains of Germany to provide deep access to mines in those locations. Sutro sized up what was happening on the Comstock when he made his first trip to the area in March 1860. He was immediately critical of the somewhat haphazard manner in which he perceived the Comstock mines were being developed, and he conceived the idea of building a tunnel to drain the mines as well as to provide better access to mills on the Carson River.

In the fall of 1864, Sutro petitioned the state legislature for a franchise to construct a tunnel to drain the Comstock mines. A site was picked low on the Flowery Range, almost at river level. A tunnel driven to the northwest from there would intersect the lode at the 1,650-level (1,650 feet below the outcrop of the lode at Virginia City); the mines would benefit by being ventilated and drained of the constant flow of hot water, and Sutro would profit by charging the mines to haul ore out of their lower levels to mills nearby on the Carson River. There was even the possibility that new, blind lodes would be cut by the tunnel, providing even more ore to haul. The tunnel seemed to be a good idea, and at first everyone, mine owners and miners alike, was in favor of it. The Nevada legislature added its official sanction, and in 1866 the United States Congress passed the "Sutro Tunnel Act" granting Adolph Sutro the right to construct the tunnel and to collect certain royalties from the Virginia City mines that would benefit from the tunnel. Sutro, however, would have to find his own financing for the venture. This is when difficulties began; the mine owners and bankers, who by now controlled much of the action on the Comstock, feared losing some of this control to Sutro and to businesses that would grow at the tunnel mouth. Lack of support from the mine owners made it almost impossible for Sutro to find financial backing and the project stalled.

In April 1869, a fire in the Yellow Jacket Mine that claimed the lives of 37 miners rallied support of the Miners Union behind Sutro's project; the miners saw the tunnel as a safe escape route—an alternative to being trapped at the bottom of a burning shaft. The Miners Union provided a small amount of funding, and digging of the tunnel began in October 1869. All was not smooth after groundbreaking, but Sutro eventually found backing from a London bank and other European sources, and in July 1878, the tunnel intersected the 1,640-foot level of the Savage Shaft. It took several more years of work to extend laterals (connections) to various other shafts on the Lode, but Sutro's idea for orderly exploitation, drainage, and ventilation of the mines was finally a reality. Unfortunately, completion came too late for the tunnel company to reap many of the benefits Sutro visualized. By 1879, the Virginia City mines had worked out most of the bonanza ore and the mines were as much as 1,500 feet below the level of the tunnel. Had the tunnel been completed in the early 1870s, millions of dollars could have been saved in pumping costs, but water still had to be pumped the 1,500 feet to the Sutro level. Ventilation of the mines was a failure; after passing through miles of the hot, wet tunnel, the air did not help the mines much. The last dream of Sutro's, that of finding new bonanzas, did not happen either. The tunnel did cut the Brunswick lode, but only low-grade ore was found and no mining was developed on it.

Despite the outcome, the Sutro Tunnel was an amazing feat of engineering. The tunnel runs in a straight line almost 4 miles (20,498 feet) from the eastern front of the Flowery Range to the Savage Shaft. The actual cost of construction through 1878 was $3.5 million, none of which was ever paid back in royalties. Adolph Sutro, however, did not go unrewarded for his efforts. He sold his stock in the tunnel company shortly after it was finished for about $900,000 and took the money back to San Francisco where he invested in real estate. He became a multimillionaire with holdings that included the Cliff House, Sutro Baths, and all of the area now known as Sutro Heights.

Water still drains from the tunnel mouth, although it has long since become impassable to travel. A mining company attempted to reopen it in the early 1980s but became mired in the mud and clay deposited in the caved tunnel over the past 100 years and walked away in defeat. The Sutro Tunnel and Drainage Co., a successor to the original company, is still in existence, and should mining ever return to the Comstock and the tunnel is still "open," the tunnel company retains the right to assess a small royalty on any ore mined above the Sutro level.

Photo: Jack Hursh

Sutro Tunnel ▲ ore car, 1999.

Portal of the Sutro ▶ Tunnel, 1999.

Photo: Jack Hursh

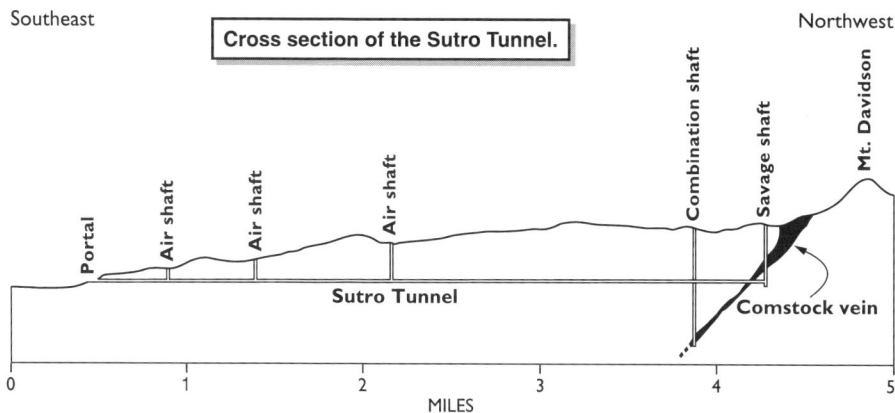

Cross section of the Sutro Tunnel.

3.8	11.6	

Sixmile Canyon road is on the left; Sixmile Canyon itself is at 7:00. This road, originally built to connect Virginia City with the emigrant routes along the Carson River and with Fort Churchill, still provides a shortcut into Virginia City. The winding mountain road is narrow with some unpaved sections. Above and beyond the canyon mouth, Virginia City's white "V" can be seen on the slope of the Virginia Range. To the right (east of U.S. 50), the road, now called Fort Churchill road, continues into the canyon cut by the Carson River across the north end of the Pine Nut Mountains and follows the river downstream about 14 miles to the site of Fort Churchill. This section of road is unpaved.

The mine visible at 10:00 provides lightweight aggregate (in this case rock, not gravel) to a Carson City firm which makes building bricks. The conical hill that hosts the mine is a Quaternary rhyolite intrusion (intruded 1.5 million years ago—this is a relatively young rock).

At about 4:00, Rawe Peak marks a high point in the north end of the Pine Nut Mountains. The old mining camp of Como lies in the hills about 2 miles south of the peak. Como, although a contemporary of Virginia City, never achieved the fame or production of its neighbor to the north.

SIDE TRIP 3, FORT CHURCHILL STATE HISTORIC PARK

The Fort Churchill road leads down river to the site of the old fort. The gravel-surface road is passable to passenger vehicles but don't be surprised if you find stretches of washboard surface. Don't travel too fast and you will enjoy the trip. Along the way, you will pass close to Susans Bluff, a rugged cliff of Quaternary basalt that forms the south face of Table Mountain and, for the entire trip, you will have peaceful views of the Carson River and old ranches on your right. Once you arrive at Fort Churchill, there is a museum and stabilized ruins of the old fort buildings to be visited. There is also a developed campground in a large grove of cottonwood trees on the river below the old fort. Just beyond the state park, the road joins U.S. Alt. 95, and from there it is only about 9 miles north to the junction with U.S. 50.

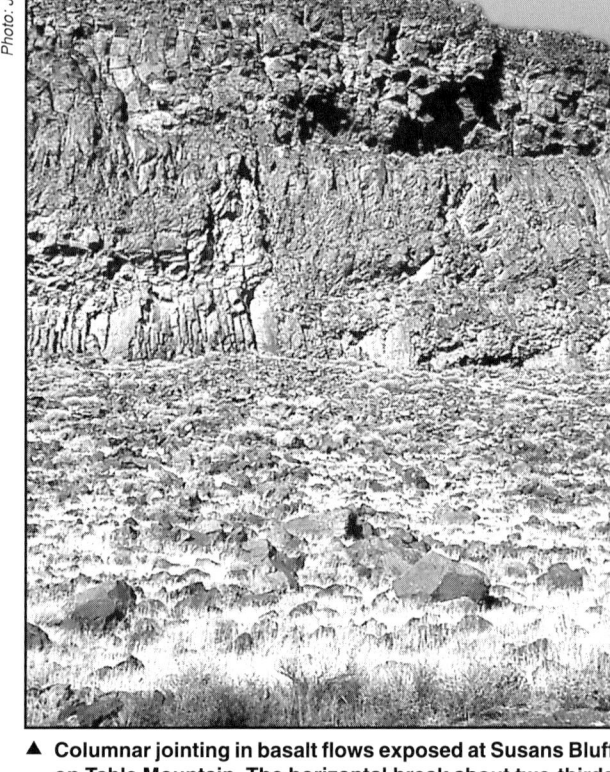

Photo: Jack Hursh

▲ Columnar jointing in basalt flows exposed at Susans Bluff on Table Mountain. The horizontal break about two-thirds of the way up the bluff is the boundary between two flows.

The Carson River on a warm winter day along Fort Churchill road. ▼

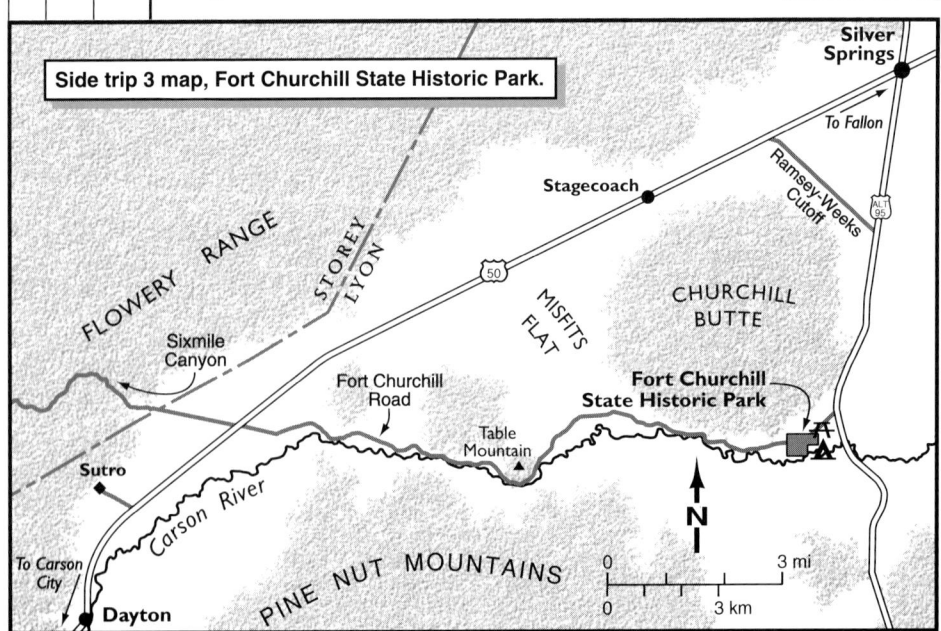

Side trip 3 map, Fort Churchill State Historic Park.

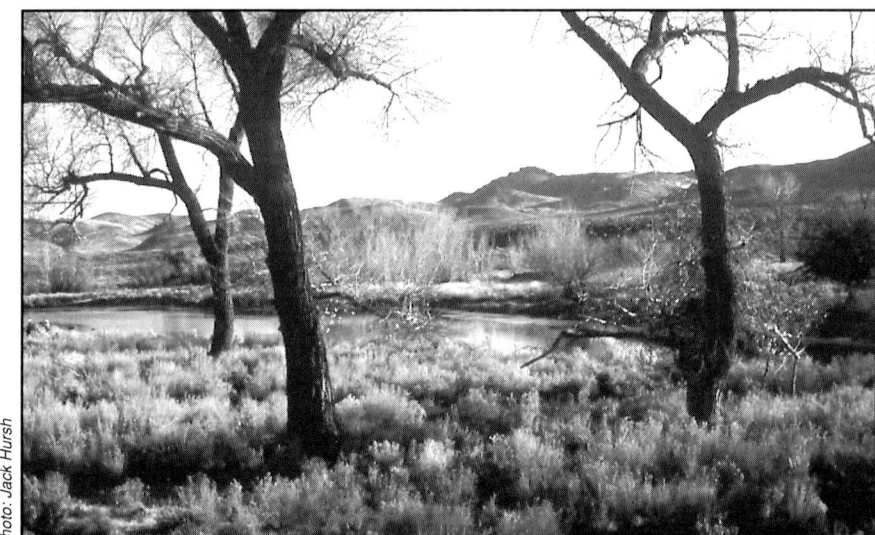

Photo: Jack Hursh

interval	cumulative	milepost	
3.4	15.0	LY 15.0	Dayton iron deposit. Although there is nothing visible except some red-stained rocks on the small hill north of the highway at about 10:00, there are about 45 million tons of iron ore buried beneath the valley and low hills to the north at this point. The deposit was discovered and first explored between 1903 and 1908, but major development drilling was not done until the 1950s. Unfortunately, the market for Nevada iron ore faded before the deposit could be developed. Housing developments in the area probably signify that the iron will remain in the ground as an untapped resource. The entrance to the area is now marked by the large "Iron Mountain Ranch" sign; the road threading through the signposts points directly to the deposit outcrop. The red-stained rocks are the surface expression of the altered zone related to the iron orebody. In the road cut on the north side of the highway, granitic rock can be seen cutting older, Triassic metamorphic rocks. The metamorphic rocks are hosts for the iron deposit to the north.
3.8	18.8		Misfits Flat. The alkali flat in the distance to the south is where portions of the movie, "The Misfits," directed by John Huston and starring Marilyn Monroe, Clark Gable, and Montgomery Cliff, was filmed in 1960.
0.3	19.1		Stagecoach, center of a sprawling bedroom community that, since the 1960s, has grown between Dayton and Silver Springs.
1.0	20.1		Churchill Butte is south of highway at 1:00 to 3:00. The west half of the butte is capped by Quaternary basalt overlying Tertiary andesite, but the lower east flanks contain one of the easternmost exposures of Mesozoic-age metamorphic rocks we will see; we are on the eastern edge of the Sierra Nevada province of granitic and metamorphic rocks, and are entering the low desert country of playa lake deposits and volcanic rocks.
5.8	25.9		Ramsey-Weeks Cutoff, a shortcut to U.S. Alt. 95 and Fort Churchill.

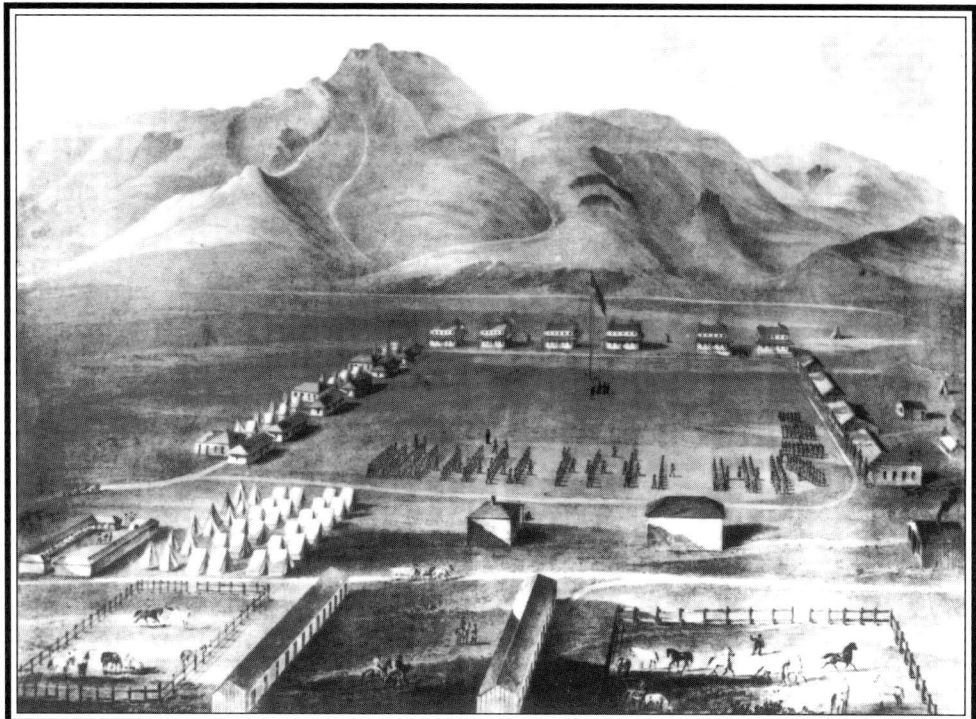

Nevada Historical Society

Fort Churchill as it appeared to an artist in the 1860s. The view is to the north, with Churchill Butte in the background.

Remains of adobe buildings at Fort Churchill, 1999.

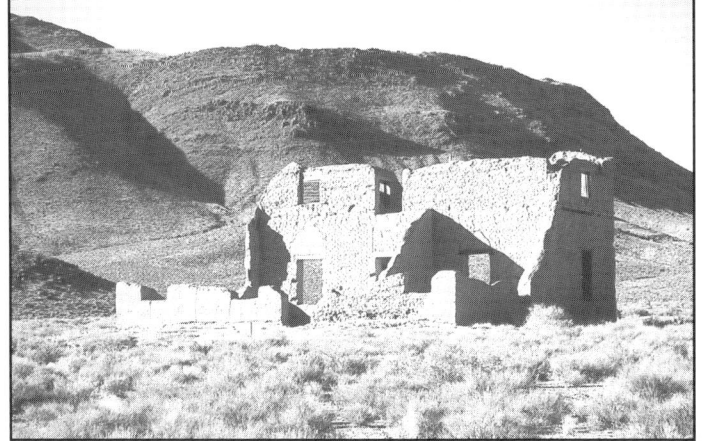

Photos: Kris Pizarro

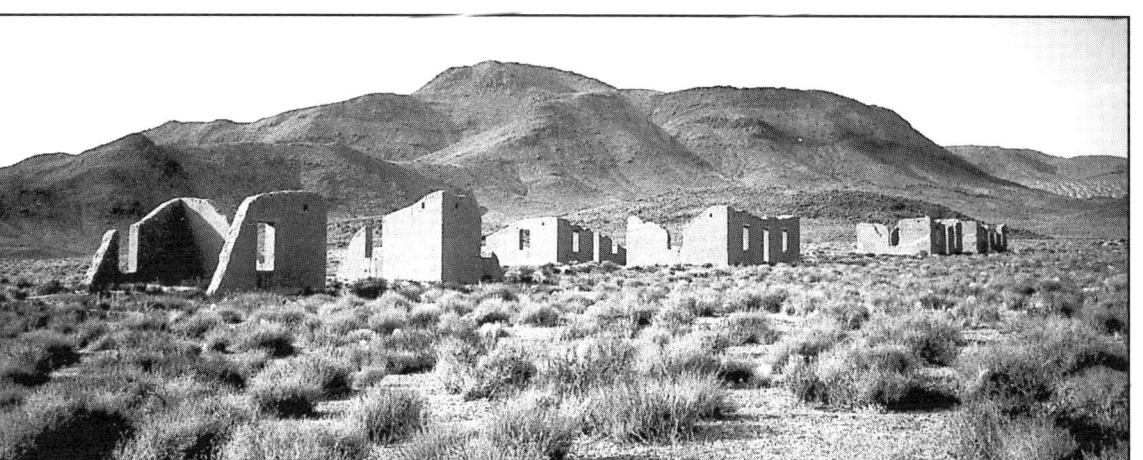

interval	cumulative	milepost
1.1	27.0	
2.4	29.4	LY 29.44

Talapoosa Mining Camp is located in the small basin at about 9:00. Discovered in 1863, deposits of gold and silver here never amounted to much. A little ore was produced in the 1930s but the district was quiet until the early 1990s, when drilling revealed a large, low-grade ore deposit. The gold price dropped about the same time plans were being made to open the mine, and Talapoosa is quiet again—for how long, no one knows. If you look closely, you can see evidence of the exploration work and old mine workings.

Silver Springs and the junction with U.S. Alt. 95.

In 1844, John Frémont camped near the present site of Silver Springs, but the present community was not established until the early 1950s. The rural lifestyle of this area attracts those who value wide-open spaces over easy access to fast-food emporiums. The community serves as a gateway to the adjacent Lahontan State Recreation Area that surrounds much of Lahontan Reservoir. There are miles of shoreline, fishing, boating, and water sports as well as camping and day-use areas. Some say that Silver Springs becomes the third largest city in northern Nevada on Memorial Day weekend when urbanites from Reno and Carson City flock to the sandy Lahontan beaches.

Fort Churchill and Fort Churchill State Historic Park are located on the Carson River about 9 miles south of here. If you took the side trip down river on the Fort Churchill road, this is where you rejoin U.S. 50.

This ends the first section of the road log. We leave the last traces of the Sierra Nevada province and move eastward into the hot lowlands of the Carson Desert.

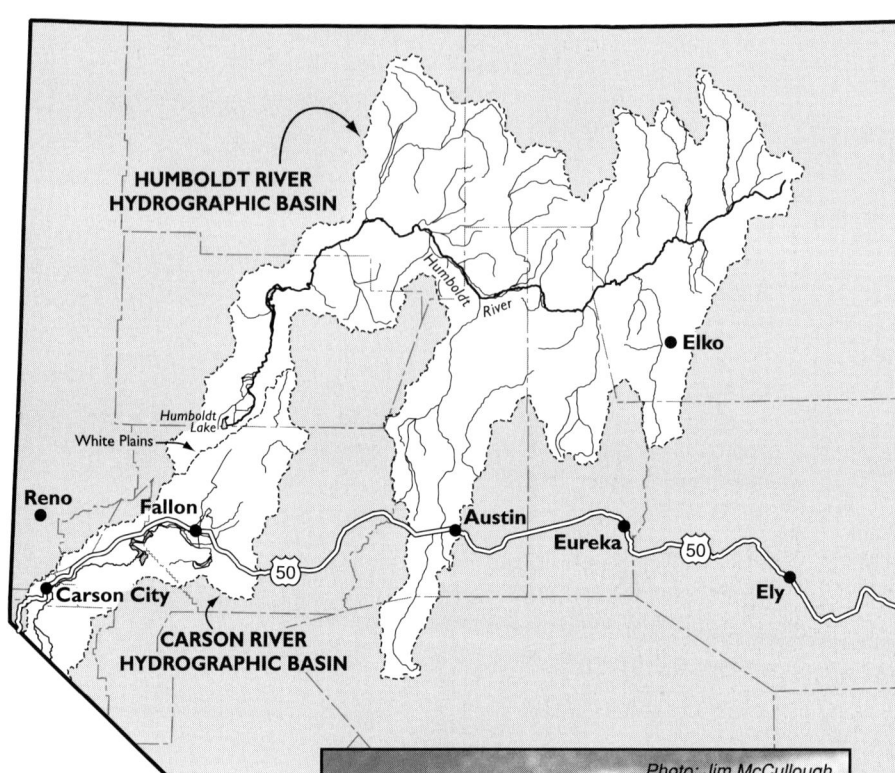

Humboldt River and Carson River hydrographic basins.

HUMBOLDT RIVER HYDROGRAPHIC BASIN

CARSON RIVER HYDROGRAPHIC BASIN

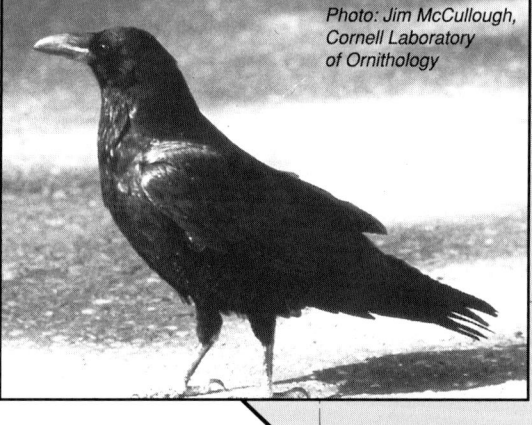

Photo: Jim McCullough, Cornell Laboratory of Ornithology

Common raven (*Corvus corax*) All black with metallic luster; heavy black bill; much larger than a crow; body 21–27" long, wingspread 45–56"; female smaller than male; has extensive vocabulary of calls and can mimic human speech; found on every continent except Antarctica.

Common ravens are conspicuous inhabitants of the Great Basin. They are usually seen one or two at a time perched on fence posts or flying along the side of the road anywhere from the hottest valleys to above the timberline. They are omnivorous and supplement a wide variety of animal food with selected plants. They hunt by sight and are more often scavenger than predator.

Ravens are amazing flyers, and loose groups can sometimes be seen engaged in dives, rolls, somersaults, and aerial combat maneuvers for no other apparent reason than fun. They mate for life and tend to remain together all year. They generally nest on cliffsides in Nevada, often in proximity to golden eagles and hawks. Young hatch in about 20 days and make their first flight in about six weeks.

Ravens have a complex social structure. They are curious and quick to learn and seem to apply reasoning to cope with entirely new situations. Not only do they profit from experience, they pass the lesson on to juveniles and other members of the group.

interval	cumulative	milepost	
6.0	35.4	CH 00	Churchill County line. Mile markers again start at zero.
0.5	0.5		Lahontan Reservoir, the body of water behind Lahontan Dam, is visible at 2:00. At least in normal or wet years it is visible; in very dry years you have to get closer to the old river channel to see water. The lake occupies part of the western arm of Pleistocene Lake Lahontan, a glacial lake that filled part of this large basin as recently as 10,000 years ago.
1.0	1.5		Wildlife viewing area

Photo: Jack Hursh

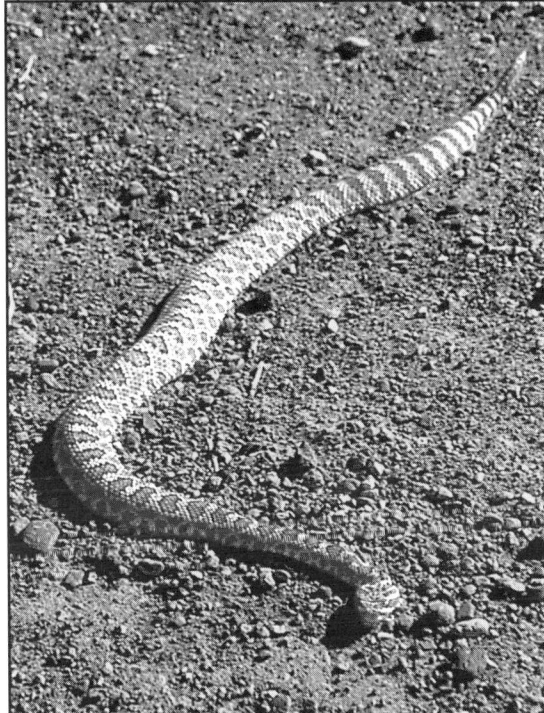

Great Basin rattlesnake (*Crotalus viridis lutosus*)

"Don't tread on me!"—This venomous resident can be found almost anywhere along the route. They are important members of the natural community. They will not attack, but if disturbed or cornered, they will defend themselves. Give them distance and respect.

SECTION II: THE CARSON DESERT

The second section of the road log, beginning at Silver Springs and extending to Frenchman Flat east of Fallon, takes us into the Carson Desert and the heartland of the Great Basin, the great interior sink of the western United States.

The Carson Desert occupies the basin surrounding Carson Sink and Carson Lake. This basin is 70 miles long in a northeastward direction and is 8 to 30 miles wide. The lowest parts of the basin are: Carson Sink (elevation about 3,870 feet), a playa 20 miles in diameter in the northern part of the basin; Carson Lake, a large, shallow lake in the southern part of the basin; and the Stillwater Marsh, a chain of small lakes, ponds, and swamps along the southeastern side of Carson Sink. Several side basins extend southeastward and southwestward from the main basin. The largest of these, Salt Wells Basin, is crossed by U.S. 50 east of Fallon.

Carson Sink is the sump for both the Humboldt and Carson Rivers. The combined drainage area of the Carson River and Humboldt River basins is 26,000 square miles.

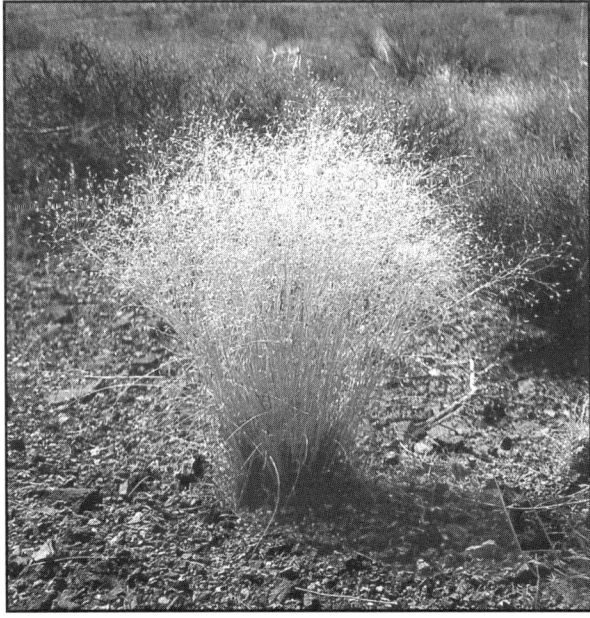

Photo: Kris Pizarro

From the north, the Humboldt River overflows Humboldt Lake only in very wet years through White Plains playa into the Carson Sink. Flow in the Humboldt River is controlled by Rye Patch Dam, about 50 miles upstream from the sink. The Carson River enters the basin at Lahontan Dam; it drains about 5,900 square miles, but most of its volume comes from only 800 square miles in the Sierra Nevada to the west. Water of the Carson River, augmented by water transported by the Truckee canal from Derby Dam on the Truckee River, is stored in Lahontan Reservoir and used to irrigate farmland surrounding Fallon. The river channel of the Carson has been modified below Lahontan Dam, but river water, plus wastewater from irrigation, still feeds Carson Lake, the Stillwater Marsh, and Carson Sink.

The Carson Desert is one of the driest and warmest parts of northern Nevada. The climate is continental in the extreme, with warm summers, cold winters, and wide fluctuations in diurnal temperatures. Summer daytime temperatures may reach 100° F and winter minimums can dive to below -5° F. The desert is in the rain shadow of the Sierra Nevada, and annual precipitation averages slightly less than 5 inches. The vegetation, which reflects the aridity, is the northern desert shrub association consisting of greasewood, shadscale, hopsage, and winterfat. *Ephedra* (Mormon tea or Indian tea) occurs locally at slightly higher elevations. Rabbitbrush is present, mainly on recently drained or disturbed ground, and desert saltgrass is found bordering the more permanent playas, ponds, and marshes. Fremont cottonwood, willow, and big sagebrush (*Artemisia tridentata*—Nevada's state flower; plates 8a to 8c) border stream courses and some lakes and ponds where the soil is relatively nonsaline. Indian ricegrass (*Oryzopsis hymenoides*—Nevada's state grass) is a common sight around Sand Mountain.

Indian ricegrass (*Oryzopsis hymenoides*) is a drought tolerant, densely tufted, perennial bunchgrass found at elevations ranging from 3,500 to 6,500 feet. Under favorable conditions it may reach heights of 18 to 24". Indian ricegrass prefers sandy soils and may form pure stands in dune areas.

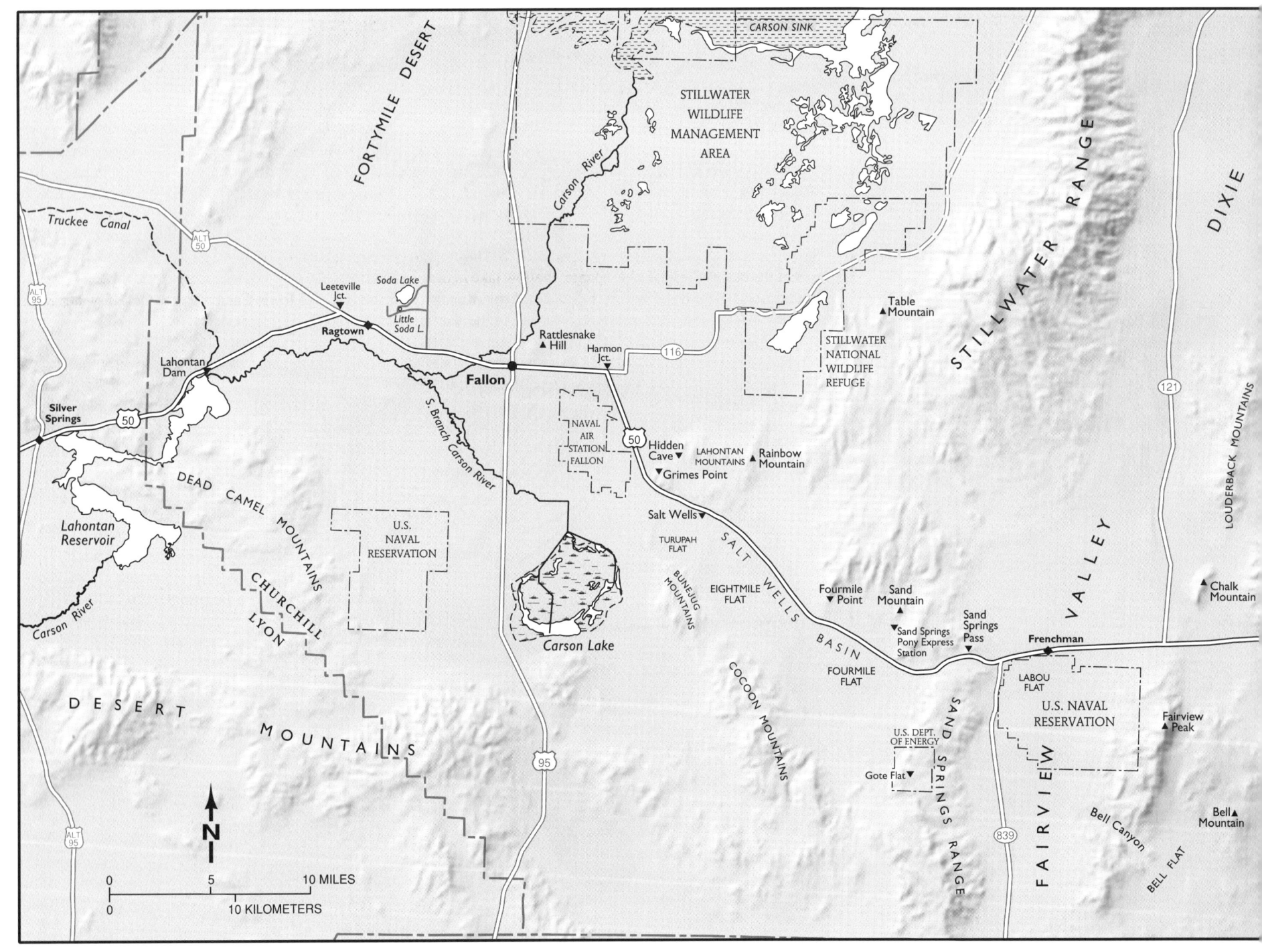

CARSON SINK

FORTYMILE DESERT

STILLWATER
WILDLIFE
MANAGEMENT
AREA

STILLWATER RANGE

DIXIE

Carson River

Truckee Canal

ALT 50

ALT 95

Leeteville Jct.

Soda Lake

Little Soda L.

Ragtown

Rattlesnake
▲ Hill

Harmon Jct.

116

Table
▲ Mountain

STILLWATER
NATIONAL
WILDLIFE
REFUGE

Lahontan
Dam

Fallon

50

Silver
Springs

50

S. Branch Carson River

Hidden
Cave ▼

LAHONTAN
MOUNTAINS

Rainbow
▲ Mountain

121

LOUDERBACK MOUNTAINS

DEAD CAMEL MOUNTAINS

CHURCHILL
LYON

U.S. NAVAL
RESERVATION

▼ Grimes Point

Salt Wells ▼

TURUPAH
FLAT

BUNEJUG
MOUNTAINS

SALT WELLS BASIN

EIGHTMILE
FLAT

Fourmile
▼ Point

Sand
▲ Mountain

Sand
Springs
Pass

Frenchman

Chalk
▲ Mountain

V
A
L
L
E
Y

Lahontan
Reservoir

Carson River

Carson Lake

COCOON MOUNTAINS

▼ Sand Springs
Pony Express
Station

FOURMILE
FLAT

LABOU
FLAT

U.S. NAVAL
RESERVATION

Fairview
▲ Peak

DESERT

MOUNTAINS

95

U.S. DEPT.
OF ENERGY

Gote Flat ▼

SAND
SPRINGS
RANGE

839

F
A
I
R
V
I
E
W

Bell Canyon

Bell ▲
Mountain

BELL FLAT

N

0   5   10 MILES

0   10 KILOMETERS

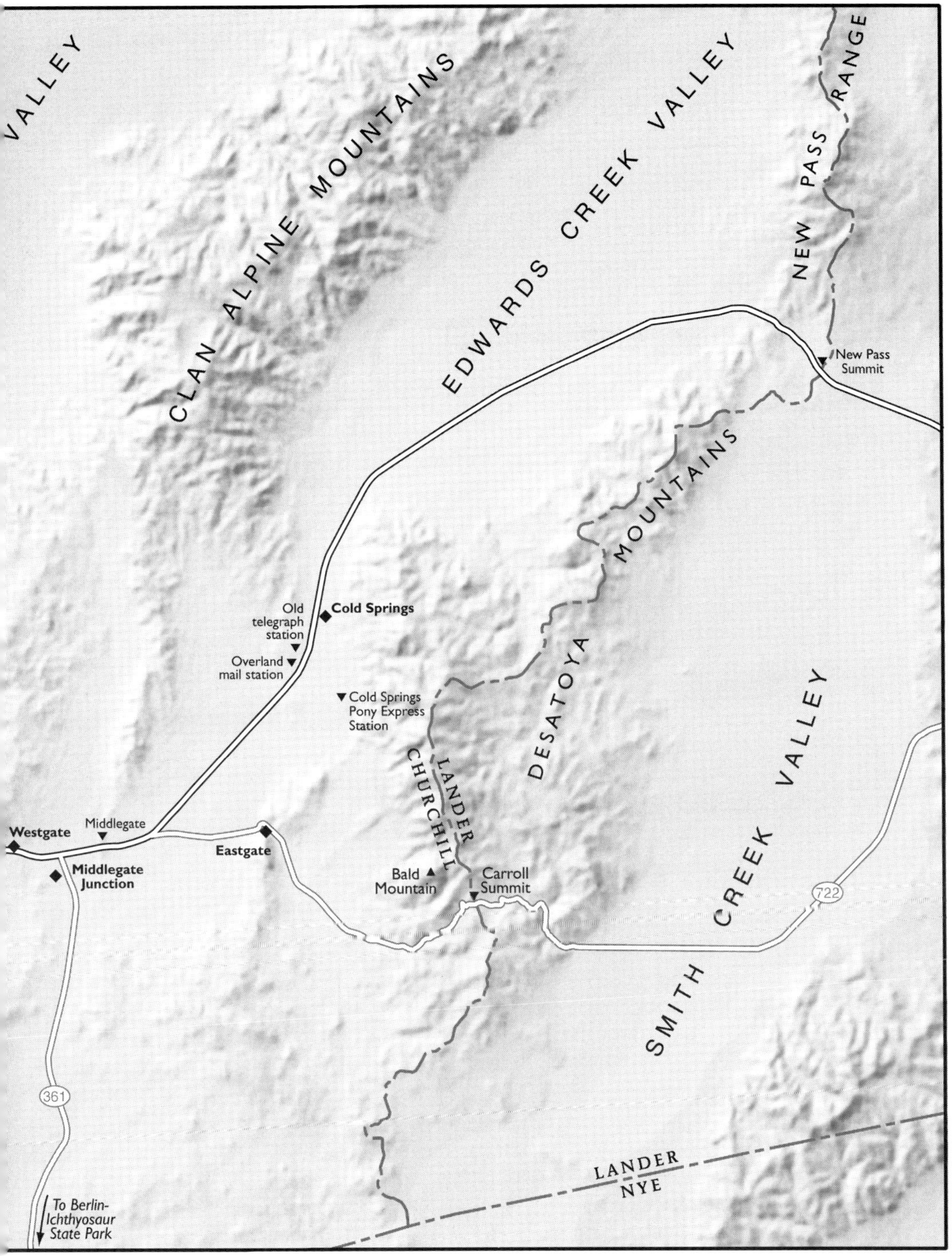

The map on the left shows a route map of Churchill County with labeled features including CLAN ALPINE MOUNTAINS, EDWARDS CREEK VALLEY, NEW PASS RANGE, DESATOYA MOUNTAINS, SMITH CREEK VALLEY, LANDER CHURCHILL, New Pass Summit, Cold Springs, Old telegraph station, Overland mail station, Cold Springs Pony Express Station, Westgate, Middlegate, Eastgate, Middlegate Junction, Bald Mountain, Carroll Summit, LANDER NYE, To Berlin-Ichthyosaur State Park, and route numbers 361 and 722.

Route map, Churchill County.

LAKE LAHONTAN

Travelers through this stretch of central Nevada need to keep the two "Lahontans" straight. Present-day Lahontan Reservoir takes its name from the extinct lake. Unless there has been a long string of dry years, water fills the lowlands to the south of the highway here and you can drive over to Lahontan Dam, see the impounded water, and if you are so inclined, even find a place to dangle your feet in it. The current lake, however, fills only a very small part of the space occupied by the glacial lake in existence here during the Pleistocene Epoch. Named for French explorer Baron de Lahontan by Clarence King during his 1870s survey of the Fortieth Parallel, ancient Lake Lahontan included the area of the present manmade lake and, through interconnections from valley to valley, extended throughout northwestern Nevada and into California and Oregon.

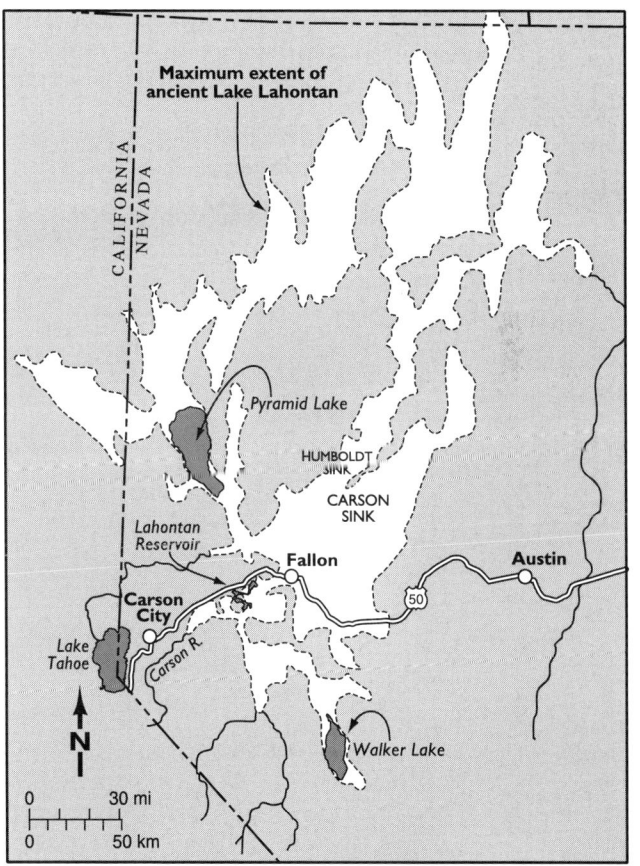

The extent of ancient Lake Lahontan within the Great Basin (after Morrison, 1964).

interval	cumulative	milepost
1.0	2.5	
1.0	3.5	CH 3.55
0.4	3.9	CH 3.9
2.0	5.9	

Lahontan Dam is visible at 2:00.

Road to Lahontan Dam.

Lahontan Dam, completed in 1915, is the key feature of the Newlands Irrigation Project, which has turned Fallon and the surrounding basin into one of Nevada's most productive farming and ranching areas. The project was one of the first authorized under the Federal Reclamation Act of 1902, and the 1903 construction contract for Derby Diversion Dam, on the Truckee River east of Reno, and the Truckee Canal was the first entered into by the U.S. Reclamation Service (later the Bureau of Reclamation). This undertaking, originally named the Truckee-Carson Project, was renamed the Newlands Project in 1919, in honor of U.S. Senator Francis G. Newlands of Nevada, an ardent supporter of federal reclamation projects and the legislation that made them possible. Operations were transferred to the Truckee-Carson Irrigation District in 1926 (NHM #215)

Crossing the Truckee Canal which brings water from the Truckee River to Lahontan Reservoir.

Note the badland topography to the left, formed by erosion of soft sands and clays of Pleistocene Lake Lahontan sediments. The small hill we are dropping down is also composed of these lake sediments.

Sediments deposited in ancient Lake Lahontan, now exposed near Lahontan Dam. Notice the layering (bedding) visible in the weathered slope.

Photo: Kris Pizarro

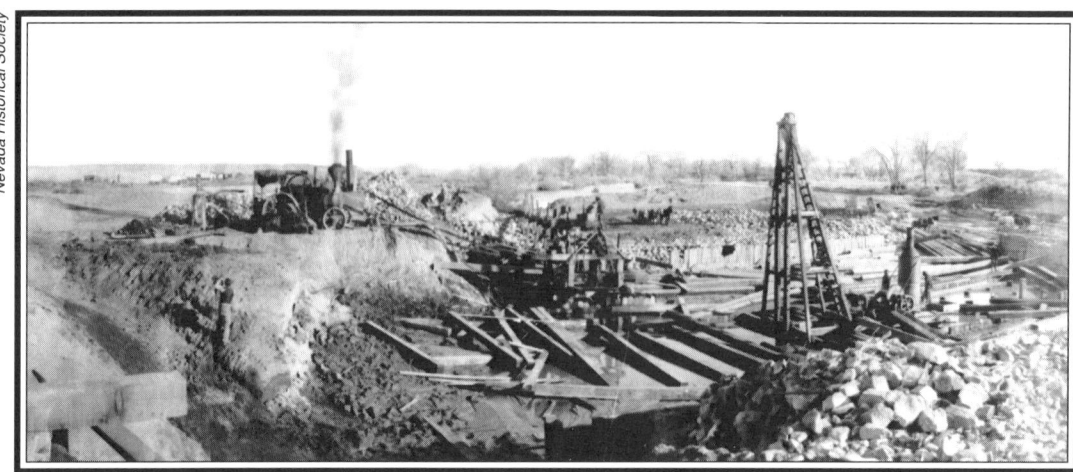

Nevada Historical Society

▲ **Truckee canal under construction, about 1910**

Lahontan Dam, 1999.

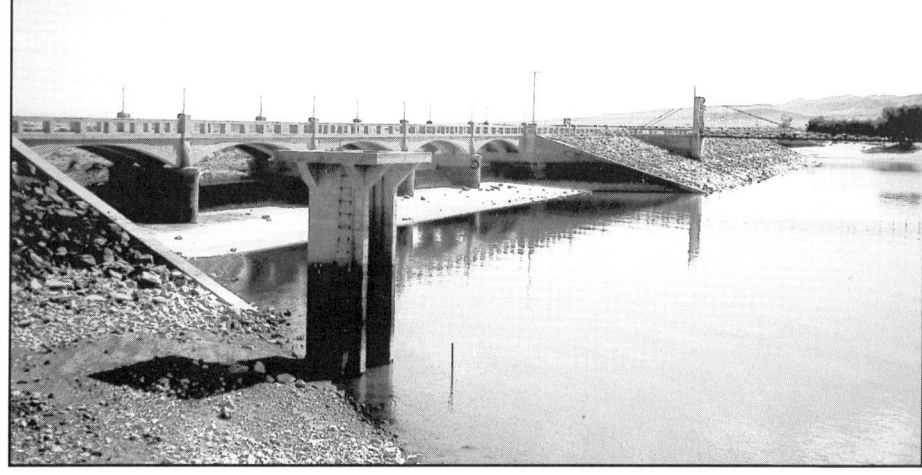

Photo: Kris Pizarro

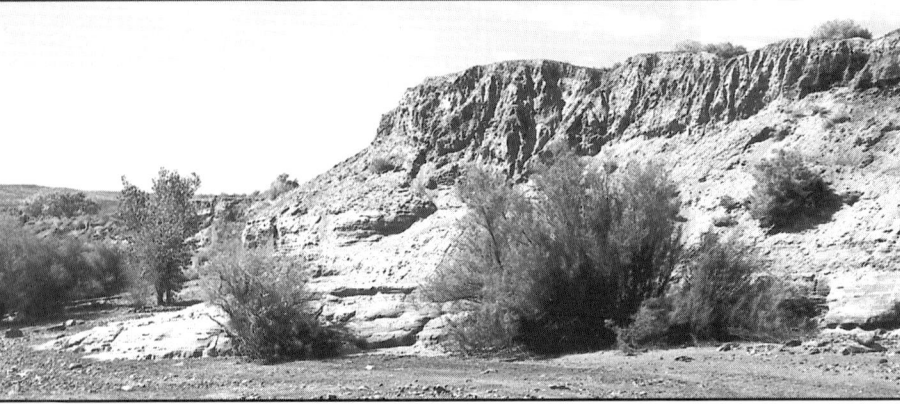

As mountain glaciers in the Sierra Nevada increased and decreased in extent due to climatic changes in the last two-thirds of the Pleistocene Epoch, the level of Lake Lahontan fluctuated. At its maximum extent during the latest glacial period, it covered an area of almost 8,500 square miles and reached a maximum depth of about 920 feet at what is now Pyramid Lake, 525 feet at Walker Lake, and 490 feet at the Carson Sink, all present-day remnants of the ancient lake. Other signs of the Pleistocene lake that remain for us to see today include extensive areas of fine-grained, powdery lake sediments, mostly silt and clay, and miles of prominent strandlines (remnant beach terraces) that rim the mountain slopes surrounding the lake's former basins. The longer the lake remained at a particular level, the more prominent the strandline. Some wave-cut terraces are as much as 300 feet wide. When the light is right, these terraces can be seen marching up most of the slopes along the highway between here and the Sand Springs Range about 50 miles ahead.

interval	cumulative	milepost	
5.1	11.0	CH 11.0	Leeteville Junction. Junction with U.S. Alt. 50.

At this point, we are on the southern margin of the Carson Sink, a major extension basin that began forming 7 to 6 million years ago. The basin is filled with about a thousand feet of late Tertiary sediments. Pleistocene Lake Lahontan sediments, mostly sand and fine clay, account for the upper several hundred feet of this section. The Carson Sink, now a large alkali flat that is the terminus of the Carson River system, extends north for over 25 miles where it merges, around the west end of the West Humboldt Range, with the Humboldt Sink, the sink of the Humboldt River system, which drains most of northeastern Nevada.

| 1.5 | 12.5 | | Ragtown, on the bank of the Carson River, was an oasis for thirsty emigrants who stopped here to recuperate after the exhausting trip over the Fortymile Desert from the last water on the Humboldt River. The Fortymile Desert was the most dreaded portion of the California Emigrant Trail, and if possible, it was traveled by night because of the great heat. Sometimes several wagon trains were camped at Ragtown at one time, recovering strength for the crossing of the Sierra Nevada ahead to the west. A burying ground near here contains more than 200 graves of early travelers who died of famine and exhaustion. The spot received its name from the common sight of pioneer laundry spread out to dry on every bush around. |

| 3.7 | 16.2 | CH 16.15 | Soda Lake Road on the left (north). |

A cluster of basalt maars (a maar is a low-relief volcanic crater that may be filled with water) with gently sloping dark cones form the low dome-like ridge about 3 miles north of the highway. The two largest maars are occupied by Soda Lake and Little Soda Lake. Both Soda Lakes are filled by groundwater. The deposits that enclose Soda Lake on three sides consist of lake sediments—sands, clays, and gravels—admixed with lesser amounts of basaltic ash and bombs that were thrown from the volcanic crater during eruption. The youngest of the eruptions that formed these maars were probably less than 7,000 years ago, making these the youngest volcanoes within Nevada. Soda Lake has a maximum depth of about 230 feet below the general surface of the desert outside of the cone. Until the early 20th century, soda (sodium carbonate) was produced from the west end of the larger lake. After an unlined canal from the Newlands project was dug around the northern end of the cone containing the lake, leakage from the canal caused the water table to rise by about 60 feet, so that the old buildings and evaporation ponds are now completely under water. With dilution, the concentration of soda in the water in the lake has decreased.

The Soda Lakes have the reputation of being the most exciting places for birding in the Lahontan Valley. The lakes are surrounded by arid desert sand hills and are a magnet for vagrant shorebirds, gulls, and a list of unusual birds including Pacific loon, oldsquaw, surf scoter, white-winged scoter, curlew sandpiper, golden plover, and Sabine's gull. The viewing areas are reached by traveling about 2 miles north on Soda Lake Road, then turning left on Cox road for 1.3 miles and finally turning right on a short stretch of dirt road to the lakes.

| 2.3 | 18.5 | CH 18.54 | Cross the Carson River. |
| 2.1 | 20.6 | CH 20.64 | Maine Street, downtown Fallon. |

FALLON

Fallon, county seat of Churchill County, is the market center for the surrounding stock-raising and irrigation-farming area. Fallon's population has increased substantially in the last few years, mostly due to the expansion of the nearby Naval Air Station Fallon, long an important center for Naval pilot training. In 1996, the Navy merged its three aviation programs, "Strike," "Topgun," and "Topdome," into the Naval Strike and Air Warfare Center, headquartered here. The most elite of Navy pilots now receive combat training here in the middle of the Nevada desert.

The old Churchill County Courthouse, constructed in 1903, is located on the northwest corner of Maine and U.S. 50. New quarters for the court house are located immediately north of the old building.

The Churchill County Museum, well worth a side trip, is located 0.7 mile south of this intersection, at 1050 S. Maine. Tours of the Bureau of Land Management's Hidden Cave leave this museum every other weekend during the summer months (description in following section).

If your timing is right, say late August into mid-September, stop in Fallon and find a stand selling Heart O'Gold cantaloupe. Once you taste one of these Fallon specialties, you'll swear off the standard supermarket melon forever.

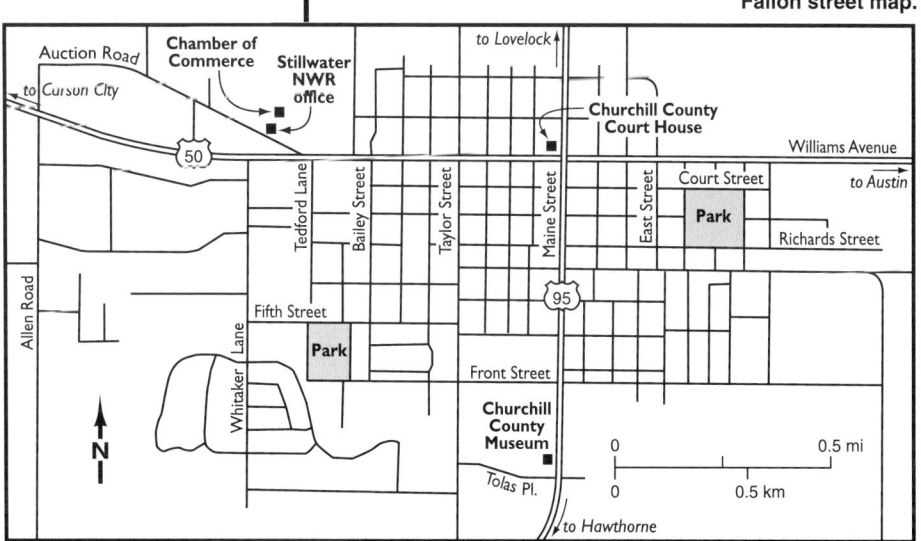

Fallon street map.

SIDE TRIP 4, STILLWATER NATIONAL WILDLIFE REFUGE

Follow the "watchable wildlife" signs to the refuge entrance. There are no structural facilities on the refuge, but there is a self-guided tour route and what will seem like an unlimited amount of area to explore. Major things to see are large numbers of waterfowl (including tundra swan), shorebirds, American white pelicans, a large dune complex, and in good years about 15,000 to 30,000 acres of wetlands. Since this is a big area, and the birding conditions vary greatly from year to year depending on water conditions, it might save a lot of fruitless driving to visit or phone the refuge office in Fallon before planning a side trip into the refuge (address: 1000 Auction Road, Fallon; phone: 775-423-5128) and ask about current "watchable" areas. During wet weather, some roads in the refuge become impassable—another good reason to check with the refuge office before heading into the area.

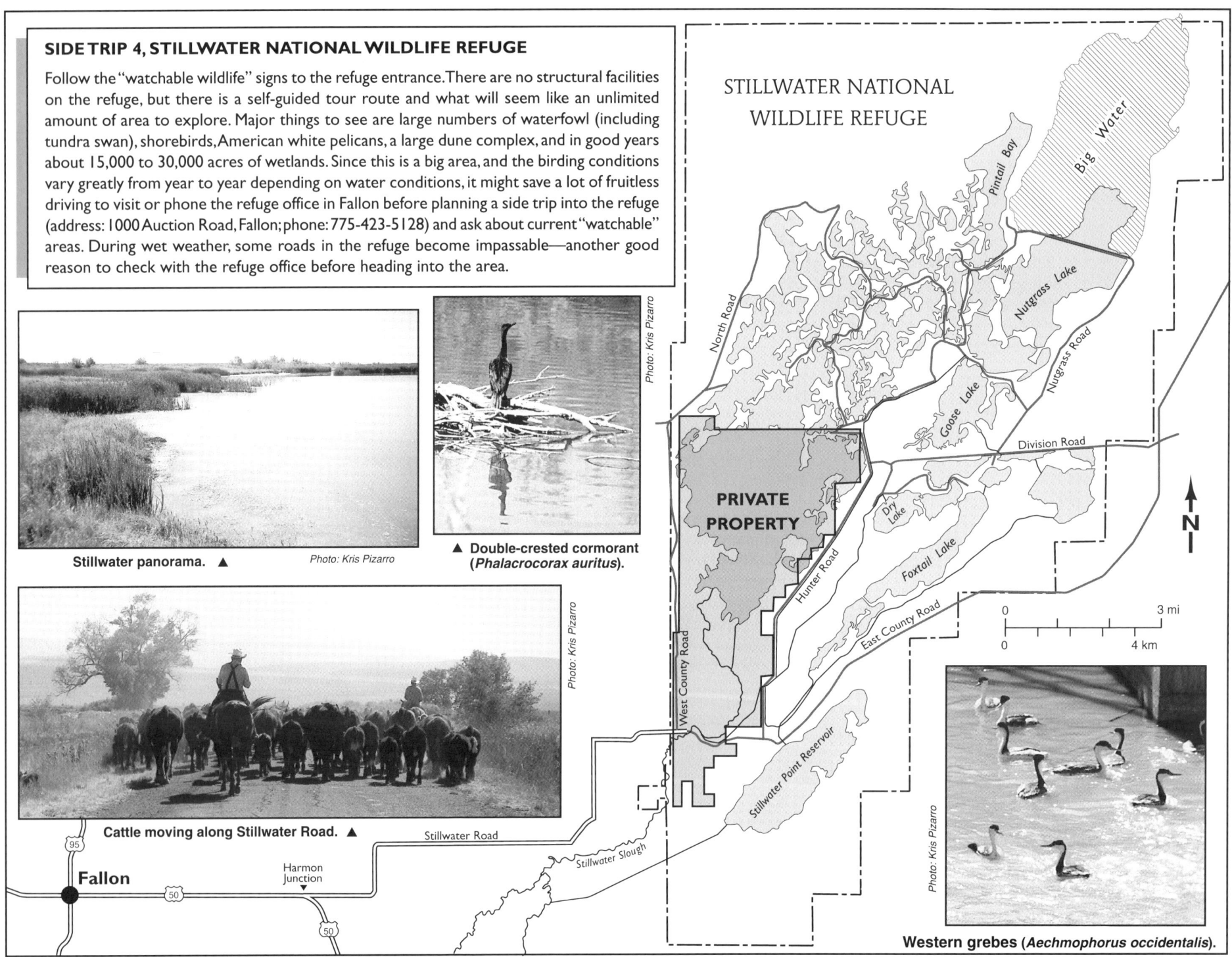

Stillwater panorama. ▲

Photo: Kris Pizarro

▲ Double-crested cormorant (*Phalacrocorax auritus*).

Photo: Kris Pizarro

Cattle moving along Stillwater Road. ▲

Photo: Kris Pizarro

STILLWATER NATIONAL WILDLIFE REFUGE

Pintail Bay

Big Water

Nutgrass Lake

Nutgrass Road

North Road

Goose Lake

Division Road

PRIVATE PROPERTY

Dry Lake

Hunter Road

Foxtail Lake

East County Road

West County Road

0 — 3 mi
0 — 4 km

Stillwater Point Reservoir

Stillwater Road

Harmon Junction

Fallon
95
50
50

Stillwater Slough

Photo: Kris Pizarro

Western grebes (*Aechmophorus occidentalis*).

0.6	21.2		Rattlesnake Hill, at 10:00 about 1 mile north of the highway, is composed of black to dark gray basalt flows formed around a shallow summit crater. Note the white "F" for Fallon on the hill.
4.1	25.3	CH 25.28	Harmon Junction, intersection with State Route 116. This road leads north to the Stillwater National Wildlife Refuge, a good side trip for birding.
1.7	27.0	CH 27.0	Naval Air Station Fallon. South of the highway at 3:00 is the runway and other facilities of the Naval Air Station. Planes can sometimes be seen taking off from here and heading out to training ranges to the south or east (we will cross one of the active ranges a few miles east of here). At 9:00 to 11:00, the lower hills are the Lahontan Mountains composed of basalts and basaltic andesites, with the flat-topped Table Mountain and the Stillwater Range in the distance. Note the plastered appearance of some of the basalts due to the deposition of Pleistocene Lake Lahontan tufa, an algal-chemical deposit of calcium carbonate. A number of Lahontan recessional shorelines are visible along the face of the hills.
3.9	30.9	CH 31.0	Grimes Point and Hidden Cave archeological sites. (mile marker CH 31.0 is just beyond—east of—the turnoff to Grimes Point)

GRIMES POINT

Grimes Point, one of the largest and most accessible petroglyph sites in northern Nevada, contains about 150 basalt boulders covered with petroglyphs. Nevada petroglyphs apparently were of religious significance in insuring the success of large game hunts and were located near seasonal game migration routes. Running east and west along the ridge, on the hill above the petroglyphs, there is evidence of an aboriginal drift fence for driving deer or antelope. The petroglyphs probably date from between 5000 BC and 1500 AD (NHM # 27).

Grimes Point Petroglyph Trail—The Bureau of Land Management has developed a 700-yard handicapped-accessible trail, winding through boulders with petroglyphs carved by prehistoric Native Americans. A Pleistocene lake once surrounded the point where animals came to drink, and human hunters waited in ambush. The oldest petroglyph style in the Great Basin, "Pit and Groove," can be found here along with curvilinear and rectilinear abstract geometric designs. Interpretive signs are placed along the trail explaining the rock art, prehistoric culture, and landscape. Photographs are encouraged, but touching or rubbing the petroglyphs will damage them. Restrooms and picnic tables have been provided at the trailhead by the BLM for visitors.

Hidden Cave—Hidden Cave was discovered in the late 1930s, and has been excavated by archaeologists three times since 1940. The cave was used as a storage site by the Lovelock Culture between 3,800 and 3,600 years ago, to safeguard their excess belongings and food supplies (people of the Lovelock Culture inhabited the area prior to the Paiutes, the inhabitants when European settlers arrived). After the last excavation in 1980, the Bureau of Land Management secured the cave behind a locked door and began offering tours to the public twice each month. On the second and fourth Saturdays of every month, the public is welcome to enter the cave with a guide and view the excavation trenches, exposed artifacts, and undisturbed soil layers dating back 20,000 years. Tours start at the Churchill County Museum in Fallon at 9:30 a.m. or at the Hidden Cave Trailhead at 10:15 a.m. on scheduled tour days.

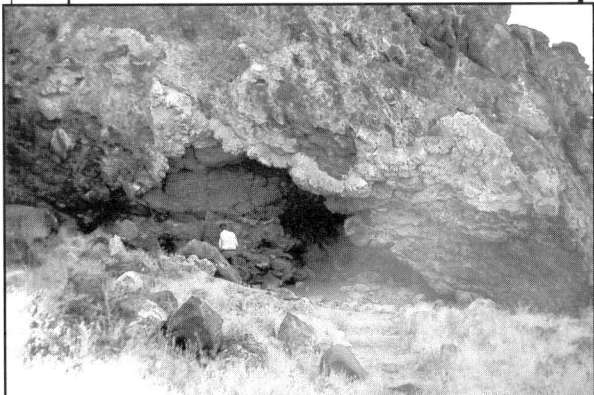

Photo: Kris Pizarro

Picnic Cave at Grimes Point. Note the tufa deposits draping the entrance to the cave.

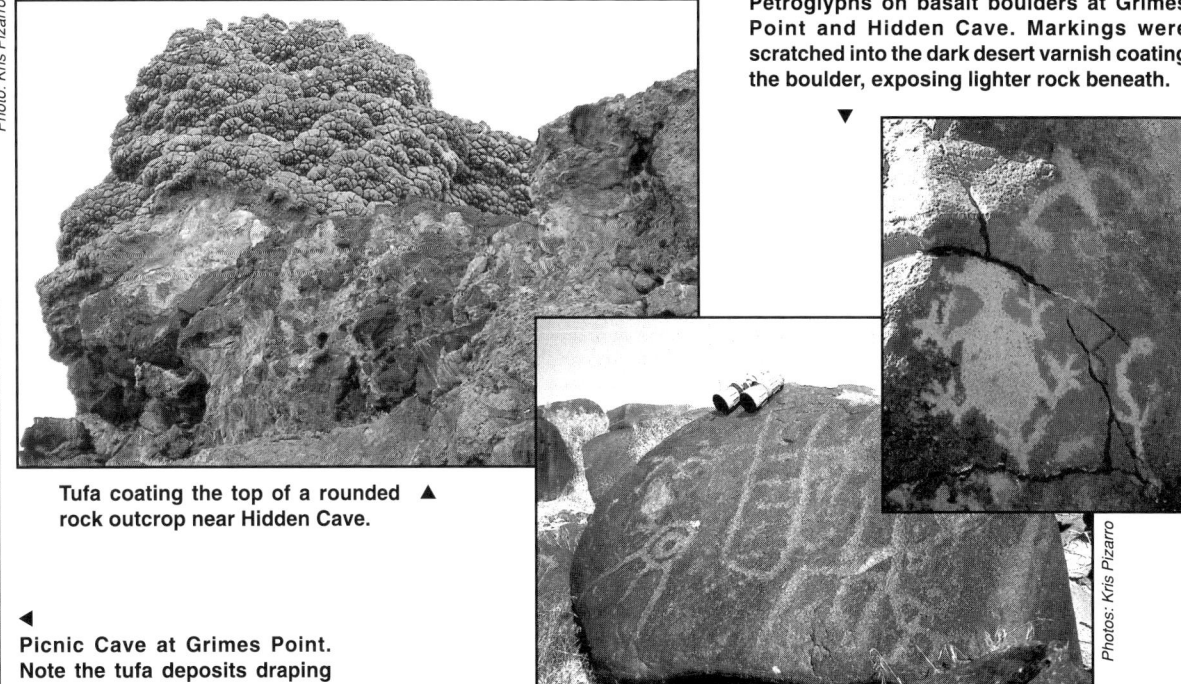

Photo: Kris Pizarro

Tufa coating the top of a rounded ▲ rock outcrop near Hidden Cave.

Petroglyphs on basalt boulders at Grimes Point and Hidden Cave. Markings were scratched into the dark desert varnish coating the boulder, exposing lighter rock beneath.
▼

Photos: Kris Pizarro

DESERT PAVEMENT

This is a good place to comment on what is called "desert pavement," the speckled, stone-littered surface that is common to flat lowlands and playa borders throughout the deserts of the west. Caused by the interaction of wind and water with silty and sandy sediments, the fine particles become swept away, leaving behind stones and pebbles. Called "deflation," because it more or less cuts the original sediment down to size, the process continues until a solid surface of stones caps the soft sediment. Now covered like some gigantic designer-built patio, the desert is paved and the process stops. Once the paving is in place, however, nature continues to add comment; the rocks in the paving become covered with desert varnish, a coating of dark minerals deposited by repeated wetting and drying cycles. Over long periods of buffeting by wind-blown sand, the stones even become faceted, allowing geologists to talk at length about how the number of faces on the stones indicates changes in wind direction and intensity.

Good examples of desert pavement can be seen along the road in many places over the next few miles. Desert varnish accounts for the almost black surface you will note on many rock outcrops in this area.

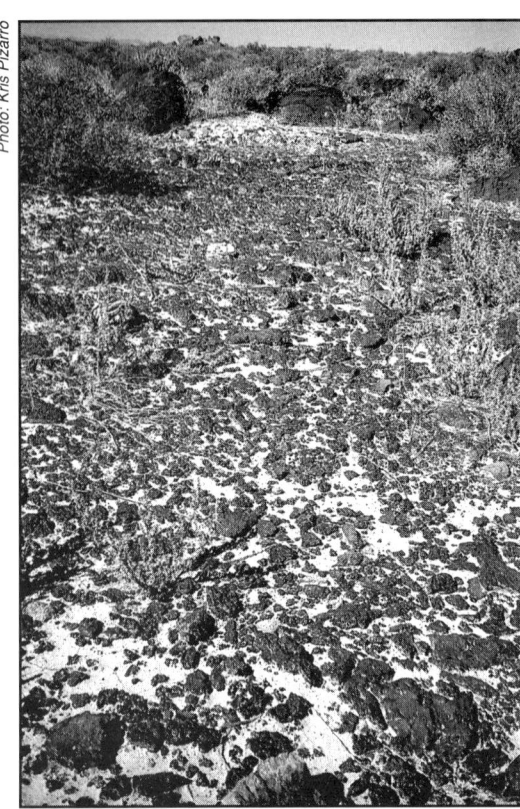

Photo: Kris Pizarro

◀ **Established desert pavement in an area near Sand Mountain.**

Short-horned lizard (*Phrynosoma platyrhinos*)—active during the day and burrows into the ground at night. It feeds primarily on ants. ▼

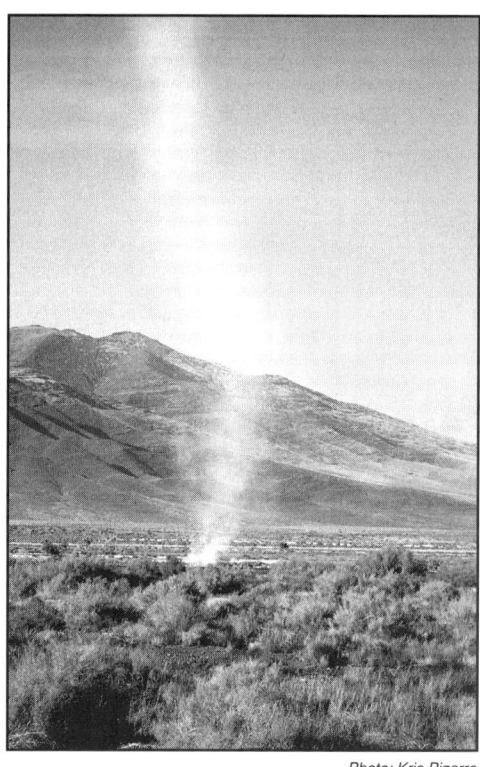

Photo: Kris Pizarro

Dust devils—resemble small tornadoes, but they're not connected to thunderstorms. They form on sunny hot days with light winds and are caused by strong convection currents rising from heated ground into cooler air above. The columns can rotate either clockwise or counterclockwise, with wind speeds that can reach 50 miles an hour or more. Most dissipate after a short time, but in exceptional cases, dust devils have spiraled to over 5,000 feet and lifted as much as 50 tons of dust and debris into the air.

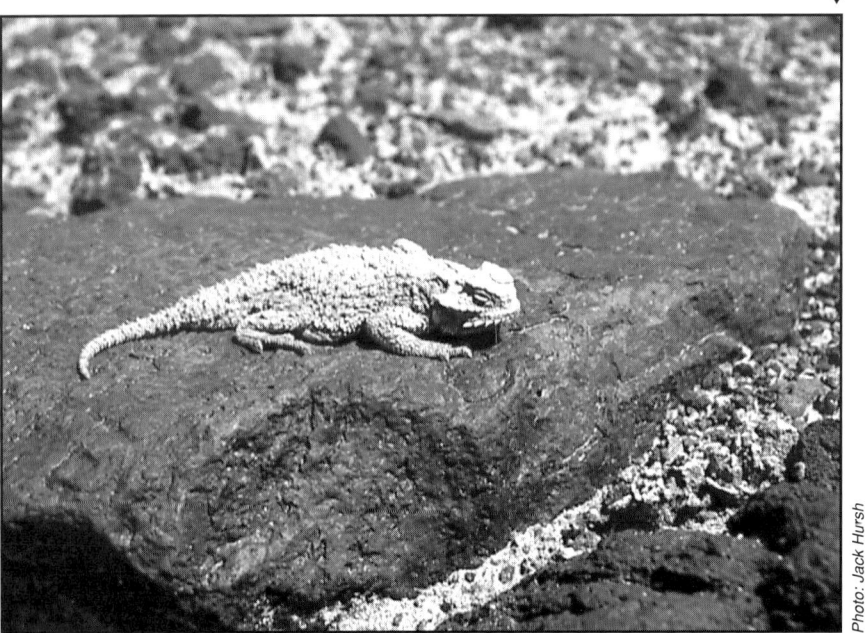

Photo: Jack Hursh

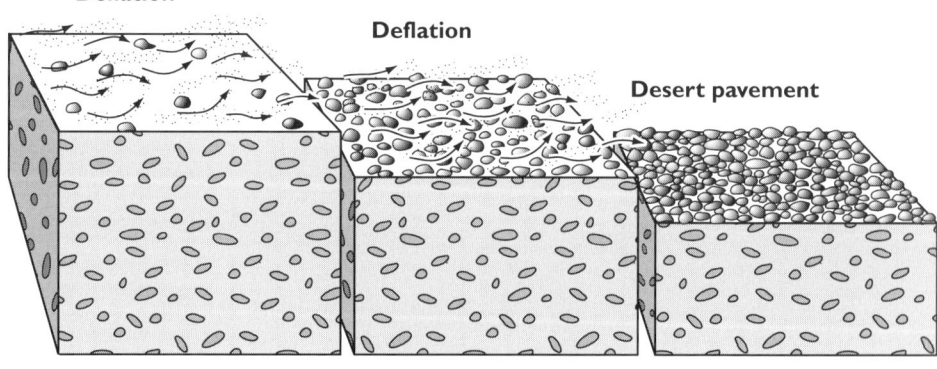

Deflation

Deflation

Desert pavement

Deflation begins

Deflation continues to remove fine particles.

Desert pavement established, deflation ends.

Development of desert pavement (after Tarbuck and Lutgens, 1999).

interval	cumulative	milepost	
4.1	34.9	CH 35.0	Salt Wells is on the right. Formerly the site of a freight and stage station dating from 1905, the location is now used for other purposes.

We are now entering Eightmile Flat, the central part of what is known as Salt Wells Basin. The floor of this flat is mainly covered with lake sediments, although some younger sand dunes and alluvial deposits partly cover the margins of the basin. Pleistocene Lake Lahontan strandlines (shorelines) can be seen on the hills to the right and left of the highway. |
| 1.1 | 36.0 | CH 36.0 | To the right, the alkali flat south of the highway is thought to be the location of the first borate discovery in Nevada. Ulexite, a borate mineral, was discovered here in 1870, and a borax plant was built by American Borax Co. the same year. This was the pioneer borax project in the Great Basin, and preceded the Death Valley "20-Mule-Team" borax industry by several years. Also to the south of the highway, the Cocoon Mountains are at 1:00–2:00 and the Bunejug Mountains are at 2:00–4:00. Bunejug Mountains were so named because, on a topographic map, the mountain resembles a June Bug. As the story goes, a small child mispronounced June Bug and called it Bune Jug, resulting in the current name.

Watch along the edge of the road here for personalized rock graffiti messages left by modern travelers in the mud of the alkali flat. Small chunks of dark basalt rock have been gathered and transformed into prose by those wishing to leave evidence that they passed this way. Please don't give in to the urge to leave your mark here, however, as it isn't safe to drive off of the pavement, and the highway shoulder is not wide enough for roadside parking. Also along this stretch, at certain times of the year, you will see a rosy, pink bloom on the thin scum of water that collects in the wide ditch between the roadbed and the playa lake mud. This is algae that makes a home in the alkaline water. |
| 6.2 | 42.2 | | Fourmile Point is on the left. This was a landmark on the old freight road between Fallon and the mining camps of Rawhide, Fairview, and Wonder.

Across the alkali flat to the west of here is the old grade of the Fallon-Sand Springs Electric Railway that crossed the flat toward the western foothills.

Fourmile Flat, the eastern part of Salt Wells Basin, is directly south (to your right) of the highway. Note that you are passing through the lowest spot in any direction. Every way you look is uphill from the alkali flat. Every drop of precipitation that enters this basin either soaks in, evaporates, or ends up in the alkaline lake that sometimes forms in the lowest part of the basin. |
| 0.5 | 42.7 | | Road to Huck Salt Co. mining facility. The salt crust that forms on the center of the playa as the thin layer of water evaporates is simply scraped off, loaded, and shipped. As more water moves to the surface and evaporates, more salt crust forms, and another layer can be mined. This is a miner's dream—a deposit that renews itself and is never mined out. Between 1863 and 1871, salt from this location was shipped to Virginia City (by camel train for awhile) for use in milling the silver ores. Salt is still produced here for use as a road de-icing chemical and for livestock feed. |

FALLON-SAND SPRINGS ELECTRIC RAILWAY

In 1914, during the boom days of the mining camps of Rawhide, Fairview, and Wonder (all to the east of here), several businessmen in Fallon conceived the idea of constructing an electric railroad between Fallon and Sand Springs, about 4 miles ahead. The idea was to do away with hauling heavy freight loads destined for the mining camps over the treacherous salt flats now crossed by our highway. Salt from the deposit in Fourmile Flat would have provided freight for a back haul. From Fallon, the railroad bed was constructed east to the northwest edge of Eightmile Flat, then it followed the foothills southward along the edge of the flat, turned west crossing the flat, and terminated about 1 mile north of Fourmile Point. By the time the roadbed construction reached this point, the mining camps had ended their "boom," also ending interest in completing the railroad (Firmin Bruner, Nevada Department of Transportation files).

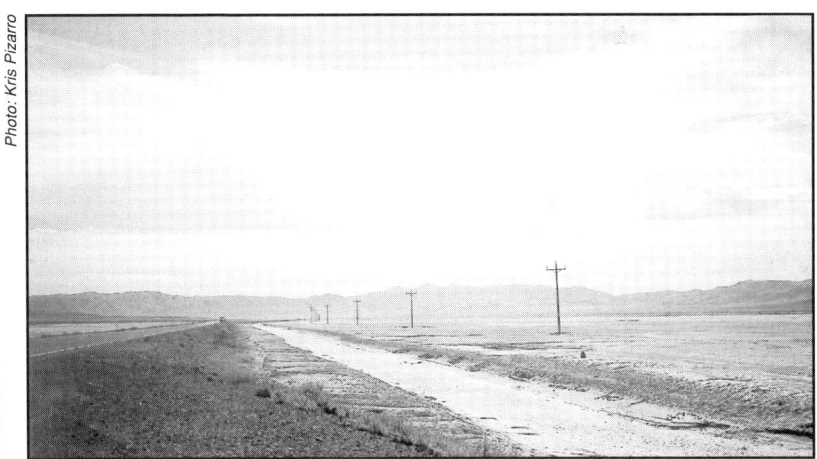

Photo: Kris Pizarro

Eightmile Flat with the Sand Springs Range in the background.

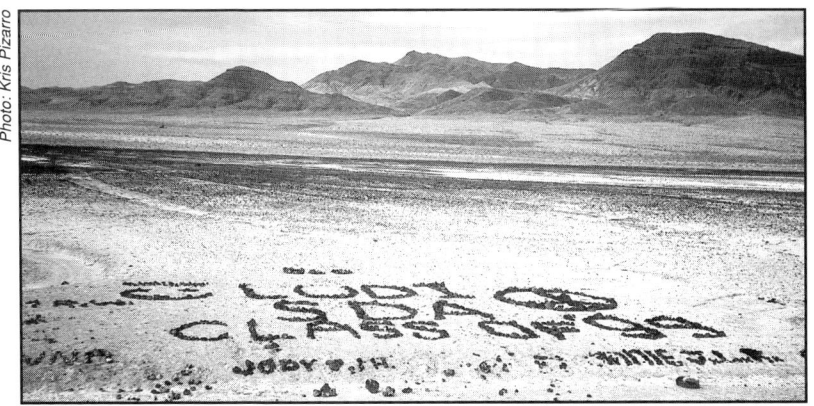

Photo: Kris Pizarro

Roadside graffiti along the edge of Eightmile Flat. Pebbles used were scavenged from the expanse of desert pavement in the background.

interval	cumulative	milepost	
0.4	43.1		Sand Mountain visible at 10:30.
1.3	44.4		LeBeau graves on left (north) mark the resting place of three young sisters who died while traveling to California in 1864. The story goes that they were very sick with diphtheria and their father left the family at this point to go ahead to find a doctor. Upon his return, he found that they had died and were buried to the northwest of here. Torrential rains washed out the remains and they were reburied at this site in 1940.
			On the right side of the highway (south), near the LeBeau graves, posts from the Overland Telegraph-Pacific Telegraph Co.'s pioneer transcontinental line can still be seen. Completed in 1861, this was the first telegraph line to cross the nation.
2.2	46.6		Road to the left leads to the Sand Mountain Recreation Area. A primitive campground is available at the foot of the mountain. The "loneliest phone," at the turnoff, is solar powered, available at anytime for the tired and sandy to order up a pizza and cold refreshments.
			This road also leads to the ruins of the Sand Springs Pony Express Station. To visit the rock ruins, watch for signs, turn left from the Sand Mountain road to a marked parking area, then follow more signs. (plate 5c)
			Wave cut terraces of Pleistocene Lake Lahontan are easily visible on the hillside directly ahead. ▼

SAND MOUNTAIN

Sand Mountain is formed from clean quartz sand, concentrated at this spot by the peculiarities of wind and mountains. The prevailing wind is from the southwest. Wind picks up fine particles from the alkali flat and surrounding area and carries them during dust storms. As the sand-laden wind begins the climb over the volcanic hills to the east, the wind energy drops and some of the sand falls out of suspension and gathers here. Over thousands of years, this action has deposited the starkly beautiful mountain of sand you see before you. Some of the sand forming this dune has traveled almost 40 miles from its source area on the Walker River delta (actually where the Walker River discharged into ancient Lake Lahontan more than 20,000 years ago). In addition to its scenic and recreational traits, Sand Mountain has another special bit of personality—it is a "booming" sand dune. The dune sands emit acoustical energy when disturbed, resulting in sounds variously described as roaring, booming, musical like a kettle drum, bass violin, or like a foghorn or low-flying aircraft. A 1904 article in the Fallon Standard said "the sand is fairly riotous, emitting a noise that sends terror to your very soul." The dune is said to sing best in the summer when it is warm and dry. Unfortunately, this corresponds with peak off-road vehicle use in the area so the sand may have competition.

The 2-mile long, 600-foot high sand dune, presents an irresistible attraction to dune buggy and off-road vehicle enthusiasts. On most weekends in good weather, the sand slopes resemble anthills from the number of dune buggies racing up and down. All it takes is a little wind to erase the tracks and leave the sand as pristine as ever, making this an ideal area for off-road vehicle play—unless you came to hear the sand sing.

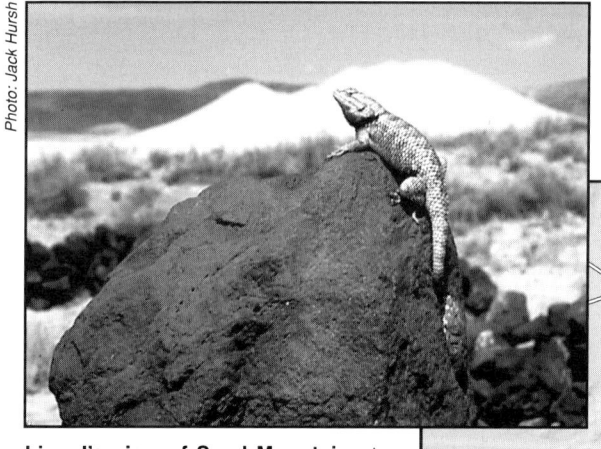

Photo: Jack Hursh

Lizard's view of Sand Mountain ▲ and the Sand Springs Pony Express Station ruins.

The sand forming Sand Mountain was carried on prevailing winds from its source, quartz ground from the Sierra Nevada by glaciers and deposited as sand in a river delta in what is now Campbell Valley. There are other dunes along the track of the wind, across the Blow Sand and Cocoon Mountains, but Sand Mountain is the end of the trail. ▼

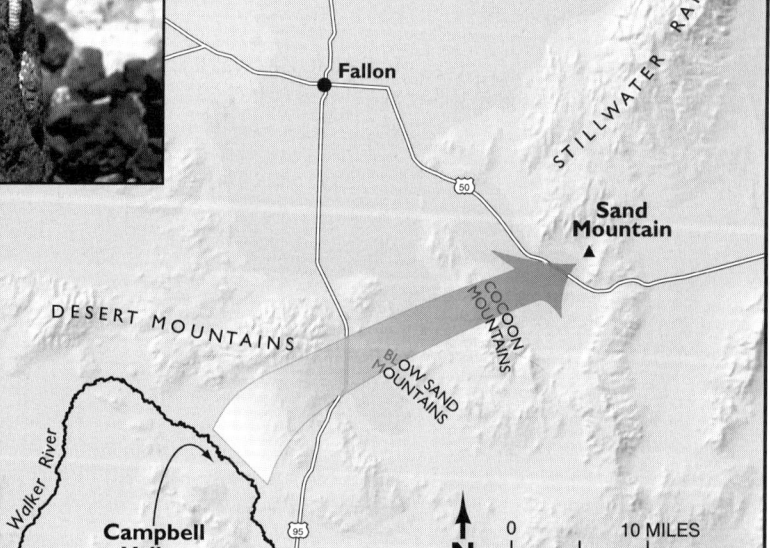

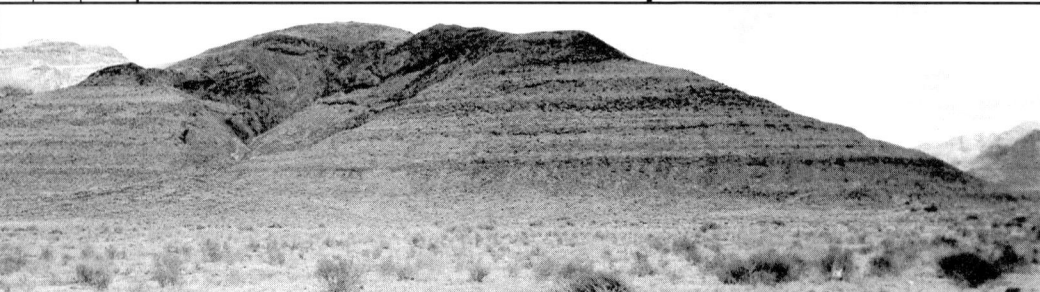

Photo: Kris Pizarro

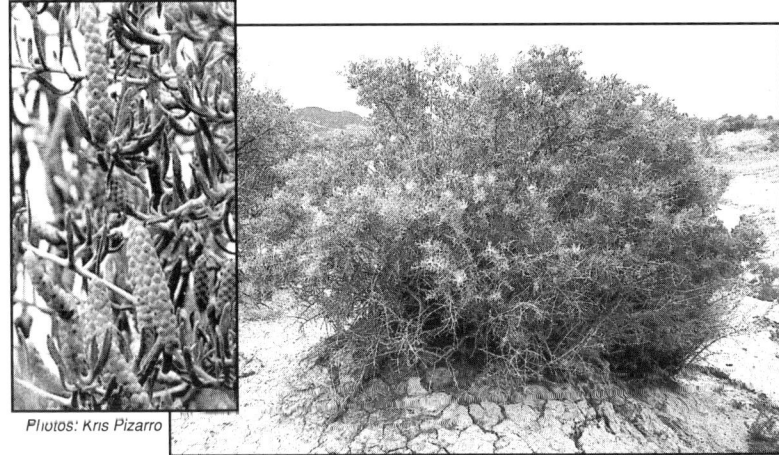

Shifting sands and greasewood at Sand Mountain.

Photo: Jack Hursh

Photos: Kris Pizarro

Greasewood (*Sarcobatus vermiculatis*) Perennial shrub, 3–8' tall; many whitish, rigid, spiny branches; leaves bright green, succulent, linear, ½–1½" long; male and female flowers (inset) on same plant mature at different times to ensure cross-pollination.

Greasewood inhabits dry valley bottoms, ephemeral stream channels, and playa margins where its conspicuous bright green color contrasts with the more muted hues of the surrounding countryside. Pure stands can often be found in highly saline areas. Though it tolerates periods of standing water, greasewood is dependent upon groundwater, and its roots may penetrate the water table 20 feet or more beneath the surface. It accumulates salts in its leaves and roots and can be toxic to sheep and cattle if eaten in large amounts.

SAND SPRINGS PONY EXPRESS STATION AND THE CHICKEN-CRAW GOLD RUSH

Covered by sand for over a hundred years, the Pony Express station was rediscovered by archaeologists in 1976, excavated, and stabilized. You can explore the ruins, walk an interpretive loop trail that describes the native plants and animals, and even see the remains of a well that supplied what passed for drinking water at this isolated station.

Sand Springs Station was the scene of an often repeated old-timer's tale of the Chicken-Craw gold rush of 1907. As the story goes, the operator of a wayside inn at the site was dressing some chickens for Sunday dinner. Fair-sized gold nuggets were found in the craws of two of the chickens. All of the remainder of the flock of chickens was processed, revealing even more gold. Convinced that the chickens were not eating gold in the neighborhood, the innkeeper set out with two companions to learn where the chickens had come from. They discovered that the chickens had been purchased near Wadsworth, 70 miles away. Confusion was added to distance when it was learned that the chickens had been owned by no fewer than four farmers before they came to Sand Springs. The area in which the chickens ate the gold nuggets for gravel remains undiscovered.

Photo: Kris Pizarro

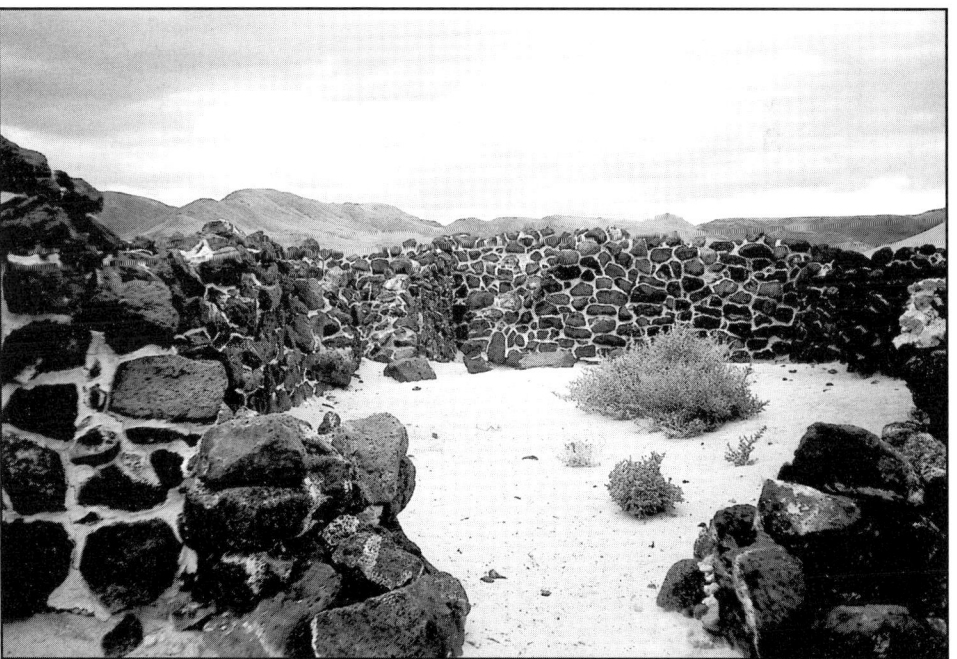

Walls of the Sand Springs Pony Express Station near Sand Mountain.

interval	cumulative	milepost	
0.8	47.4		Ahead to the right (south of the highway) lies the Sand Springs Range. Most of the central part of this range is composed of Cretaceous granitic rock, but there are small patches of Jurassic-Triassic metasedimentary rocks at its south end and along the highway at the north end. Deep beneath Gote Flat, located a few miles to south, on the east side near the range crest, the Department of Energy (then the Atomic Energy Commission) detonated an underground nuclear device in 1963. This is one of two locations in Nevada where nuclear tests were conducted outside of the boundary of the Nevada Test Site.
1.3	48.7		To the left, north of the road, note the light-colored, altered Tertiary-age rhyolite dikes cutting dark-colored Triassic slate and phyllite. The best exposure is in the old road cut; blocks of dike rock can be seen on the dump on the west end of the cut.
0.5	49.2		Another good exposure of the light-colored dike in dark slate and phyllite, on the left. (plate 5a)
0.8	50.0	CH 50.0	To the right (south of the highway) is the Summit King (Dan Tucker) Mine, where more than $2 million in silver and gold was mined from west-trending quartz veins cutting both Jurassic-Triassic metasedimentary rocks and Tertiary volcanic rocks. The deposit was discovered in 1905, but most of production was between 1940 and 1951, making this one of the last active "boom camps" in the state.

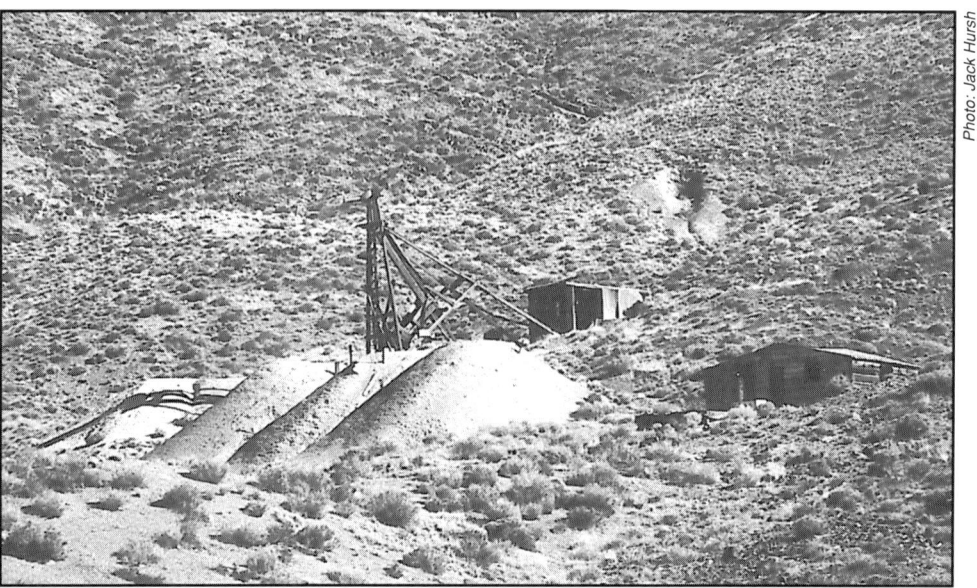

Photo: Jack Hursh

The Summit King (Dan Tucker) Mine.

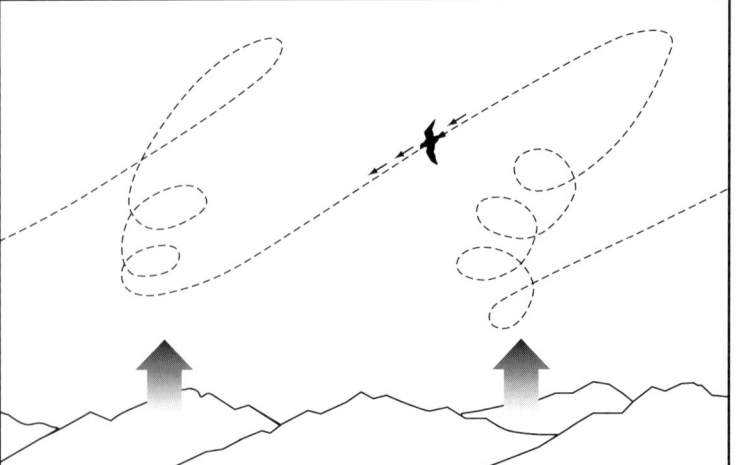

Turkey vulture (*Cathartes aura*) 26–32" long; wingspan to 72"; brownish black with unfeathered red head (immature head is black) and yellow feet; usually silent; migrates south in autumn and return in early spring.

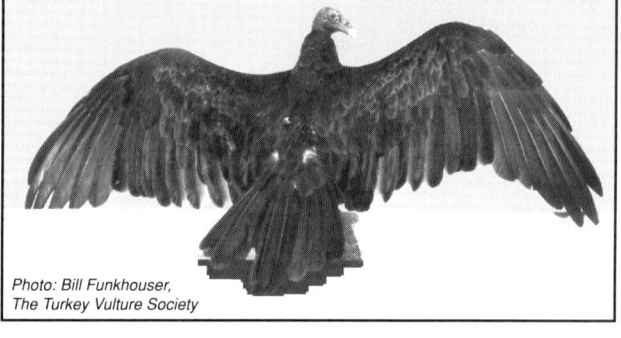

Photo: Bill Funkhouser, The Turkey Vulture Society

The turkey vulture excels at two things— soaring and clearing the countryside of the dead. The vulture and other carrion eaters dispose of millions of carcasses each year, thereby eliminating a source for the spread of infection and disease, not to mention helping to keep our roads and highways scenic. Its name is from the Greek *kathartes*, meaning "purifier" or "cleanser." Its featherless head is easy to keep clean, its feet are adapted for walking and holding food in position, and its digestive tract detoxifies materials as they pass through.

The turkey vulture rides rising thermals in large circles and spirals, using both its keen eyesight and acute sense of smell to locate food. It flaps its wings only as a last resort to remain in the air. In flight it is easily distinguished from other large birds, particularly the golden eagle, because it holds its wings in a shallow V-shape (dihedral) and tips and rocks from side to side while soaring.

Turkey vultures don't bother much with nest building. Two eggs are deposited on the floor of a crevice or cliffside cave, and both parents share incubating duties. Young vultures hatch in 38–41 days and make their first flight in about 2½ months. When not nesting, turkey vultures are gregarious birds that roost together, often soar together, and if there's enough, eat together.

Hawks, eagles, and vultures soar by taking advantage of rising currents of warm air called thermals (modified from Terres, 1980).

interval	cumulative	milepost	
1.7	50.5		Summit of Sand Springs Pass (elevation 4,631 feet).
1.7	52.2		We are leaving the Sand Springs Range and dropping into Fairview Valley. As soon as we clear the range, the panorama includes the southern end of the Stillwater Range at 9:00, Dixie Valley in the distance at about 10:30 bordered by the Louderback Mountains with the Clan Alpine Mountains beyond at 10:30 to 12:00. Chalk Mountain is at 11:00. Labou Flat occupies the center of Fairview Valley, south of the highway, and Fairview Peak is at 1:30 beyond Labou Flat.
0.4	52.6	CH 52.54	Intersection with State Route 839 (Scheelite Mine road).

This road was paved years ago to provide access to the now abandoned Nevada Scheelite Co. tungsten mine near the south end of the Sand Springs Range. Today, the road is one of the main access roads to the Rawhide gold-silver mine, an open-pit mine at the site of the old camp of Rawhide.

About 4.9 miles south on State Route 839, a well-maintained gravel road leads up to Gote Flat, mentioned earlier as the site of an underground nuclear detonation. This test, named Project Shoal, was conducted 1,211 feet below the surface and was considered a relatively small yield device at 12 kilotons. There isn't much to see; only a concrete pad and one small sign mark the location.

Note the observation towers and targets in the center of Fairview Valley. This is one of the Navy's active bombing ranges, and it is common to see planes making runs at the targets. Planes usually approach high from the south, dive, then rapidly climb out to the east or west. Practice bombs (no explosives) are used on the run close to the highway, but real explosives are dropped on targets across the valley on the lower flank of Fairview Peak.

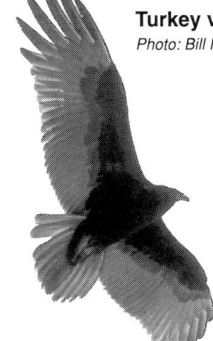

Turkey vulture in flight.
Photo: Bill Funkhouser, The Turkey Vulture Society

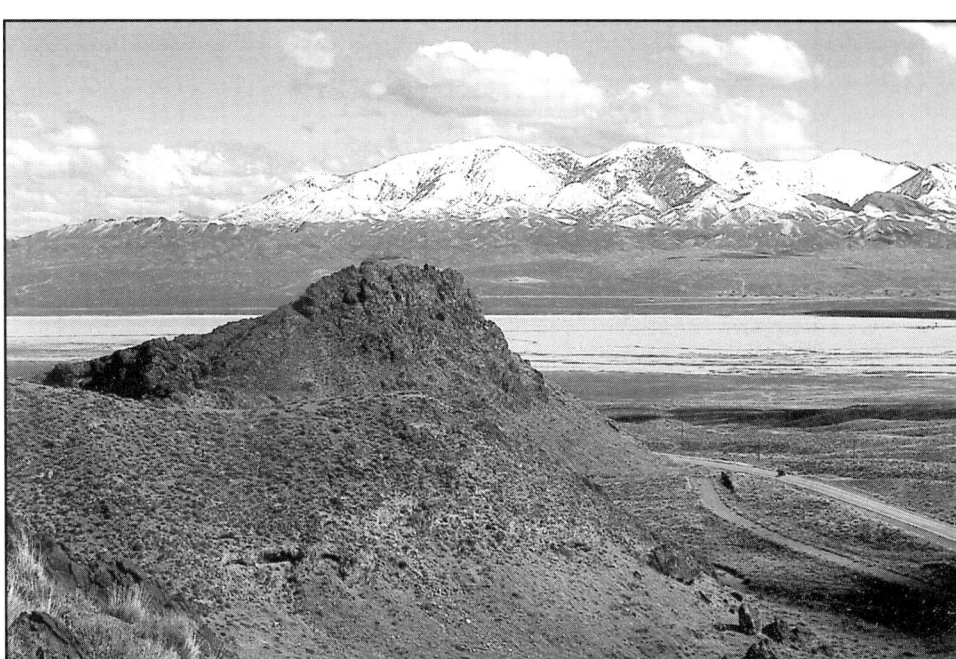

Photo: Jack Hursh

Sand Springs Pass with U.S. 50 in the right foreground descending into Fairview Valley. Fairview Peak is covered with a typical blanket of March snow. The hills in the foreground are capped by rhyolite welded ash-flow tuff.

RAWHIDE

The site of the mining camp of Rawhide is 20 miles south of U.S. 50 on State Route 839, then about 3 miles further on gravel-surface road.

Seeing Rawhide as a last chance to hit it big in a mining boom, promoters from Goldfield and other faded Nevada boom camps descended on Rawhide in 1906 and whipped up a frenzy of mining stock activity. By June 1908, the camp had grown into a city of 8,000 with a full complement of saloons (over 40), hotels (30), barbershops (10), brokerage houses (125), churches (2), a school, and a red light district (one-quarter mile long). In September of that year a disastrous fire wiped out much of the town. About the same time, investors noted that the mines were not producing as anticipated, and the bubble began to burst. The camp produced about $2 million total and was soon reduced to a ghost camp.

Prospectors (mining company geologists this time) returned to Rawhide in the 1960s and 1970s when interest in precious metals heightened. In 1982, Kennecott Exploration began a program that led to discovery of over 29 million tons of gold-silver ore in the area of the old townsite. Production began in 1990 and, in 1997 alone, the mine produced 120,000 ounces of gold and over 1.1 million ounces of silver—far more than the total previous production of the camp.

interval	cumulative	milepost	
2.1	54.7	CH 55.0	
1.0	55.7		

Site of Frenchman on the right, south of the highway. Starting about 1906 as a freight station on the road to the camp of Fairview, this site was the location of a gas station, restaurant, and bar until a few years ago. The buildings were just beyond the end of a line of targets on the Navy bombing range in the valley to the south. The Navy eventually bought the place and leveled it, leaving only a dead tree to mark the site.

Drag marks are reported to have been seen on the surface of the playa along the highway in this area. Drag marks, made by stones or other objects, are common on Nevada playas. They escaped general attention until the publicity given the Racetrack Playa in Death Valley caused Nevada playas to be investigated. Drag marks on this playa were first observed in the spring of 1956, probably having been formed the previous winter. They gradually disappeared during the dry years following 1956, only to reappear sometime between 1960 and 1962. Many of the stones leave concentric or parallel tracks, suggesting that they were locked in place by a thin sheet of ice during movement while the playa was covered with water.

This marks the end of the second section of the road log; the wide Carson Sink and other alkali flats filled with sediments deposited by ancient Lake Lahontan now give way to a landscape dominated by volcanic rocks and calderas. Fairview Peak, ahead on the right, is the first of a number of volcanic centers that we will pass in the next part of the trip.

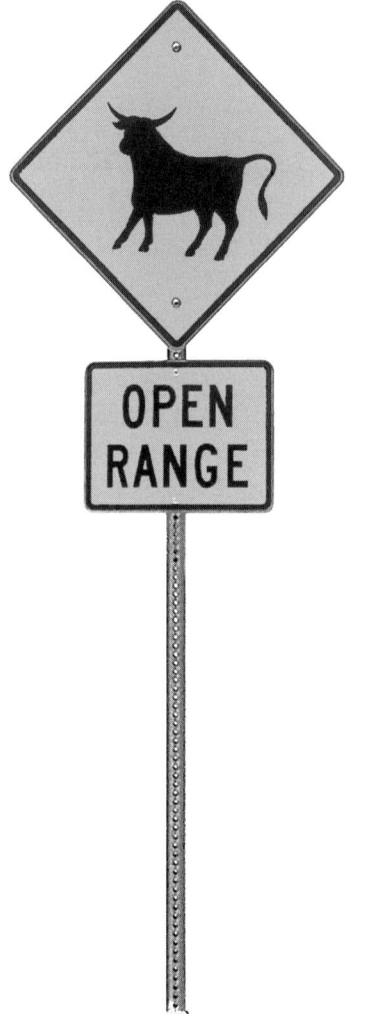

SECTION III: VOLCANOES, LAVA FLOWS, ASH-FLOW TUFFS, AND CALDERAS

After crossing the desert lake and sink features surrounding Fallon, the "loneliest road" now enters a section of Nevada dominated by thick flows of volcanic rock—the third part of the log. Much of this material came from calderas, a type of volcano that collapses during eruption of massive amounts of ash. This section of U.S. 50 crosses a small part of the area of volcanic activity, and passes by several major calderas.

Volcanic rock is formed from molten material erupted from pockets of magma within the crust of the Earth. Central eruptions create the most familiar volcanic topography—the cone-shaped mountain. These discharge lava and volcanic ash through a pipe-like opening to the Earth's surface. Successive flows from non-explosive eruptions result in the buildup of broad aprons of volcanic rock around the central vent. These aprons coalesce to form a composite cone, resulting in the familiar mountain form we call a volcano. When conditions occur that cause water vapor and other gases to build up in a magma chamber, a sudden release of pressure, perhaps caused by earthquakes, can cause an explosion. The rock froth, mostly ash, that results from this explosion is rapidly ejected from the vent as a pyroclastic flow. These flows, fiery clouds of rock and dust, blast away from the source vent and destroy everything in their path. Such flows from Mount Pelée destroyed the city of St. Pierre in Martinique 1902, and similar flows wiped out much of the forests on the flanks of Mount St. Helens in Washington in 1980. Pyroclastic flows and other debris can extend great distances from the source vent and, when solidified and cooled, form sheets of rock called ash-flow tuff. During and following eruption of the pyroclastic material, the roofs of empty magma chambers, no longer able to support themselves, collapse into the void leaving a steep-walled, basin-shaped depression much larger than the original crater. These basins are called calderas and can range from a mile to more than 10 miles in diameter. The process may not end there; magma can again move into the deflated chamber, forcing the collapsed roof and overlying rock back upward like a huge plug, resulting in a resurgent caldera. Magma can also flow through cracks around the collapsed plug and fill the caldera basin with thick volumes of volcanic rock. Present-day landforms of calderas range from water-filled depressions, like Crater Lake in Oregon, to rugged mountain masses formed from resurgent central cores.

Ash-flow tuffs of the Great Basin in Nevada and adjacent states, formed during mid- to late-Tertiary time, cover an area of about 80,000 square miles; the next 100 miles or so of U.S. 50 will take us across a small part of this large area. There are no caldera lakes in central Nevada, but our route takes us by calderas (Fairview Peak and the central Desatoya Mountains), and sheets of tuff deposited outside calderas (Hickison Summit).

interval	cumulative	milepost	

1.0 | 56.7 — Fairview Peak caldera.

Fairview Peak, visible between about 2:00 and 3:00, occupies the central part of the Fairview Peak caldera. The highway ahead roughly coincides with the north margin of the caldera. Bell Canyon, cutting the range at about 3:00, marks the south margin. The mountain front is the topographic west wall. Fairview Peak is underlain by rhyolitic ash-flow tuff that filled the caldera.

1.0 | 57.7 — NHM #202, Fairview, is on the right. The site of Fairview is in the foothills about 1 mile to the southeast (large concrete object visible from the highway is the remains of a safe from one of the town buildings). The Fairview deposits, located out of view in a basin another mile or so to the south on the northwest flank of Fairview Peak, were discovered in 1905. The area was the scene of a frantic boom in 1906–1907, and a substantial town that boasted 27 saloons, hotels, banks, assay offices, a newspaper, post office and a miners' union hall soon came into being. Most production occurred between 1911 and 1917, when about $1 million in gold and $2.8 million in silver were produced.

Although mining property in the old camp is still privately owned, much of the surrounding area, including the access road to the camp, is within the Navy bombing range and is off limits to the public.

2.9 | 60.6 | CH 60.52 — Intersection with State Route 121 north to Dixie Valley.

Historical Marker (NHM #201) commemorates Wonder, another 1906-era mining camp located 13 miles to the north. Like Fairview, Wonder's boom was brief but spectacular. Between 1907 and 1920, the Nevada Wonder Mining Co. produced some $5.8 million in silver and gold. In the 1980s and 1990s, several attempts were made to mine low-grade ore remaining at the site, but none were successful.

A prominent Wonder native, Eva Adams, was administrative assistant to Senator Patrick A. McCarran for many years and was director of the U.S. Mint during the 1960s.

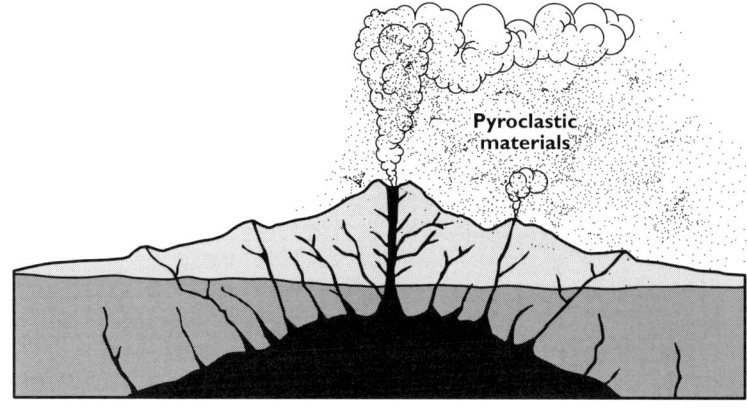

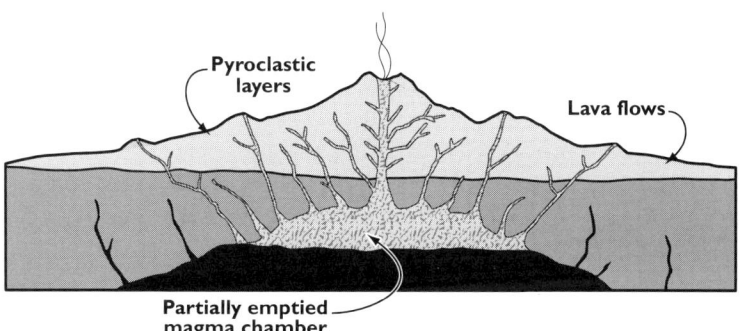

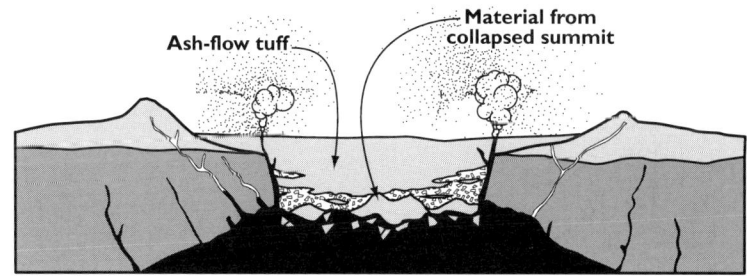

Caldera formation begins when magma forces its way to the surface through fissures and vents, resulting in explosive volcanic eruptions that deposit pyroclastic materials (volcanic ash and ash-flow tuff) around the vent areas. The expulsion of material to the surface leaves a void in the underlying magma chamber which then collapses, forming the caldera. More pyroclastic material erupts from vents around the caldera, eventually filling it with thick layers of ash-flow tuff. Magma continues to push its way toward the surface, pushing the older collapsed material and newer ash-flow tuff fill into a dome marking the site of the caldera.

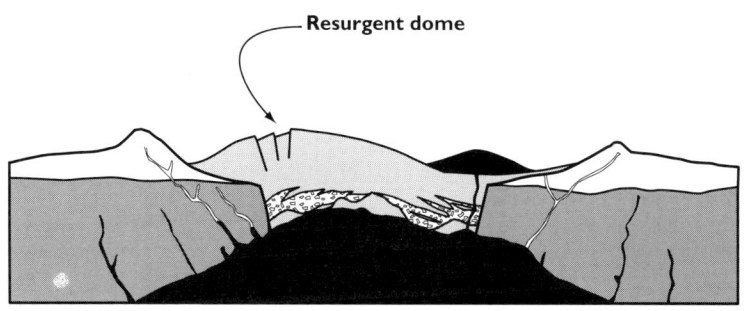

		CH 60.52

Chalk Mountain is ahead on the left at 10:00. Chalk Mountain is composed of Mesozoic dolomite (light color) that has been intruded by Cretaceous granitic rocks (dark area on the west flank). Lead and zinc have been mined from replacement deposits in the dolomite on the east side of the mountain. A wide variety of oxide minerals of lead, zinc, iron, and other metals can be found on mine dumps on the southeast side of Chalk Mountain (access by four-wheel drive vehicles only). Chalk Mountain is bounded by steep normal faults on both the east and west sides; it is north of the margin of the Fairview Peak caldera. (plates 6b and 6e)

1.7 62.3

Fairview Peak fault, gravel road to the right.

Informational signs are placed along this gravel road by the BLM to explain good exposures of the Fairview Peak fault. The first sign is 0.4 mile off the asphalt. The most dramatic exposure of the fault break is 5.8 miles south of the highway up a steep (but passable by passenger cars) section of road where up to 15 feet of vertical exposure is still evident.

4.0 66.3

Windmill on the left, Westgate is just ahead (narrowest point between cliffs on right and cliff north of creek bed on left).

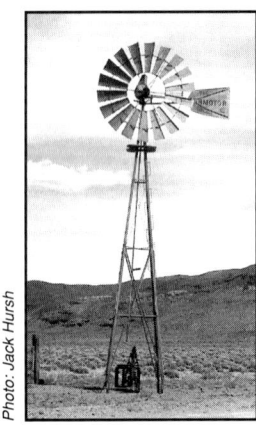

Photo: Jack Hursh

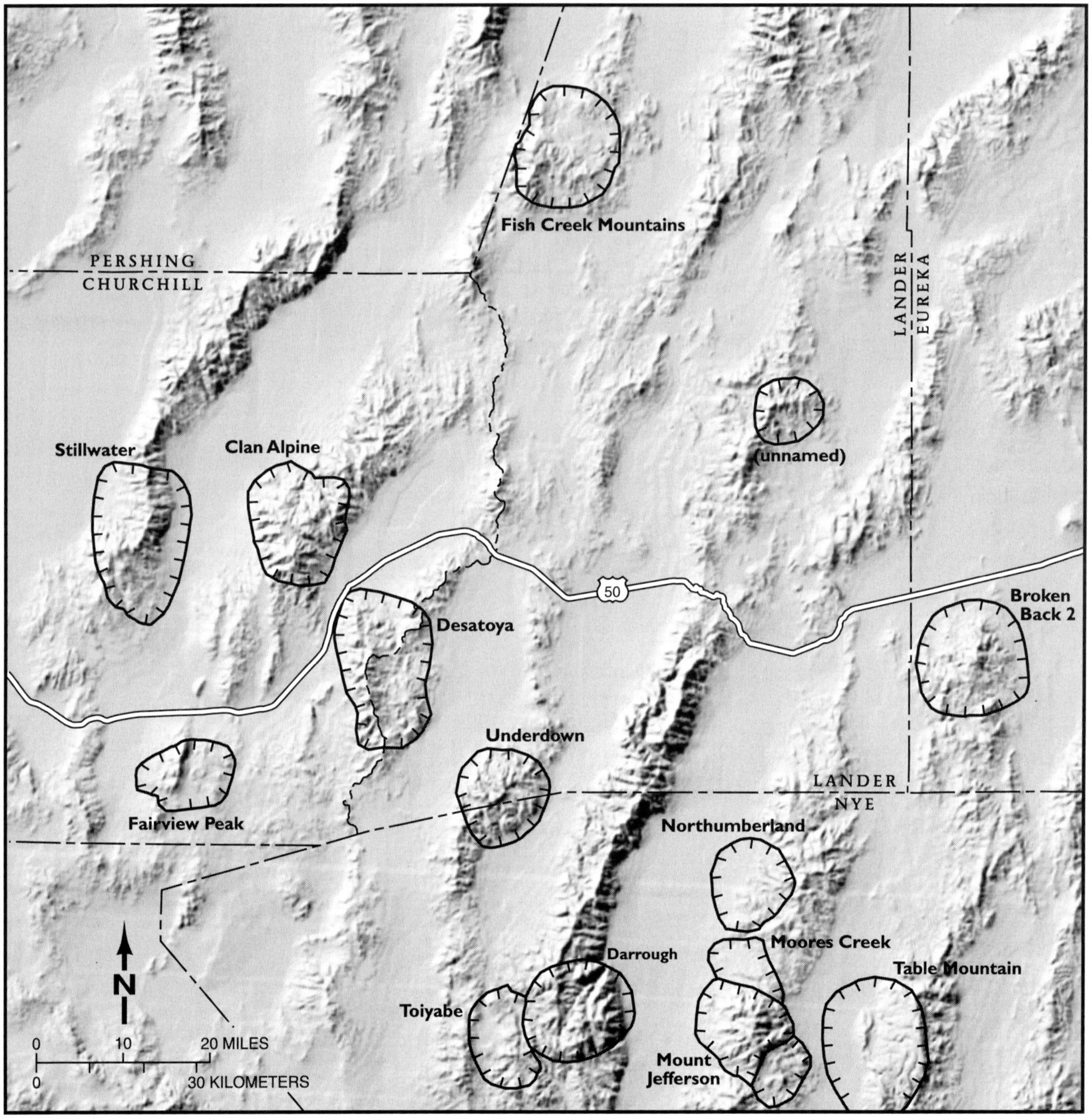

Locations of known calderas in the mountain ranges flanking U.S. Highway 50. Most of the boundaries shown are approximate, and some include overlapping calderas of slightly different ages (map by Chris Henry, NBMG).

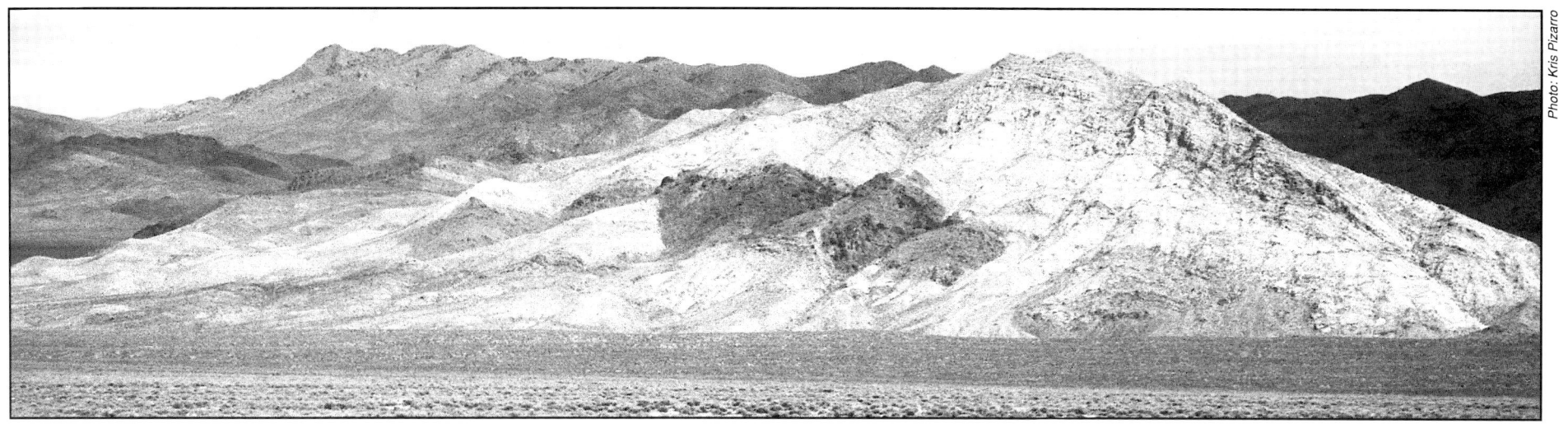

Photo: Kris Pizarro

Chalk Mountain. The elongate dark spot on the mountain face is granodiorite that has intruded normally gray dolomite, altering it to the white, recrystallized rock you now see.

Central Nevada scenery viewed from the east slope of Fairview Peak. Barren, rolling hills to the east are composed of rhyolite ash-flow tuff filling the Fairview Peak caldera. The Gold Basin mining district, marked by faint bulldozer scrapings and small dumps, is to the right of center. The Desatoya Mountains are in the background.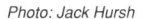

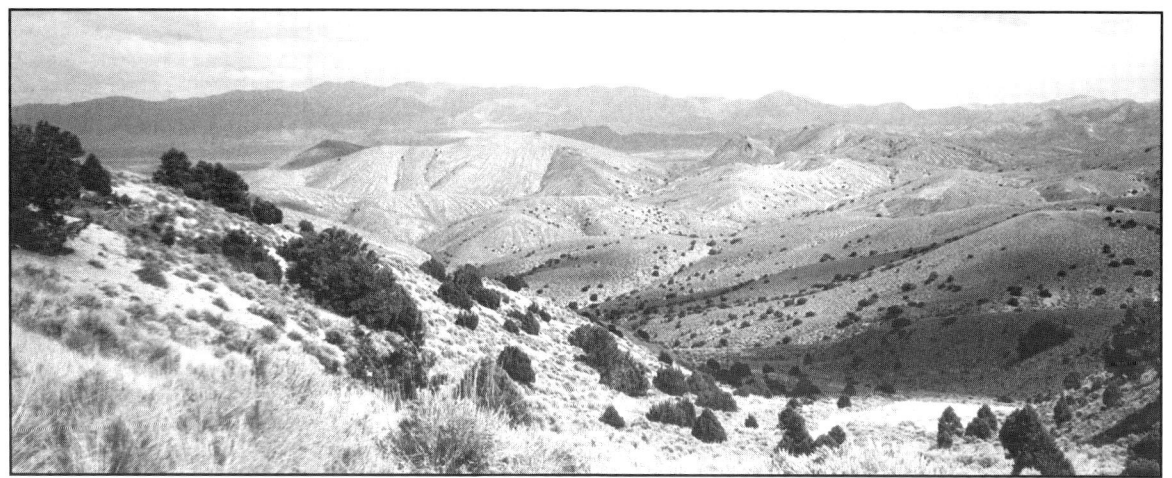

Photo: Kris Pizarro

Trace of the Fairview Peak fault. Ground to the left moved down along the fault relative to the ground to the right, creating the white scar on the hillside.

Photo: Jack Hursh

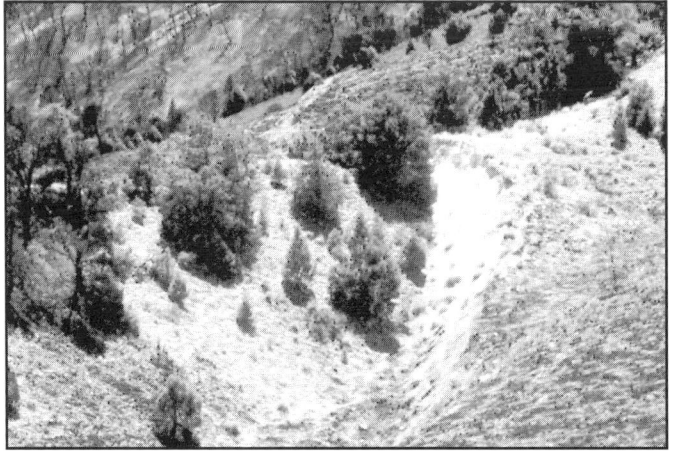

DIXIE VALLEY – FAIRVIEW PEAK EARTHQUAKE AREAS

The fault break exposed on the east side of Fairview Peak to the south of here is the surface expression of the first of two severe earthquakes that rocked this area in 1954. This earthquake, of Richter magnitude 7.1, took place at about 3 AM on December 16, 1954. A second earthquake, of magnitude 6.9, followed about 4½ minutes later. The second earthquake was centered along the northwest edge of Dixie Valley, about 35 miles to the north. The Fairview Peak fault has been described as a "contagious" fault. The earthquake that caused it may have triggered the second earthquake in Dixie Valley. The two earthquakes occurred within a zone of surface faulting about 50 miles long and 10 miles wide that is marked by historical as well as prehistorical fault scarps of very fresh appearance.

SIDE TRIP 5, BERLIN-ICHTHYOSAUR STATE PARK

The Berlin-Ichthyosaur State Park was established in 1957 to protect and display North America's most abundant concentration of ichthyosaur fossils. Ichthyosaurs were prehistoric marine reptiles that lived at the same time as the dinosaurs. Ichthyosaur fossils are found on all continents except Antarctica, but the fossils found at this State Park are the largest specimens known, most being about 50 feet long. The fossil site was discovered in 1928 and excavations continued through 1960. About 40 ichthyosaurs have been discovered in various locations throughout the park in Triassic mudstone. The ichthyosaur is the Nevada state fossil.

The park also preserves the 1896–1911 mining town of Berlin, established by miners working at the Berlin gold mine and mill. During its heyday, Berlin and its suburbs supported a population of 200 to 250. Protected by a watchman for many years before formation of the park, many buildings remain to be viewed. Visitors can walk on their own through the town, and there is a guided tour of a portion of the old Diana Mine. For a more detailed story of Berlin-Ichthyosaur State Park, refer to Nevada Bureau of Mines and Geology Special Publication 5, *Child of the Rocks*.

In addition to all of the things to see, there is a day-use picnic area with tables, grills, drinking water, and restrooms nearby. There is also a well-equipped campground with some spaces suitable for RVs to 25 feet.

From Berlin, a loop can be made through the [almost ghost] town of Ione to the Reese River Valley, then north to intersect the Carroll Summit road, and, eventually rejoin U.S. 50 just west of Austin. Since the road from Ione north is largely unpaved, we recommend that you retrace your steps back to Middlegate Junction instead.

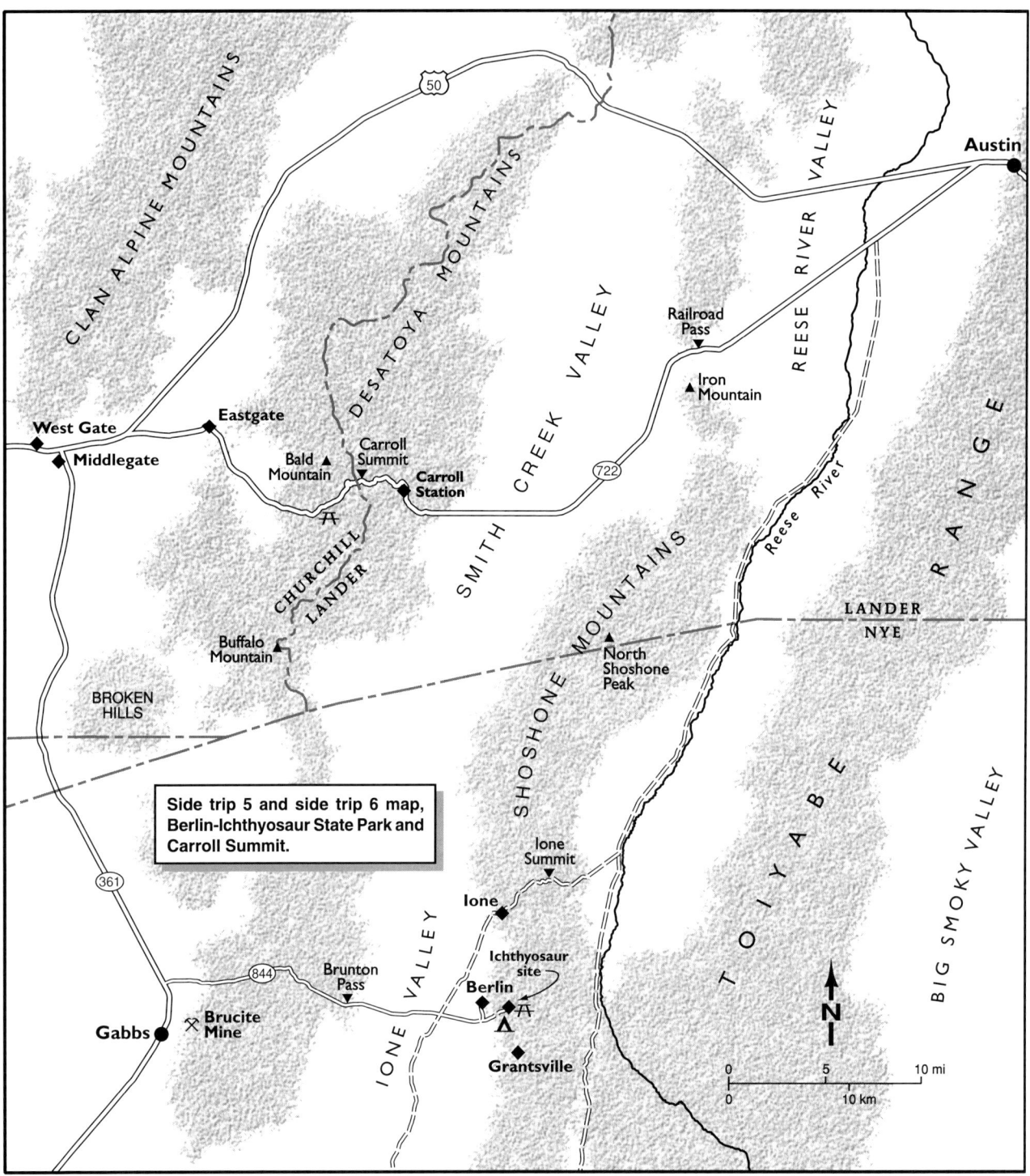

Side trip 5 and side trip 6 map, Berlin-Ichthyosaur State Park and Carroll Summit.

interval	cumulative	milepost	
1.4	67.7	CH 67.76	Middlegate Junction, intersection with State Route 361. Middlegate Bar, site of part of the Shoe Tree story (ahead), is to the right.
			This is the starting point for an optional side trip: the Berlin-Ichthyosaur State Park is about 51 miles southeast of here via a good paved highway (State Route 844). The park offers a preserved ghost town, an interpretive center at the ichthyosaur fossil locality, and a well-maintained campground.
2.2	69.9		Middlegate. The highway passes through a split in a low ridge, called Middlegate by the locals. Bighorn sheep have been seen on these rocky bluffs—keep a sharp eye!
0.1	70.0	CH 70.0	Shoe Tree to the left, north of the highway. It's billed as the largest shoe tree in the world, at least in Nevada. The large cottonwood at the edge of the steep-walled wash was, until a few years ago just another Nevada native, adding its own type of color to the desert landscape. Exactly how the tree earned its shoes is still open to debate but, according to Don Cox in a September 4, 1998 article in the Reno Gazette-Journal, it all started with a fight between a guy and his new bride, or maybe his girlfriend, who were passing by. He threw her shoes into the tree and repaired to the Middlegate Bar to think about it. The two made up and happily went on their way. Shoe-tossing seems to have caught on, and now hundreds of old shoes hang from the branches of the cottonwood. ▼

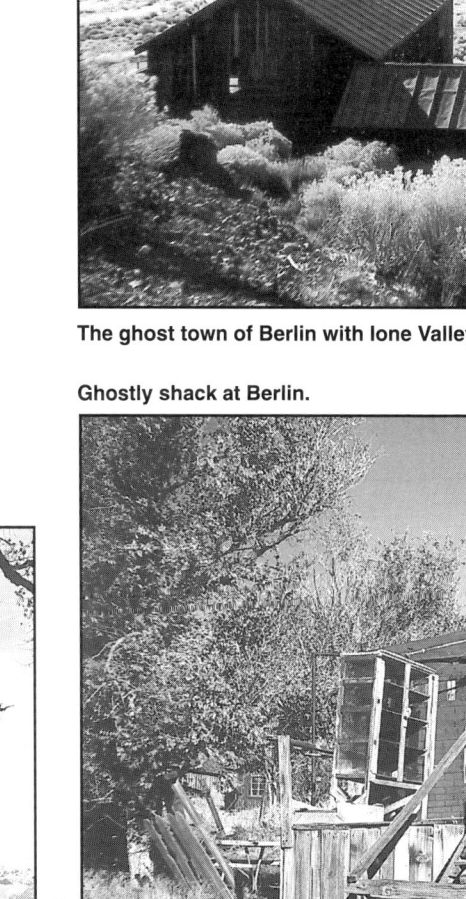

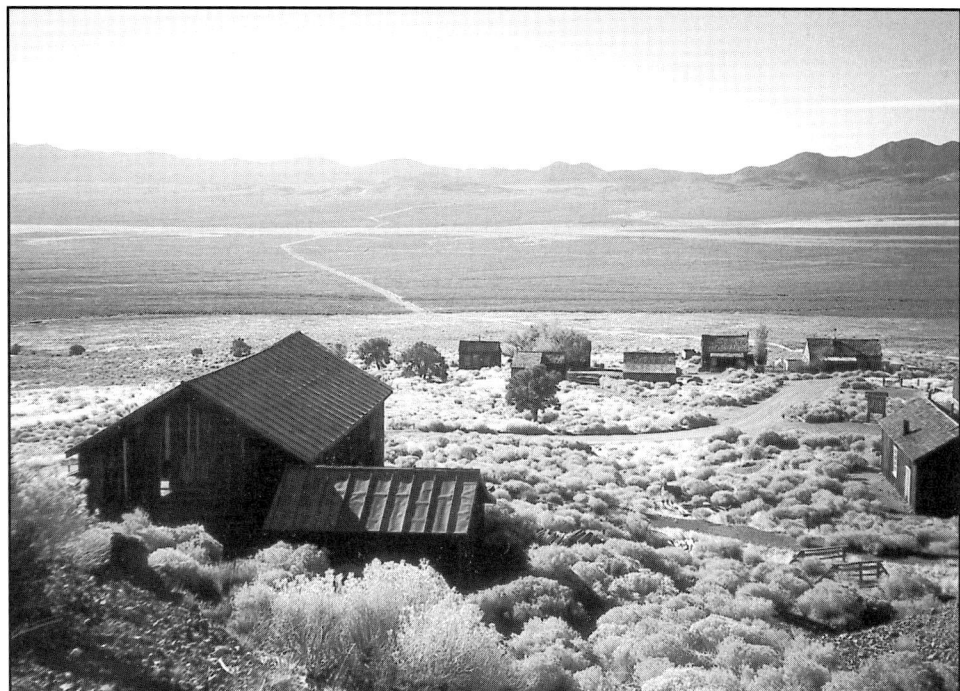

Photos: Kris Pizarro

The ghost town of Berlin with Ione Valley in the background.

Ghostly shack at Berlin.

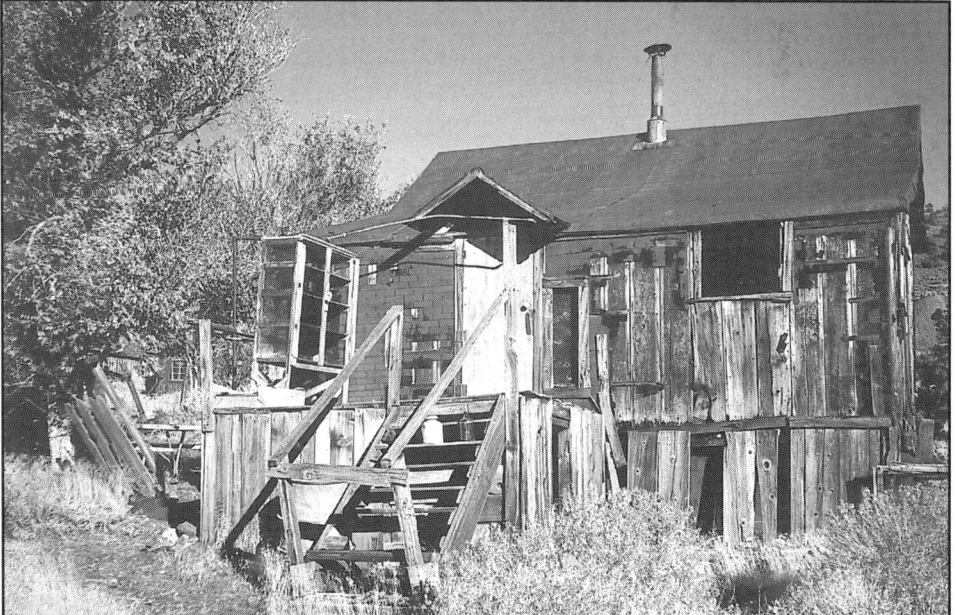

Photos: Kris Pizarro

interval	cumulative	milepost
1.0	71.0	CH 71.0

Junction with State Route 722 to Eastgate, stay on U.S. 50 (main highway to left) unless you are taking side trip 6 to Carroll Summit.

Along the right side of the highway, cuts and trenches have been dug to explore a deposit of zeolite. The zeolite occurs in Miocene lakebeds that crop out in the area between the two highways. Zeolite minerals were used as filters in many industrial processes, but have largely been replaced by man-made materials that have similar properties. Small amounts are mined from this area for use as kitty litter, one market still covered by the natural material.

interval	cumulative	milepost
1.7	72.7	

Good exposures of gently dipping Miocene lakebeds can be seen in the low hills to the left of the highway.

Rugged outcrops of welded ash-flow tuff at Eastgate. ▼

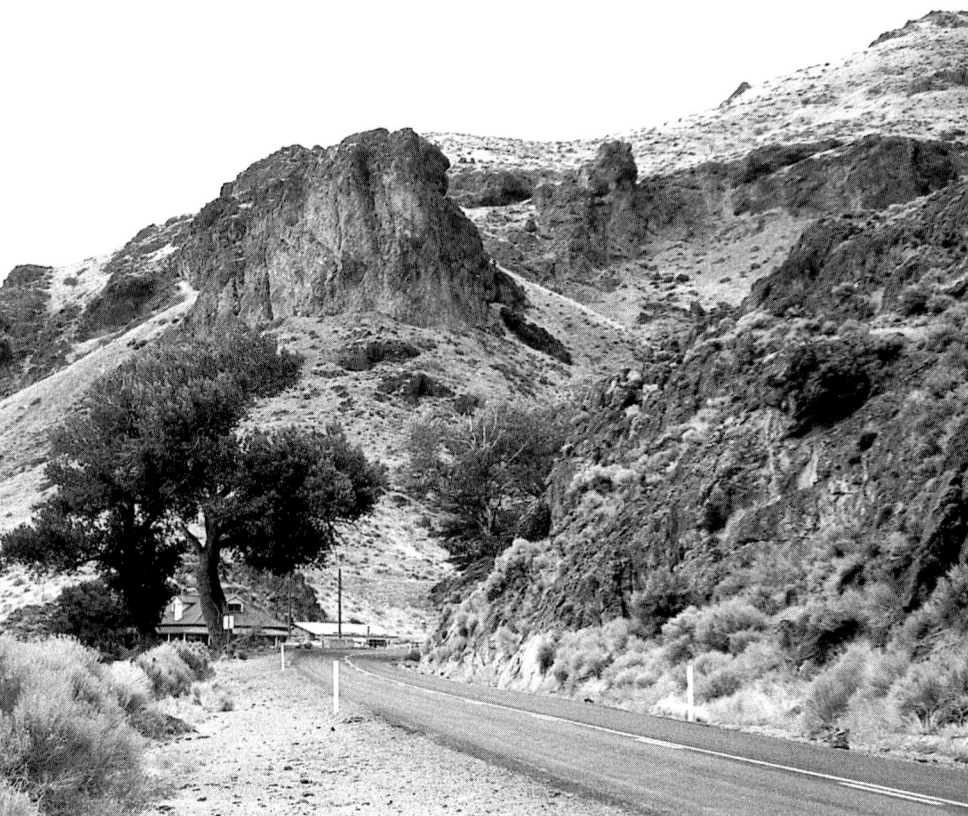

SIDE TRIP 6, CARROLL SUMMIT

State Route 722 continues east to Eastgate, about 5 miles ahead at the cleft through the foothill range, then continues over the Desatoya Mountains at Carroll Summit, crosses Smith Creek Valley and the Shoshone Mountains, and reconnects with U.S. 50 west of Austin. When the Lincoln Highway, later to become U.S. 50, was established through this area in 1913, it deviated from the historical Simpson Survey-Pony Express route and followed this southern route over Carroll Summit. In the early 1960s the road was realigned to follow the historical northern route. The southern route over Carroll Summit is a scenic, alternate way to reach Austin. It does require a little more mountain driving, but about 12½ miles from the junction there is a tree-shaded rest area (with a picnic table), on a small stream with (usually) running water, which provides a good lunch stop.

Things to see on this side trip include good views of rugged cliffs of rhyolitic ash-flow tuff as the road enters the canyon east of the Eastgate ranch buildings, and an area where Buffalo Creek (the drainage on the right of the highway) has cut deep, vertical cliffs in the soft stream-deposited sandy silt that forms its banks. This area provides opportunity for birding, as several varieties of birds like to nest in cavities in the cliff faces (plates 7d and 7e). The bottom land along the creek provides habitat for small mammals and other creatures (watch out for rattlesnakes here). Since the route climbs fairly quickly into the Desatoya Mountains, you will pass from low desert vegetation dominated by shadscale into sagebrush and grass, then into piñon-juniper woodland, then back into sagebrush as the road drops into Smith Creek Valley to the east. Also, there are some spectacular outcrops of columnar-jointed ash-flow tuff that can be seen in the canyon walls just before the road breaks free of the eastern front of the Desatoya Mountains. As the road approaches the Shoshone Mountains, Iron Mountain can be seen first on the east (right), then on the south (still right) as the road swings east to cross the mountains at Railroad Pass. Iron Mountain is a small plug of rhyolite that cuts slightly older ash-flow tuff; it is distinctively darker than the surrounding tuff. The last 15 or so miles of State Route 722 crosses the Reese River Valley, passes the Austin Airport, then connects with U.S. 50 about 1 mile west of Austin. **See map on page 54**.

Wagonjack Shelter, natural cave in ash-flow tuff near Eastgate. ▶

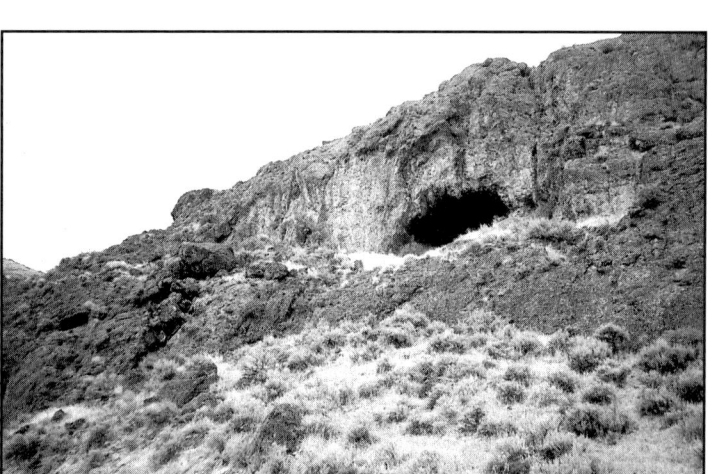

Photo: Kris Pizarro

Photo: Kris Pizarro

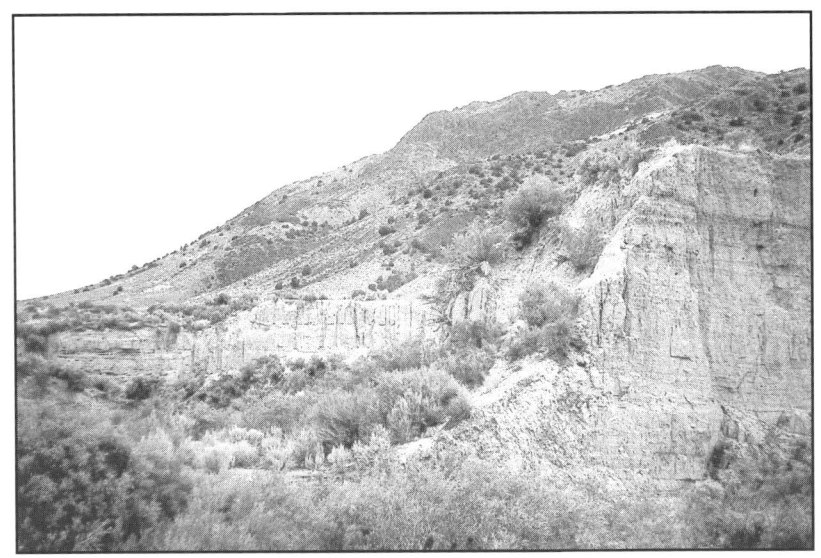

Photos: Kris Pizarro

Buffalo Creek Canyon, east of Eastgate. The near-vertical cliffs formed as Buffalo Creek cut its way through older stream sediment deposits.

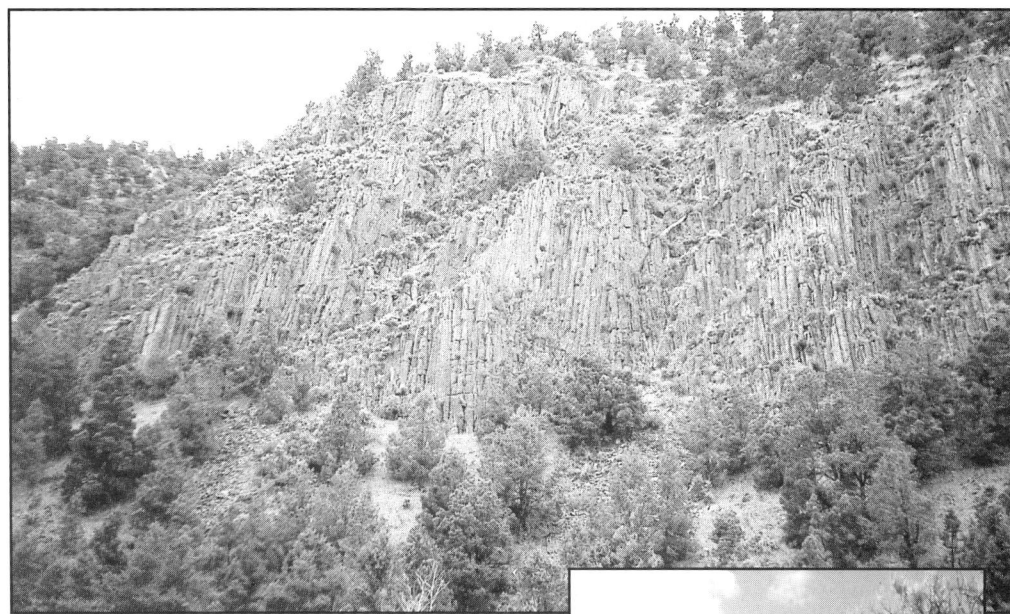

Columnar jointing in rhyolitic ash-flow tuff, Carroll Summit.

Photo: Kris Pizarro

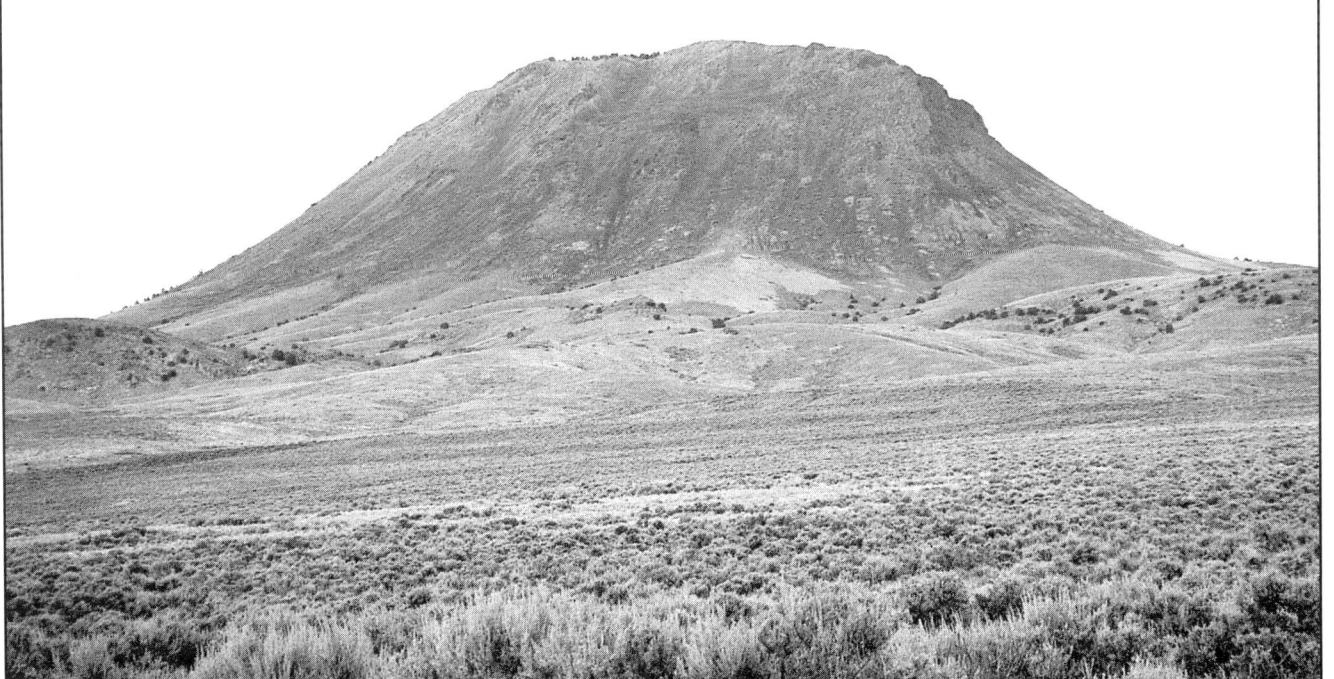

Photo: Jack Hursh

Close-up of columnar jointing in rhyolitic ash-flow tuff.

Iron Mountain in the northern Shoshone Mountains.

interval	cumulative	milepost
3.3	76.0	CH 76.0

The view to the right, at 1:00 to 3:00 is of the thick pile of ash-flow tuff that fills the Desatoya Mountains caldera. We are looking at the steep western wall of the caldera.

To the left of the highway, we can still see light-colored lakebed sediments in the low hills in the foreground.

As we travel this next stretch of highway, envision what this place would have looked like in 1861—no highways or power lines, no fences. But there were roads, poles with wires, corrals filled with horses, and a flood of people heading west. Without doubt, this was a busy and sometimes dangerous place. There was substantial activity with wagon trains heading west to the gold country of California and the Comstock Lode, commercial stage lines taking passengers from points east to the Comstock Lode and San Francisco, and the daily Pony Express riders hauling First Class postage at $5.00 per ounce between St. Louis and Sacramento.

interval	cumulative	milepost
4.0	80.0	CH 80.0

Ruins of an Overland Stage station are on the left (NHM #83).

In its day, an important stagecoach stop on the Butterfield (1861–1866) and Wells, Fargo & Co. (1866–1869) stage route. Fresh horses, blacksmith services, and wagon repair facilities were available here. (plate 7c)

The Pony Express Cold Springs Station was constructed in 1860 on the sagebrush-covered bench a short distance east (to your right) of the highway. Pony Bob Haslum, a Pony Express rider famous for his very long ride, extended by the outbreak of the Paiute War, rode from Friday's Station, at Lake Tahoe, all the way to Smith Creek Station (ahead on the Pony Express route). On his return trip through Cold Springs, he found the station attendants killed by Paiutes. He kept going to Sand Springs Station where he was able to get a fresh horse to continue back to Friday's Station. This was a 380-mile trip on horseback in only two days.

To the north are the ruins of a telegraph repeater and maintenance station that serviced this segment of the Overland Telegraph-Pacific Telegraph Co.'s pioneer transcontinental line, completed between Sacramento and Omaha in 1861. The coming of the transcontinental railroad and its parallel telegraph line along the Humboldt River to the north spelled the demise of both the telegraph line and the stage route here. The telegraph line was abandoned in August 1869.

interval	cumulative
0.2	80.2

Pony Express informational board (turn right).

interval	cumulative
0.4	80.6

More ruins.

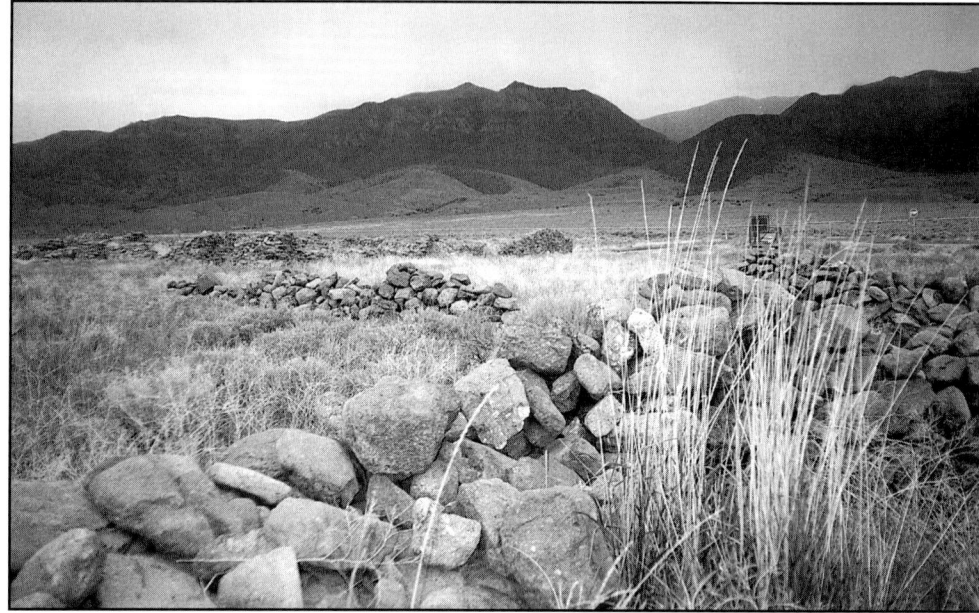

Rock walls remaining at the site of the Overland Stage station near Cold Springs with the Desatoya Mountains in the background.

Photo: Kris Pizarro

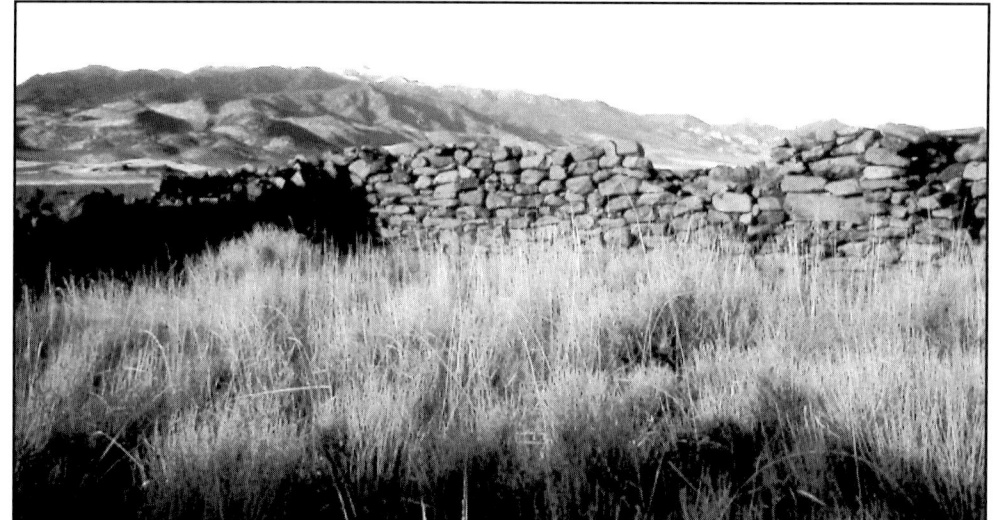

Cold Springs Pony Express ruins with the Clan Alpine Mountains in the background.

Photo: Jack Hursh

interval	cumulative	milepost	
1.1	81.7		Cold Springs, gas station and Nevada State Highway Maintenance Station.
0.4	82.1		From this section of highway, we have a good view of the massive caldron-filling tuffs that form the southern Clan Alpine Mountains. This range has not been studied in detail, so we don't know as much about the volcanic history as at Fairview Peak.
2.0	84.1		The rocks forming the low hills at the range fronts on both the Clan Alpine Mountains to the northwest and the Desatoya Mountains to the southeast flowed out from the calderas in the central parts of these ranges.
2.0	86.1		The caldera margin in the Desatoya Mountains can be seen at 3:00. Note the massive nature of the tuffs inside of the caldera (to the right of the margin), compared to the thinner layered ash-flow tuffs to the north (left) that flowed outside of the caldera margin.

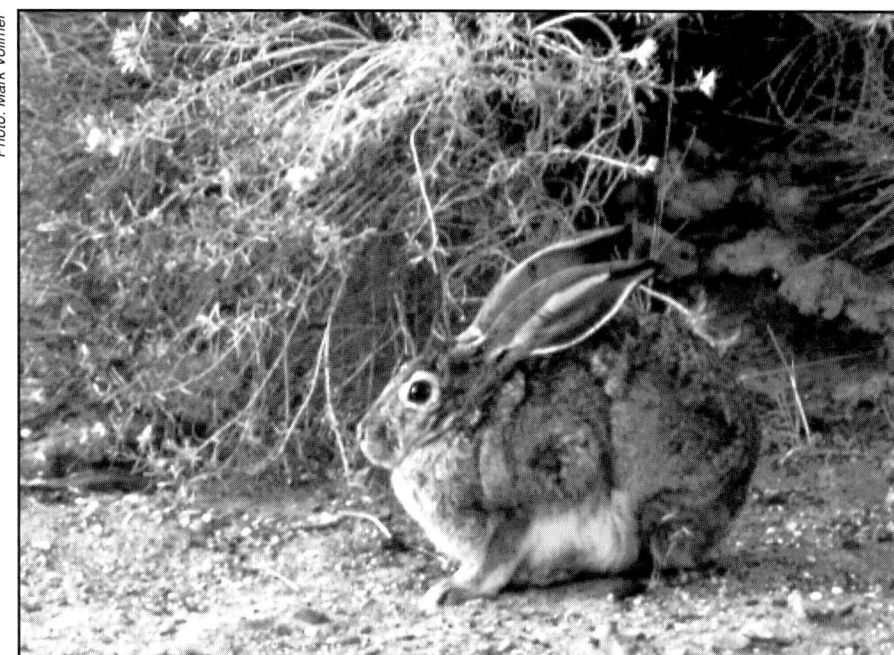

Photo: Mark Vollmer

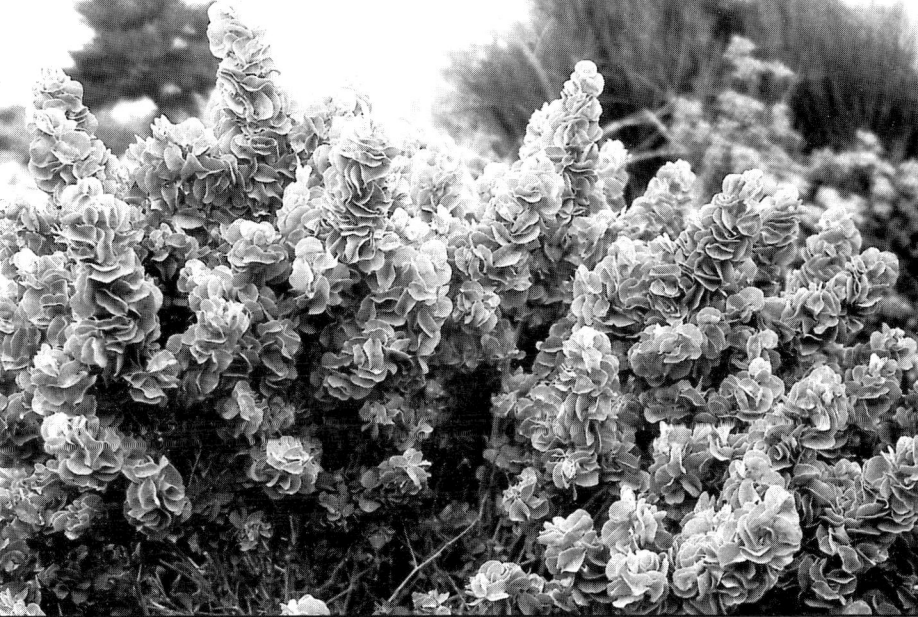

Photo: Kris Pizarro

Spiny hopsage (*Grayia spinosa*) medium-sized shrub, 1–3' tall; spiny, grayish stems, elliptical leaves about 1" long.

Hopsage grows in association with sagebrush, and on talus slopes and alkaline flats. Flowers are inconspicuous. Seeds of female plants (photo) become enclosed by two large bracts whose colors vary from whitish green to shades of red.

Black-tailed jackrabbit (*Lepus californicus*) Body 18–25" long; large ears up to 6.5" long; weight 3–7 pounds; grayish to sandy colored above, ticked with black; buff to white below; black-tipped ears; tail has black stripe above; very large hind feet.

The black-tailed jackrabbit is not a rabbit, but a hare. Its young are born fully furred with eyes open, unlike true rabbits whose young are born blind and nearly hairless. It was originally named "jackass rabbit" by the pioneers, who marked the obvious resemblance to its namesake. The jackrabbit relies on its oversized ears both to warn it of approaching danger and to radiate body heat during hot summer days.

The black-tail becomes active at dusk and feeds mostly at night on grasses, herbs, and stems and leaves of shrubs. It spends the day resting in a shallow depression or "form" under a bush with a good view of approaching predators, particularly coyotes, eagles, and large hawks. When disturbed it takes off at great speed, easily reaching 35 mph or more. It often bounds high in the air every few jumps to keep track of its pursuer.

The female may mate several times between early spring and midsummer. Over a period of years there is an ever expanding breeding population and numbers eventually skyrocket. Hunting by predators, disease, and/or lack of adequate food eventually cause a rapid decline in numbers. The rise and fall in population follows a 9–11 year cycle.

COLOR PHOTO CAPTIONS

PLATE 5

5a **Rhyolite dike** (left) that has bleached and altered (center) originally dark **Triassic sedimentary rocks** (right). *Photo: Kris Pizarro*

5b **Basalt boulders** coated with **desert varnish**, near Grimes Point. *Photo: Kris Pizarro*

5c **Sand Mountain** with rock walls of the **Sand Springs Pony Express Station** in the foreground. *Photo: Jack Hursh*

5d **Desert bighorn sheep** (*Ovis canadensis*) **Rams 5¼–6' long, ewes 4¼–5½' long; sturdy, muscular build; brown to tan with white belly, rump, muzzle, and eye patches. (Ram, ewe, and yearling ram shown)**

Rams have massive horns that grow over and behind the ears in a C-shaped "curl." Horns are permanent and grow incrementally year by year. It may take 7 to 8 years for a ram to acquire a full curl. Ewes have shorter, more slender horns that curve slightly back. Pictured from left to right are a mature male, mature female, and a male yearling.

Bighorn sheep occupy rough, precipitous terrain near sources of permanent water. Bands follow regular feeding routes and maintain bedding grounds that may be used for years. The Desert bighorn sheep is Nevada's state mammal. *Photo: Jim Yoakum, Nevada Chapter, The Wildlife Society.*

5e **Prince's plume** (*Stanleya pinnata*) A member of the mustard family, prince's plume is found in sandy soils and talus slopes, often in association with sagebrush. *Photo: Jack Hursh*

5f **Black-collared lizard** (*Crotaphytus insularis*) *Photo: Jack Hursh*

PLATE 6

6a **Desert evening primrose** (*Oenothera deltiodes* var. *piperi*) A May bloomer seen near the Sand Springs Pony Express ruins. *Photo: Jack Hursh*

6b **Chalk Mountain** seen from the Fairview Peak road. The **Augusta Mountains** are to the north, beyond the cloud shadow. *Photo: Kris Pizarro*

6c **Yellow prickly pear cactus** (*Opuntia polycantha*) forms low, spiny mounds seldom more than 6 inches high. It prefers sandy soils and can be found in desert areas, as well as piñon-juniper woodlands. Prickly pear flowers may also be magenta. *Photo: Jack Hursh*

6d **Blazing star** (*Mentzelia laeviculmis*) blooms in mid to late summer and can commonly be seen growing alongside roads and highways. *Photo: Jack Hursh*

6e Sunset on **Chalk Mountain**. *Photo: Jack Hursh*

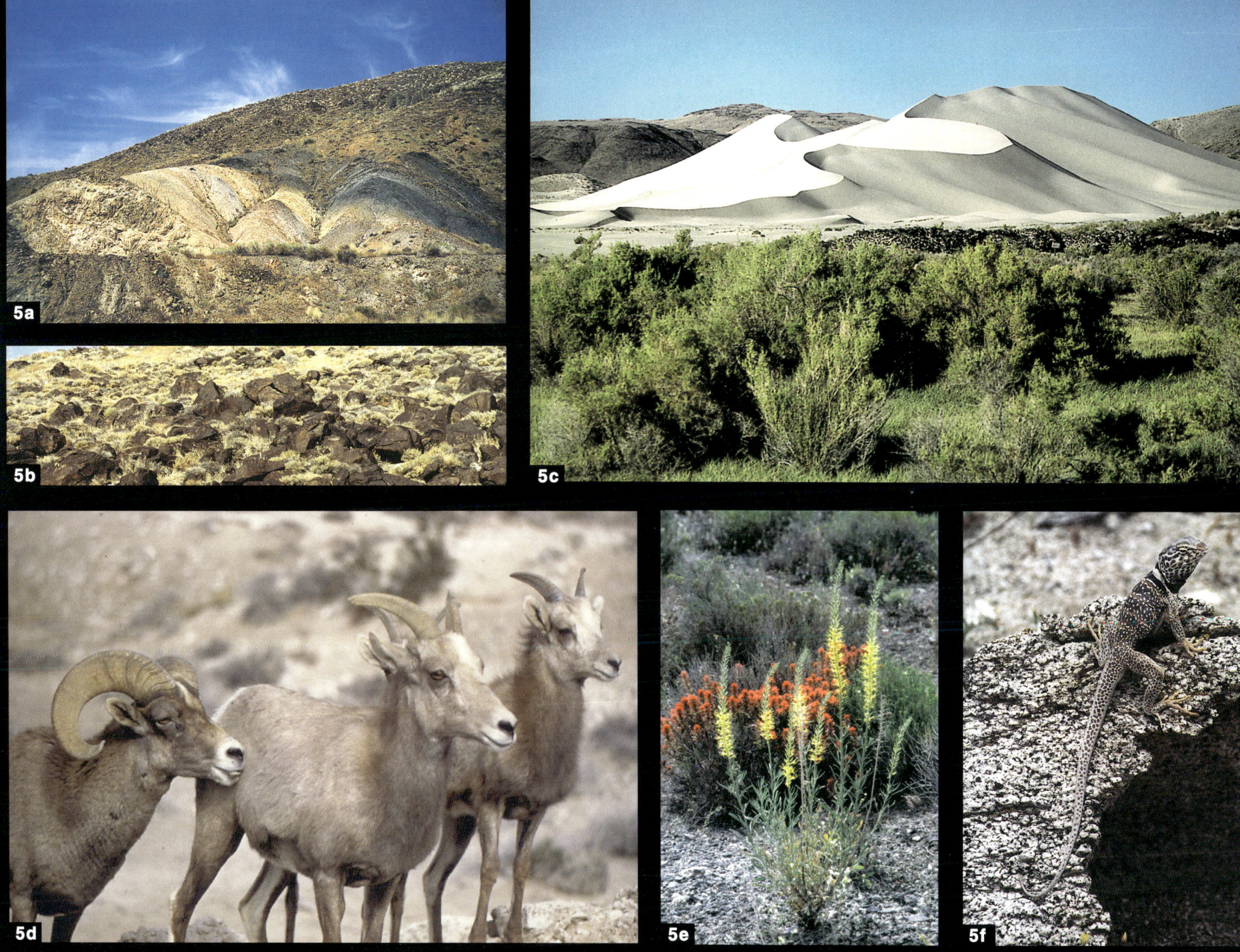

5a

5b

5c

5d

5e

5f

PLATE 5

61

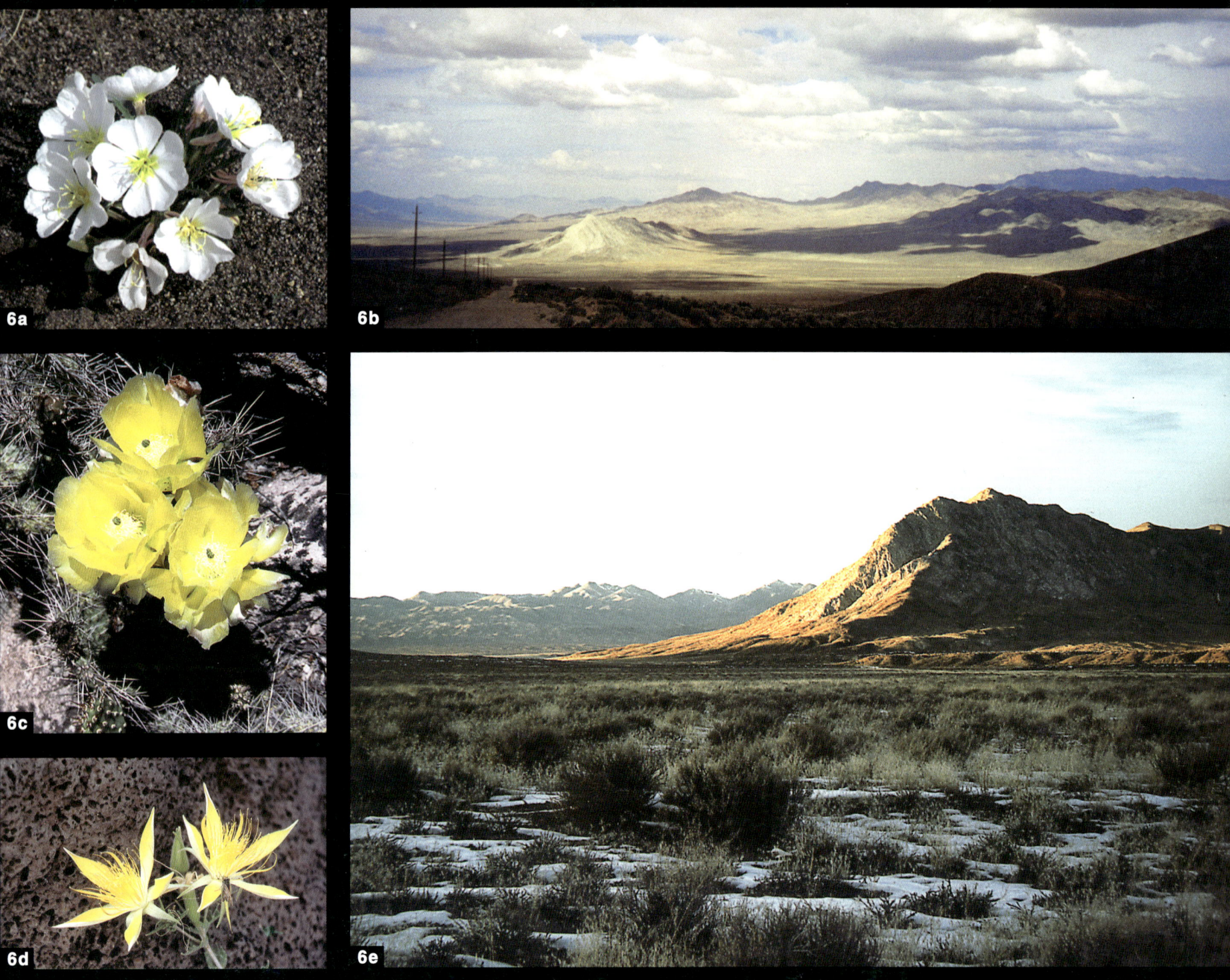

6a

6b

6c

6d

6e

PLATE 6

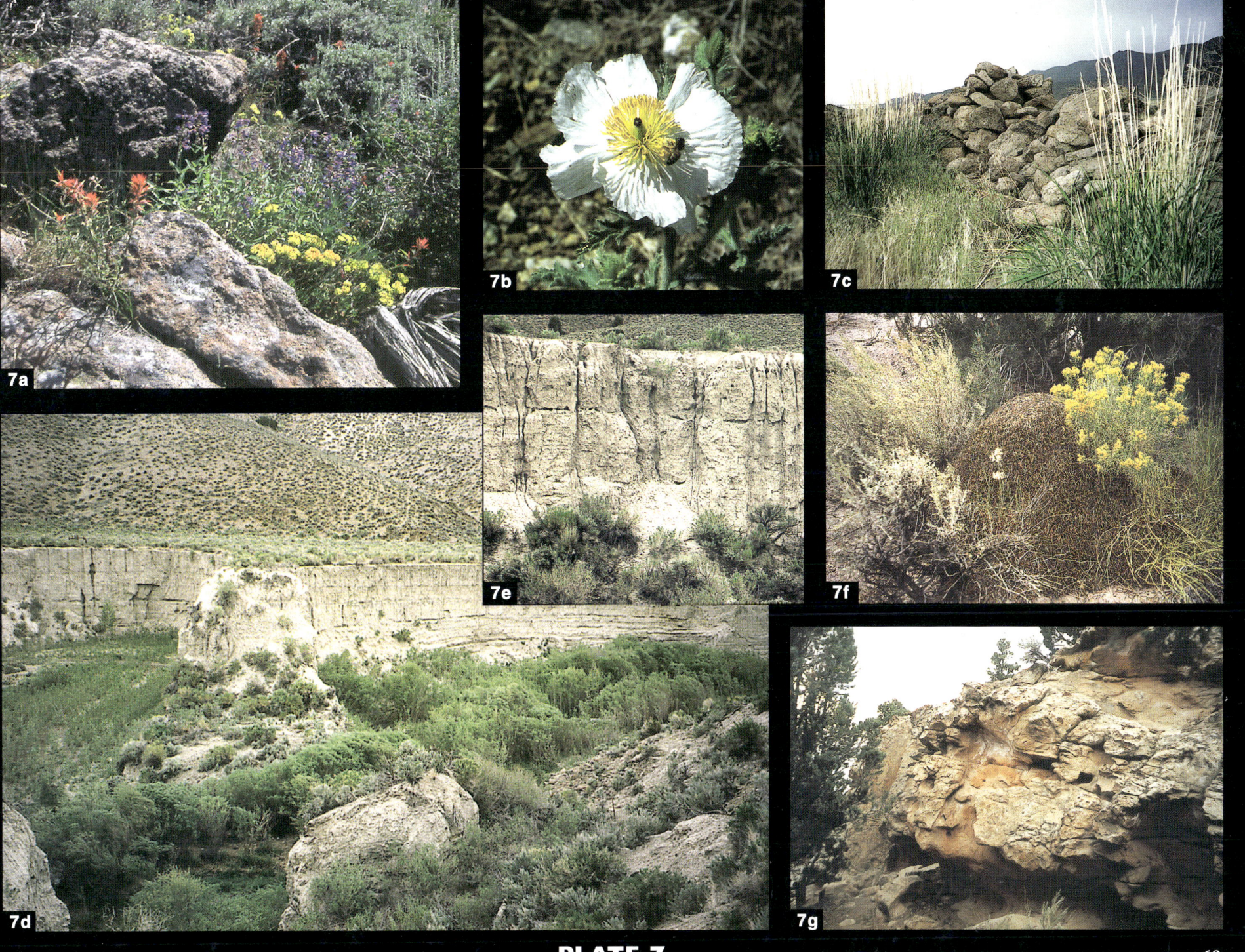

7a

7b

7c

7d

7e

7f

7g

PLATE 7

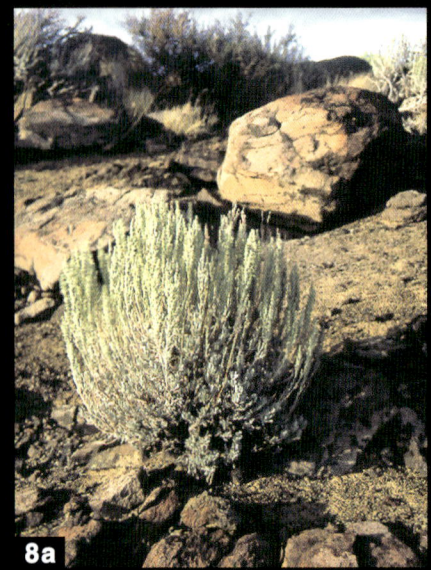

8a

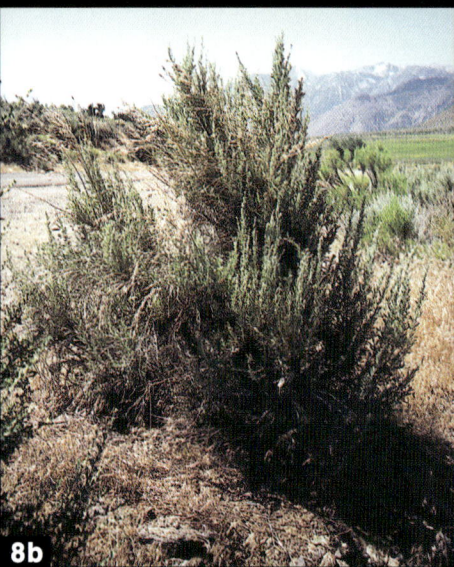

8b

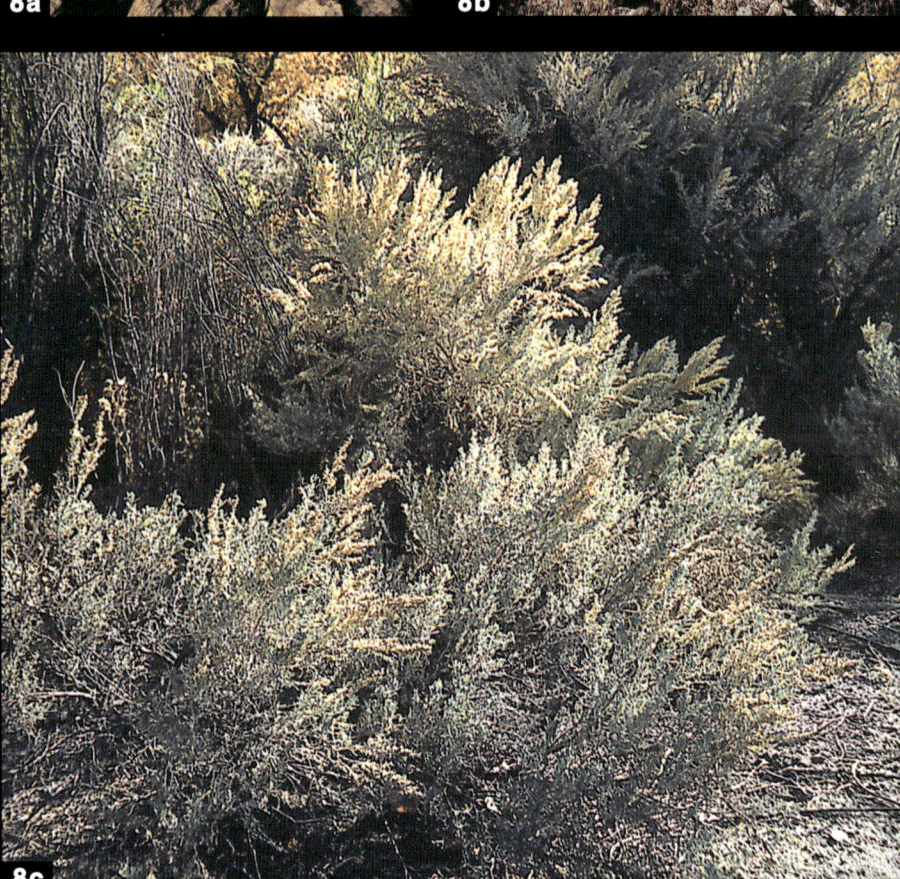

8c

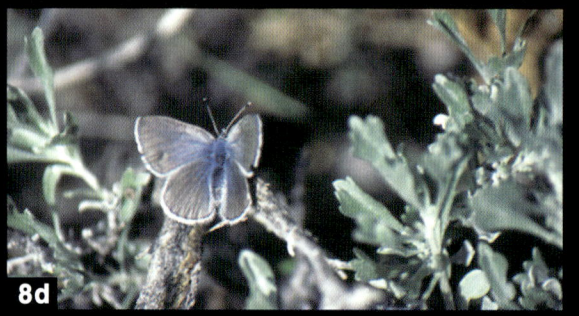

8d

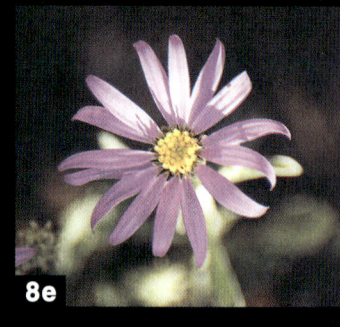

8e

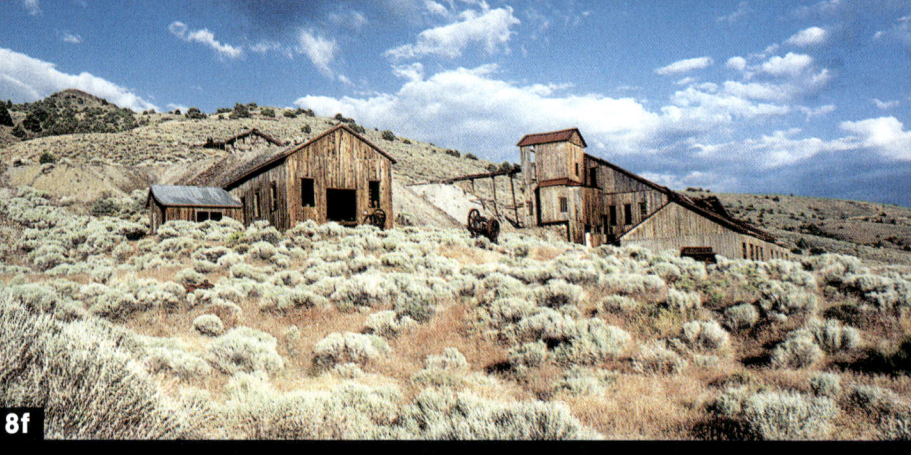

8f

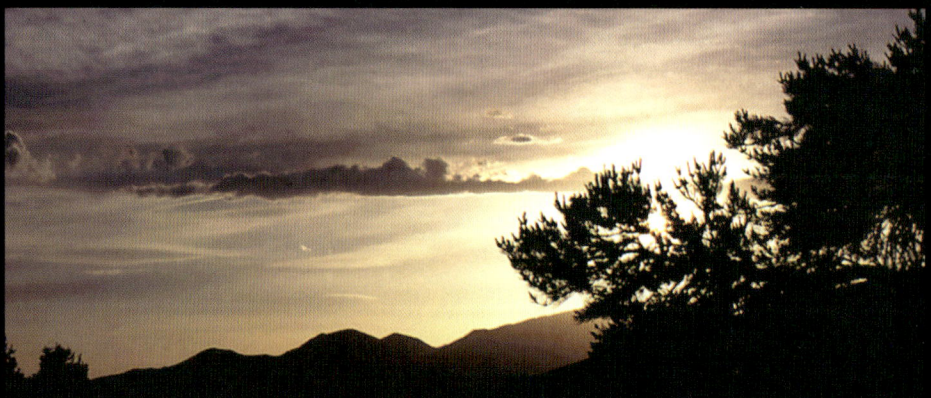

8g

PLATE 8

COLOR PHOTO CAPTIONS

PLATE 7

7a **Paintbrush**, **penstemon**, and **sulfur flower** (desert buckwheat) add color to the landscape. *Photo: Jack Hursh*

7b The **prickly poppy** (*Argemone platyceras*) grows in dry, sandy and gravelly soils and is commonly seen along roadsides. It grows to 3 feet tall with stems and leaves covered with sharp spines. *Photo: Jack Hursh*

7c Rock ruins at the **Cold Springs stage station**. *Photo: Kris Pizarro*

7d Steep banks and meanders of **Buffalo Creek**, cut in soft, **stream deposited sediments**. *Photo: Kris Pizarro*

7e A variety of birds and small mammals make their homes in the walls of the wash. **American kestrels** and **scrub jays** are often observed here. *Photo: Kris Pizarro*

7f **Ant hill** at Hickison Petroglyph Recreation Area. *Photo: Kris Pizarro*

7g Cavities in a weathered outcrop of **welded ash-flow tuff** within the Hickison Petroglyph Recreation Area. *Photo: Kris Pizarro*

PLATE 8

8a–c **Big sagebrush (*Artemisia tridentata*) Many-branched shrub with fibrous, shredded-looking bark; three-lobed, densely hairy, grayish-green leaves; aromatic; very small yellowish flowers on dense panicles appear August through October.**

Big sagebrush has a broad ecological tolerance and can be found from the floors of valleys up to 10,000 feet in elevation. This is the shrub that gives so much of the West its characteristic gray-green color. Under ideal conditions it can grow to 12 feet tall, yet in arid regions with poor soil it may average less than a foot in height. The shrub produces a huge quantity of tiny seeds that are spread primarily by the wind. It has been estimated that one shrub, about 3 feet in diameter, can have around 450 flowering branches that produce about 350,000 seeds. It is commonly found in association with rabbitbrush, green ephedra, spiny hopsage, and bitterbrush.

Big sagebrush uses several mechanisms to increase its absorption and retention of water. It grows two types of leaves: longer, three-lobed leaves that remain on the plant all year, and smaller, nonlobed leaves that appear in early winter, enabling the plant to take advantage of moister growing conditions. The latter drop off during dry conditions the following summer, and the plant becomes somewhat dormant. Dense hairs on the leaves reflect sunlight and aid in slowing water loss. The plant has a two-part root system. Widely spread, shallow roots take in water from passing storms while a very long tap root extends downward to take advantage of underground reservoirs. Big sagebrush supplies most of the diet of sage grouse and is a primary browse for deer and antelope. *Photos: 8a-Jack Hursh, 8b and 8c-Kris Pizarro*

8d Here a **Blue** (about 1 inch wingspan) has lit on sagebrush. Numerous species of Blue butterflies make Nevada home. *Photo: Jack Hursh*

8e Numerous species of **asters** and **daisies** are found throughout the Great Basin. *Photo: Roy W. Cazier*

8f Mill and associated buildings at the **Berlin Mine**, Berlin Ichthyosaur State Park. *Photo: Aleta Hursh*

8g Sunset at **Berlin**. *Photo: Jack Hursh*

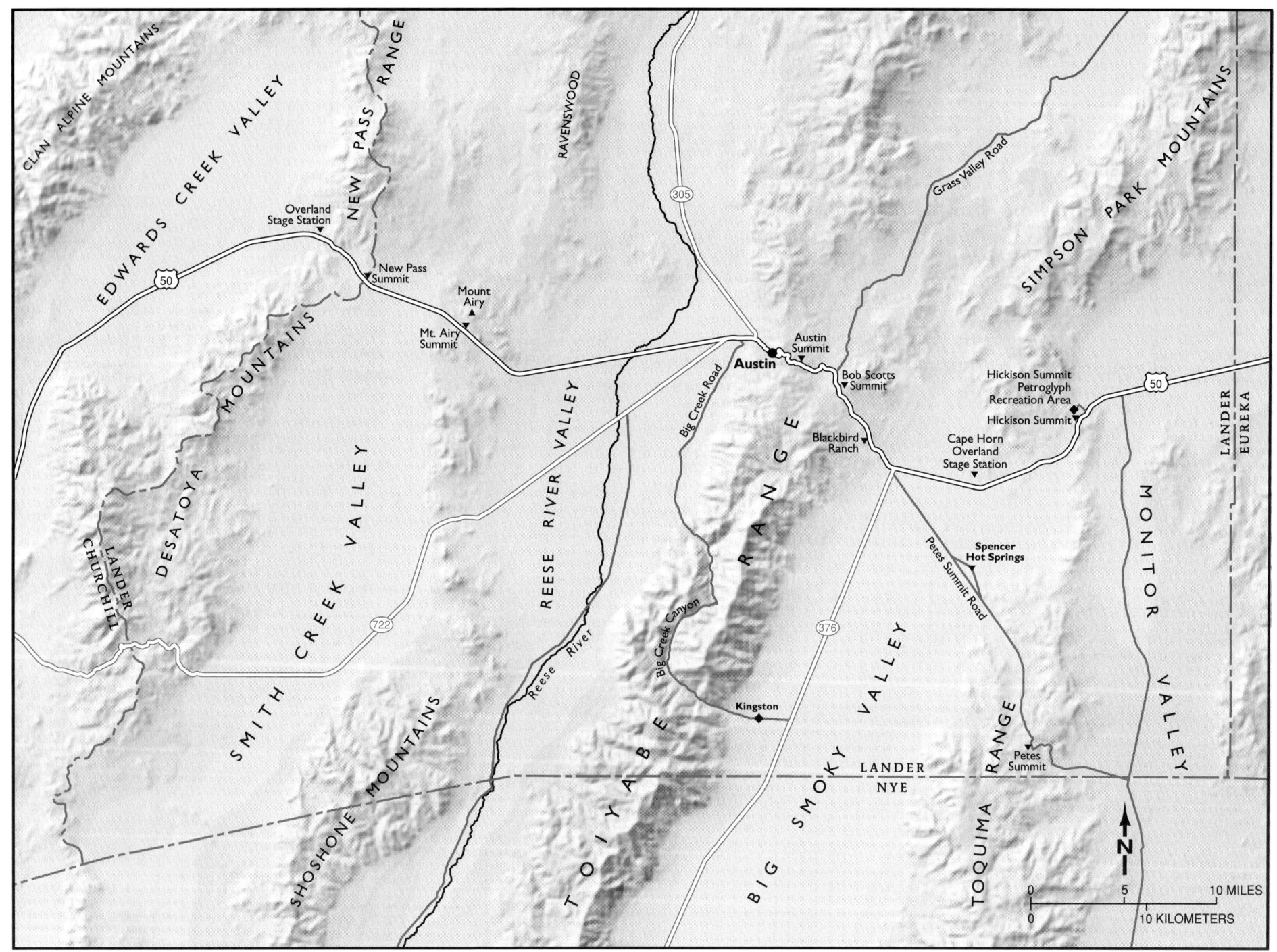

Route map, Lander County.

interval	cumulative	milepost	
17.0	103.1		Entering the canyon that separates the New Pass Range, to the left (north) from the Desatoya Mountains (south). Exposures of Oligocene New Pass Tuff can be seen in the road cuts on both sides of the highway. This tuff unit, a crystal-rich rhyolite, is believed to be outflow material from a caldera in the southern Stillwater Range 50 or so miles west of here.
1.9	105.0		Ruins of Overland Stage station on the left. (NHM #135)
1.9	106.9	LA 00	Lander County line, Highway markers again start at zero.
0.3	0.3		New Pass Summit (elevation 6,348 feet)
0.2	0.5		New Pass Mine road. The New Pass gold mine is located on the Churchill-Lander County line about 2½ miles up this road. Gold was discovered there in 1864, but not much happened in the district until after 1900. Mining has been carried out intermittently on narrow, gold-bearing veins since that time, but total gold production has been very small.

Also north of the highway a few miles is the location of the now inactive Shoshone Turquoise Mine. Last mined in the 1980s, raw turquoise from this area found its way into finished jewelry sold all over the southwest. Just to the east of the summit, New Pass road, to the right, leads south into Smith Creek Valley. The Shoshone Mountains are east of the valley.

Photo: Kris Pizarro

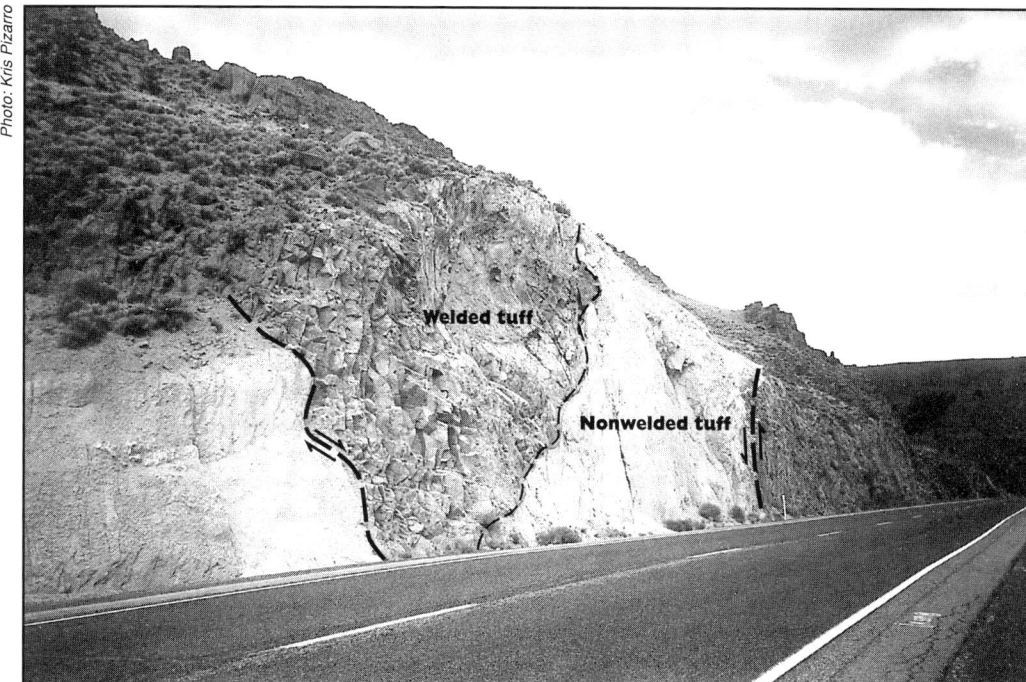

Photo: Kris Pizarro

Road cut west of New Pass. Nonwelded ash-flow tuff (white) is overlain by welded, crystal-rich welded ash-flow tuff (buff-colored, harder unit). The rocks are cut and offset by normal faults, causing the white unit to be moved up into view again at the left of the photo.

◄ **Contact between nonwelded ash-flow tuff (white unit at lower right) and welded ash-flow tuff (blocky unit to left).**

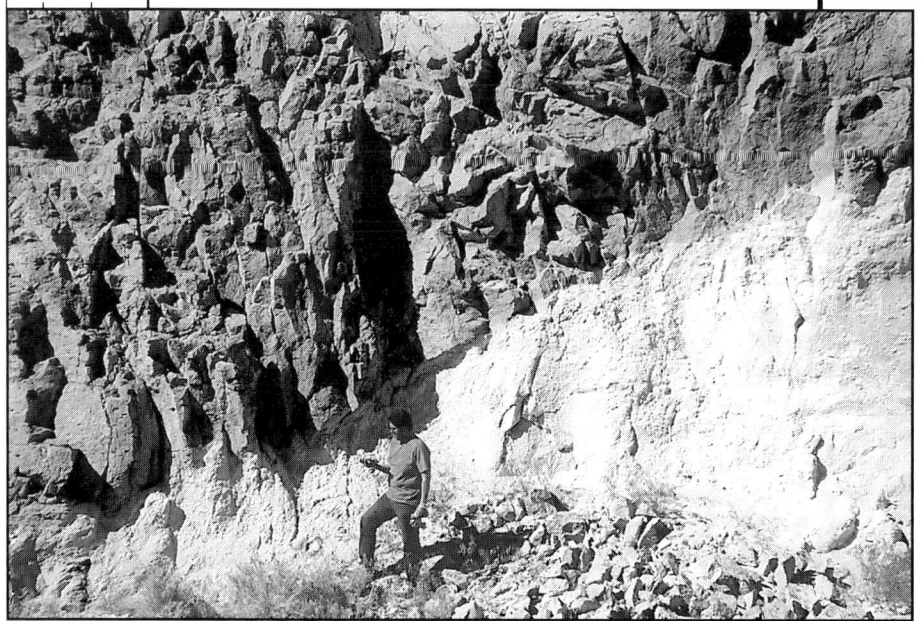

NEVADA TURQUOISE

Nevada has a fairly well established history as the Silver State, and has now created a new niche as one of the most important gold-producing regions in the world. Few realize, however, the important role Nevada turquoise has played in the gemstone world.

Other than obsidian and other hard, flinty stones used for primitive tools and weapons, turquoise was probably the first mineral product to be mined in what is now Nevada. Evidence has been found in southern Nevada of turquoise mining by Native Americans dating back to 300 A.D. to 500 A.D. In more recent history, Nevada has been an important supplier to the southwest Indian turquoise industry. Most of Nevada's turquoise mines fall along a broad, north-south-trending belt that passes through the center of the state. We are now starting to cross that belt, and the Shoshone Mine north of New Pass Summit is one of the newer mines. There are several mines to the north, south, and east of Austin, but the most famous of the turquoise producers are near Battle Mountain and Tonopah.

interval	cumulative	milepost
5.9	6.4	
3.6	10.0	

Mount Airy Summit (elevation 6,679 feet). Site of another Overland Stage stop, on the left. A station was maintained here from 1861 to 1869.

Ash-flow tuff is exposed along the highway for the next several miles. Notice the prominent columnar jointing in unit that caps the cliff on the left, just beyond the summit. The rugged, columnar-jointed unit overlies softer pale green to white volcanic ash deposits that form the gentle slope below the cliff and make up the low hills between the cliff and the highway.

We are now entering the Reese River Valley. Ahead and to the right is the Toiyabe Range, the highest mountain range in central Nevada. The gap in the range straight ahead at 12:00 is Big Creek Canyon.

Reese River, which has its headwaters in the high peaks of the Toiyabe Range, flows north to join the Humboldt River at Battle Mountain. A cold, clear trout stream as it leaves the mountains, Reese River is usually little more than a cloudy trickle at the point our highway crosses it a few miles ahead.

During the boom years of the Reese River mining district and the "Rush to the Reese" in the early 1860s, entrepreneurs floated stocks in the Reese River Navigation Co. to eastern investors. The scheme was that rich ores could be taken by paddle wheel steamships down river to the Central Pacific railhead at what is now Battle Mountain. Shipping by steamships would be far more economical than by either freight wagons or railroad, and investors could reap fortunes.

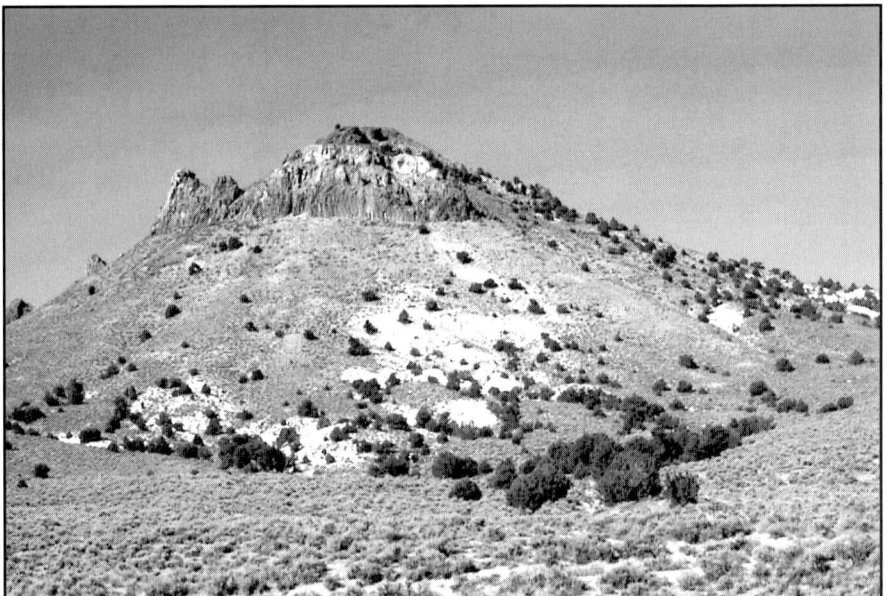

Photo: Jack Hursh

Hill to the north of the highway at Mount Airy Summit composed of white, bedded volcanic ash capped by a more resistant welded ash-flow tuff unit displaying prominent columnar jointing.

The west flank of the Toiyabe Range, south of Austin. Big Creek Canyon is to the right.

Photo: Kris Pizarro

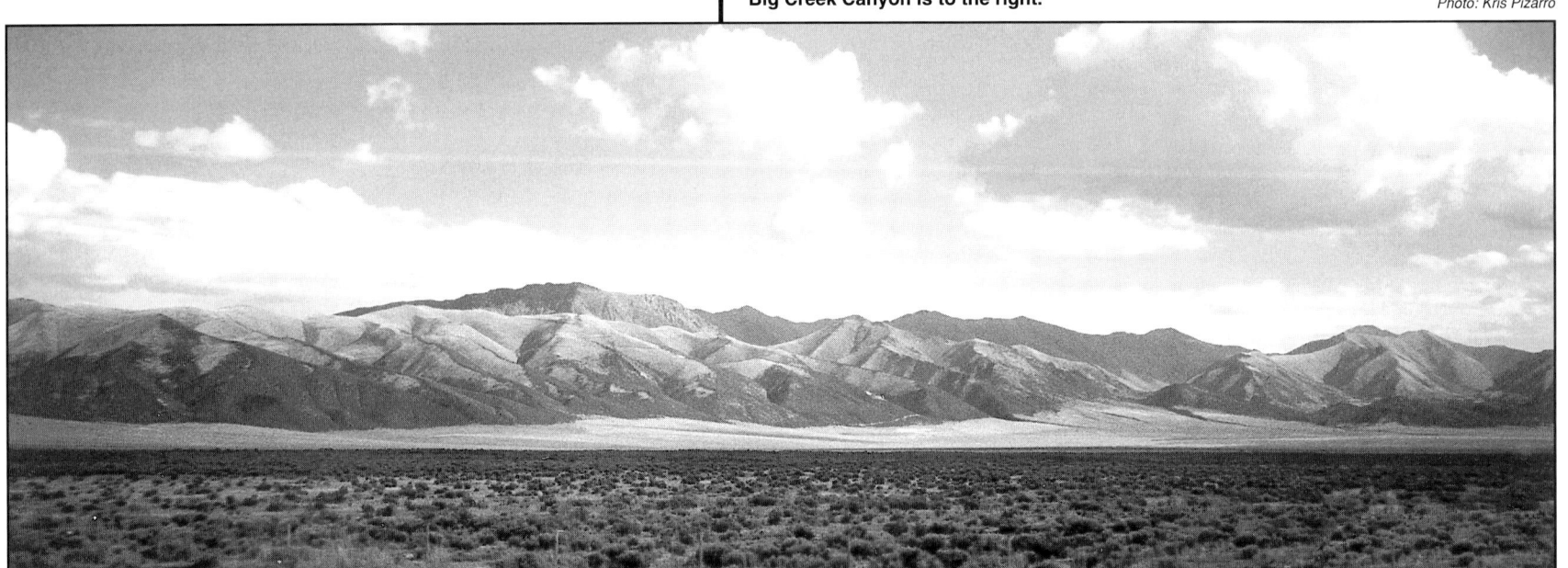

interval	cumulative	milepost	
5.7	15.7		Historical route of Pony Express trail crosses the highway.
0.6	16.3		Long anticipated crossing of the Reese River. As you cross, envision Mississippi River style paddle wheel steamships plying the muddy waters. Don't blink or you might miss the river.
4.7	21.0		Junction with State Route 722, the east end of the Carroll Summit road.
1.3	22.3		Junction with Big Creek road. Gravel road (usually well maintained) enters Big Creek Canyon about 9 miles south of here. A U.S. Forest Service campground about 1 mile up the canyon provides a pleasant place to spend a night or two. Beyond the campground, the road quickly becomes a four-wheel-drive adventure, but it continues over the Toiyabe Range at Big Creek summit (+8,500 feet) and descends the eastern slope via Kingston Canyon. Kingston Creek has stretches with good fishing, and there is another campground on the creek just above the village of Kingston. The road is paved from Kingston to State Route 376, which connects with U.S. 50 about 12 miles ahead.

This is the end of the third section of the road log. We leave a terrain dominated by volcanic rocks and begin our traverse of fault-block mountains composed mainly of Paleozoic-age sedimentary rocks.

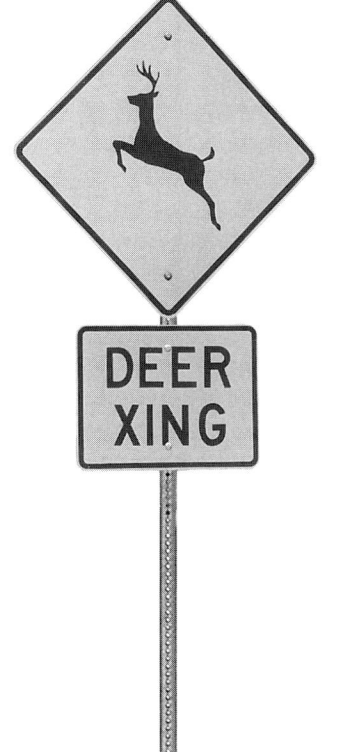

SECTION IV: FAULT BLOCK MOUNTAINS, THE ROBERTS MOUNTAINS THRUST BELT, AND NEVADA GOLD

Although most of Nevada's U.S. Highway 50 is within the Basin and Range province, the tell-tale features of this province are best exposed in this fourth and last section of our road log. This segment begins as the route climbs into the Toiyabe Range west of Austin. For the remainder of the journey across Nevada, U.S. 50 climbs over one tilted, fault-block range after another, each separated by yet another alluvial valley.

We are still in caldera country, and we will see several more major calderas east of the Toiyabe Range. However, from here to the end of our log at the Utah state line, impressive outcrops of Paleozoic-age sedimentary rocks exposed in fault-block mountains dominate the geology. U.S. 50 now begins crossing the Roberts Mountains thrust belt where low-angle faulting has pushed upper-plate, or hanging-wall, rocks as much as 90 miles eastward, overriding more-stationary lower-plate, or footwall, rocks. This is the home of Nevada's famous "Carlin-type" gold, and there are hundreds of known gold deposits within a broad area extending about 150 miles north, 90 miles south, and 170 miles east from here. The belt contains (depending somewhat on the current price of gold) about 50 operating gold mines. Most of these mines are far to the north of us, but two major mine areas near Eureka are visible from the highway. The log ends as U.S. 50 crosses into Utah in Snake Valley, east of the Great Basin National Park.

Photo: Roy W. Cazier

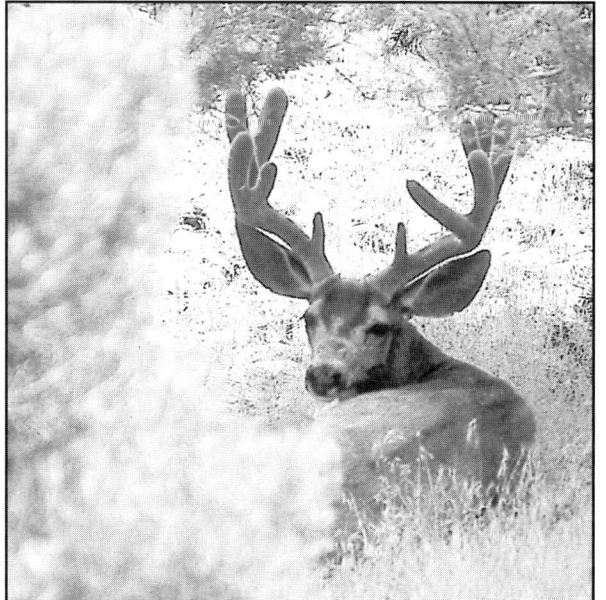

Mule deer (*Odocoileus hemionus*) Adults 32–42" high at shoulder; large male can weigh 200 pounds; ears very large and move independently; adults reddish- or yellowish-brown in summer, grayish in winter; rump and throat white; males larger than females.

Mule deer are found in mountains and foothills across the Great Basin. They often migrate up and down seasonally to avoid heavy snows. They prefer mixed habitats and forest edges and are mostly active mornings, evenings, and moonlit nights. They eat grasses and herbs when they are available but depend mainly on a variety of shrubs. They are also fond of mushrooms and fruit, particularly apples.

In April bucks begin to grow antlers. They are spike-like in the second summer and branched in older deer and can attain a spread of four feet. The antler's covering of skin or "velvet" dries and is rubbed off by September. The mating season, or rut, begins in late fall. Antlers are dropped by the following March.

Fawns, usually twins, are born around June and are kept concealed for the first month. Their spots disappear with their first molt into a winter coat. Does and fawns often stay together until the next young are due.

0.8	23.1	

As we start up the grade into Austin, we pass through Austin's historical cemetery; the older section is on the right.

| 0.1 | 23.2 | |

Intersection with State Route 305 (on left), which follows the Reese River Valley north for about 90 miles to Battle Mountain where the Reese River enters the west-flowing Humboldt River. Battle Mountain is located on Interstate 80, which follows the old emigrant route.

| 0.2 | 23.4 | |

Stokes Castle is visible at about 2:00 to 2:45 south of the mouth of Pony Canyon (the canyon our highway is climbing into). This tall stone structure was built in 1897 by Anson Stokes, owner of the narrow gauge railroad that carried ore and supplies between the Austin and the Central Pacific Railroad at Battle Mountain. The three-story castle was built as a summer home for the Stokes family, but was only used for a short time, and then abandoned.

Beyond the cemetery, as the road curves through the first cuts in the mountainside, notice the many iron-oxide-stained fractures and thin quartz veins that cut the exposed granitic rock.

| 0.7 | 24.1 | |

Entering Austin, center of the Reese River mining district. The graded road to the right, around the hill, leads to Stokes Castle. The road is usually in good shape and, although there are no facilities, the Castle is a good place for a picnic—there is even an old mine dump to examine.

The silver mines were located on the north and south sides of the steep canyon above the town, but most were within an area of about one-half square mile on Lander Hill, directly ahead of us as U.S. 50 swings to the left and climbs out of the first hairpin turn leaving upper Austin.

Photo: Kris Pizarro

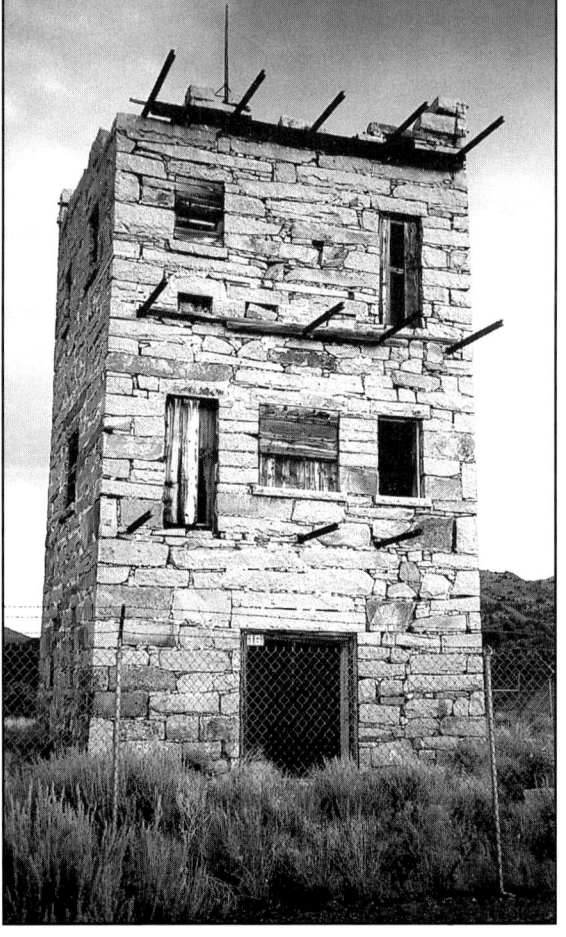

Stokes Castle.

AUSTIN AND THE REESE RIVER MINING DISTRICT

Silver was discovered in Austin in 1862, touching off a "Rush to Reese River" that was second only to the Comstock rush a year or so earlier. Austin became a boomtown with a population of almost 10,000. Production of $25 to $65 million was reported up to 1886 by which time most of the rich ore was mined out. Arrival of the Nevada Central Railroad in 1897 did little to help the town's fortunes, and Austin slid into the semi-ghost town status that it has enjoyed ever since. Austin was the county seat of Lander County until 1980, when Battle Mountain finally carried the county government north. With this move, one of the town's few remaining reasons for existence ceased, and Austin's population has continued to dwindle.

Many historical buildings can be seen on both sides of the main street, including the old courthouse, the International Hotel, and the Gridley Store. The Gridley Store's claim to fame revolves around a lost election bet and a sack of flour. It was a Civil War era election. R.C. Gridley, the store's proprietor, bet on a southern sympathizer who lost to a Unionist. The bet required Gridley to carry a 50-pound sack of flour from upper Austin to Clifton, a small settlement that was at the mouth of Pony Canyon, below the present cemetery. Gridley made good his bet and carried the sack; then loser, winner, and followers repaired to the nearest saloon for refreshments. The idea then arose to auction the sack and donate the proceeds to the Sanitary Fund, a Civil War precursor to today's Red Cross. The idea took hold, the sack was taken to Virginia City to auction, then to other cities around the country including San Francisco and New York. Before it was over, Austin's sack of flour had raised over $175,000 (a lot of money in 1864) for the benefit of the Sanitary Fund.

Austin is one of those rare places that impresses, in one way or another, everyone who passes through, and usually provides each with a unique "Austin Story" to tell. One such story, overheard here a few years ago, involves one of the three impressive brick churches you will spot as you climb up through town. These churches attest to Austin's attempt to temper the wilder elements of a frontier mining community. Only one still has regular services and another is undergoing renovation. Based on the story told, however, one traveling minister may have permanently crossed this stop off his list. It seems that, during a thunderstorm, lighting struck the steeple of one of the churches and the charge passed through the church to the ground, welding the bed usually occupied by the minister to a refrigerator standing next to it.

In another incident, two visitors spent the evening strolling along the main street admiring historical buildings. Next day on their way to breakfast the travelers found that two of the more picturesque structures had collapsed overnight, leaving behind only piles of rubble.

interval	cumulative	milepost	
0.9	24.8		Gridley Store, rock building on the left.
0.4	25.2		Many large mine dumps can be seen ahead and on both sides of the highway. Silver minerals in these mines were found in thin quartz veins that cut through a Jurassic-age granitic pluton. While an occasional specimen of ruby silver (pyrargyrite, a silver-antimony sulfide with a dark, ruby-red color) can be found on the old dumps, most are on private land and are off limits to the public.
0.7	25.9	LA 26.0	Wide road cut on the left exposes a typical (but barren of silver) quartz vein of the type mined during Austin's boom days. You can see the iron- and manganese-oxide-stained quartz in granitic wall rock. The wall rock is slightly altered (note the white, chalky appearance of some of the minerals in the rock) and is also stained by iron oxides (rust).
1.1	27.0	LA 27.0	Austin Summit (elevation 7,484 feet). The rock along the highway at the summit is Jurassic quartz monzonite, the same as in the mining district to the west, but it is a harder rock (not altered as it was near the mines) and is a little more resistant to weathering. A fire in 1983 burned the already sparse piñon-juniper forest from the summit and western slope of the range here, exposing the rounded knobs of granitic rock to view. Note the vertical jointing (cracks) in the granitic outcrops.
1.4	28.6		Picnic table on the left
0.7	29.3		Intersection with Grass Valley-Cortez road. Gravel road to the left travels 75 miles northerly through some even lonelier country before it connects with paved State Route 306. A side trip over this road would lead you past isolated ranches, along the Simpson Park Mountains, past the ghost town of Cortez, and finally to Crescent Valley where several large, open pit gold mines are currently active. This trip is recommended only for those with sturdy vehicles, adequate water, and emergency supplies (you probably won't see anyone for the entire 75 miles), and it is definitely not recommended in bad weather.
1.3	30.9		Bob Scotts Summit (elevation 7,267 feet) This is a good place to take note of the piñon-juniper association that is so common across this region. These two species of bushy and more-friendly-than-noble trees enjoy each other's company so much that they commonly occur together.

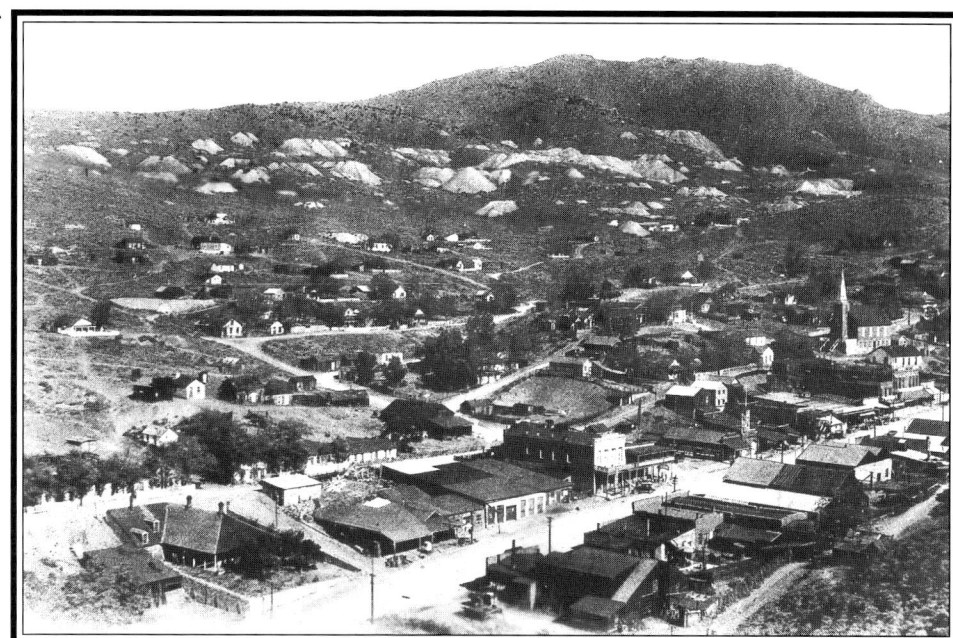

Nevada Historical Society

Austin, probably around 1915. The mine dumps are on Lander Hill, Mount Prometheus forms the skyline to the right.

Photo: Kris Pizarro

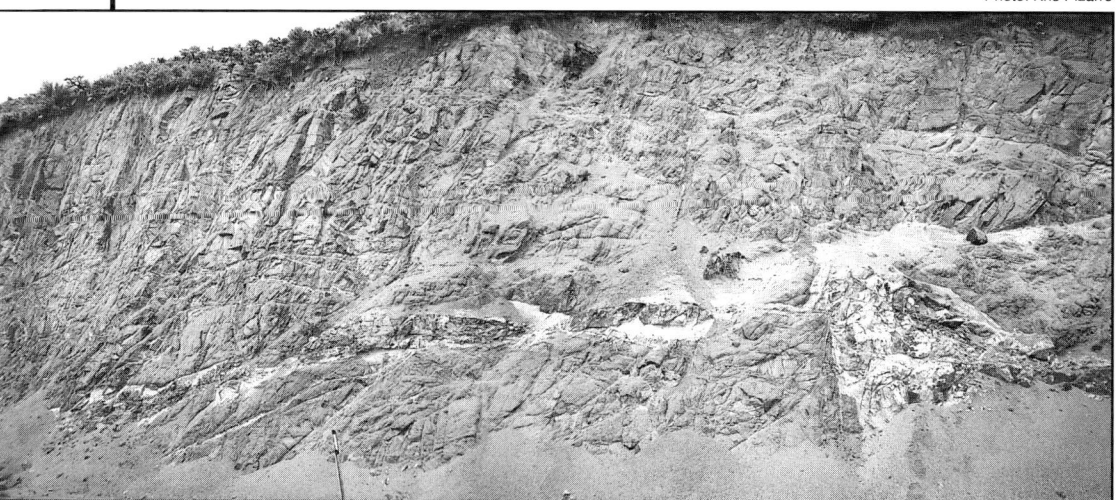

Narrow quartz vein (white streak across lower part of photo) cutting highly fractured granitic rock, road cut west of Austin Summit.

PIÑON COUNTRY

Except in the foothills of the Sierra Nevada and in a few isolated canyons in some of the higher ranges of the state, "forest" in Nevada usually means piñon—after all, singleleaf piñon is one of Nevada's two state trees (the other is the bristlecone pine).

Nevada is included within over 75,000 square miles of southwestern United States landscape that is dominated by a single woodland community. This is the piñon-juniper woodland, the characteristic vegetation type of the southern Rocky Mountains, the mesas of the Colorado Plateau, and the mountains of the Great Basin. The piñon grows on the middle to low slopes of the ranges, at elevations between 5,000 and 8,000 feet, where the mean annual precipitation is between 12 and 18 inches. Eleven species of pine make up the group generally known as piñon pine. In Nevada, and only a few other isolated spots in Arizona, New Mexico, southern Utah, and the Mohave Desert of California, we have the singleleaf piñon (*Pinus monophylla*). This piñon is unique among pines around the world—it only has one needle in its needle bundle. Stop, look, and enjoy the aroma of a real Nevada tree—but don't brush too close to one or you will quickly learn about pine pitch (rubbing alcohol is useful if you don't heed this warning). The piñon's companion in Nevada is the Utah juniper (*Juniperus osteosperma*). These hardy trees, sometimes incorrectly called cedars, are almost always associated with the piñon. They, however, can tolerate dryer conditions and sometimes form pure stands, extending even to the margins of the alkali flats.

Pine nuts from the piñon, high in fat and protein, provided the Native American residents with a valuable and perennial food source. Miners and white settlers in the region, with a more short-term outlook, viewed the nut groves as a wood source—wood for cabins, wood for heat, and wood to make charcoal to fuel boilers and smelters at the mines. These two cultural viewpoints soon clashed and destruction of the pine nut "orchards" was possibly one of the factors that led to the "Battle of Pyramid Lake" in 1860. This skirmish, fought along the Truckee River between Wadsworth and Pyramid Lake northeast of Reno, led to the deaths of some 46 soldiers and civilian volunteers from the Virginia City area. The Indians suffered three wounded and lost two horses in what has been described as the most disastrous conflict the whites ever waged in what is now the state of Nevada. It was this confrontation that prompted the attack on the Cold Springs Pony Express station, resulting in the death of the attendants and Pony Bob Haslum's long ride back to Friday's Station and safety.

Meanwhile, the whites found that piñon made very poor lumber. It was knotty and, because of the stubby nature of the trees, it came only in short lengths. During the boom days of the 1860s, piñon pine, known as "Reese River lumber," sold for only about half the price of imported top-grade lumber. Piñon, however, made very good charcoal for the mine furnaces. By the late 1870s, most of the piñon forests within as much as a 50-mile radius of Austin, Eureka, and other mining towns were completely stripped. The forests you now see are second-growth trees that have returned since mining activity ceased.

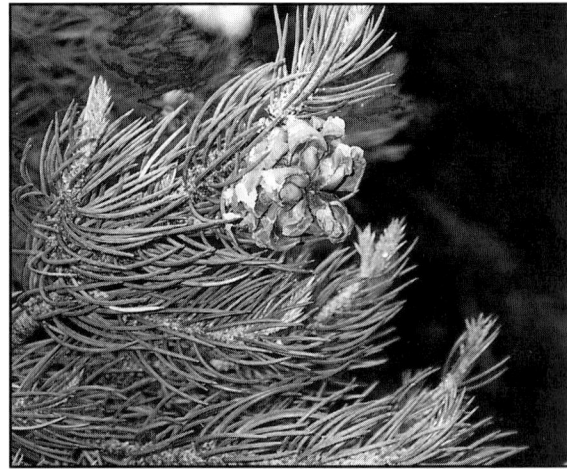

Photo: Kris Pizarro

Photo: Jack Hursh

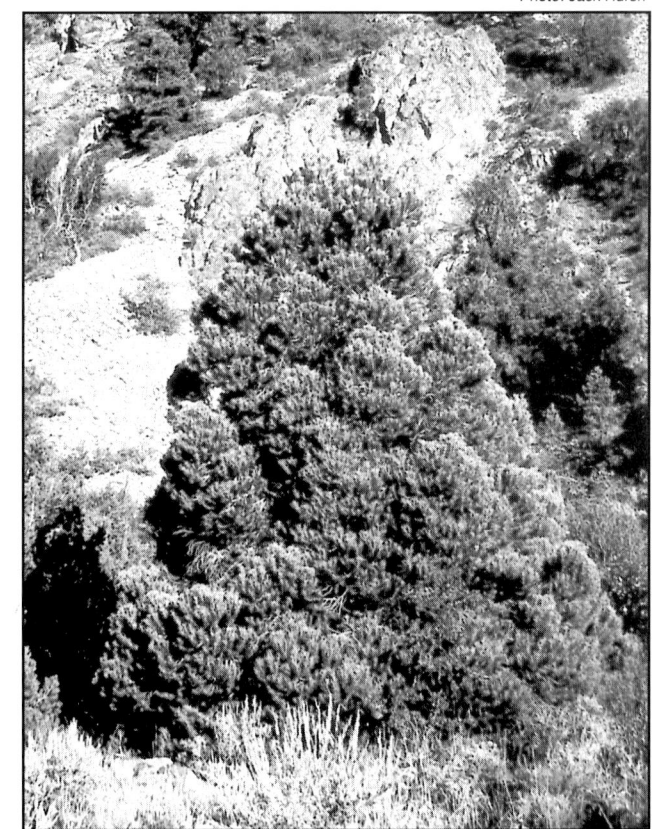

Singleleaf piñon (*Pinus monophylla*) Evergreen; height to 50', typically 11–20': trunk 6–14" in diameter, gray to dark brown; needles one in bundle, 1–2¼" long, gray-green, resinous; cones 2–3" long, egg-shaped or rounded, yellow-brown.

The singleleaf piñon is a long-lived, slow-growing, small pine commonly found in association with Utah juniper at elevations between 5,000 and 8,000 feet. Young trees are compact and pyramidal and become rounded and irregular in shape with maturity. Typical life span is 350 to 450 years.

Trees produce a significant number of cones by the time they are 75 to 100 years old. Cones mature in fall every other year and carry an average of 6 to 9 large seeds. The seeds, or pine nuts, are high in nutrition, comparable to peanuts, walnuts, and pecans.

Piñon seeds must be buried ¾ to 1¼" deep to germinate successfully. Numerous rodents and birds such as the piñon jay and Clark's nutcracker cache seeds underground or in leaf litter for winter consumption, in the process providing suitable placement for uneaten seeds to begin growth in spring.

Utah juniper (*Juniperus osteosperma*) Typically 7–20' tall; short trunk with low spreading branches; leaves yellow-green, scale-like and usually opposite in twos or threes; bark reddish-brown to gray-brown and fibrous; berries ¼–⅝" diameter, bluish.

Utah juniper is a small, long-lived tree commonly occurring in association with singleleaf piñon at elevations between 5,000 and 8,000 feet. Young trees have rounded or conical crowns that become more ragged and flattened with age. The species is more drought tolerant than piñon and may form pure stands at lower elevations and on more arid sites.

Most trees begin to bear seed at from 10 to 30 years of age. Every two years trees produce abundant crops of berries, each containing one or two seeds. Juniper berries remain on the tree throughout the winter and provide a vital source of food to many birds and animals. The tree is served in turn by the wildlife that eat the berries and disperse the undigested seeds as they move about the countryside.

Utah juniper provides cover for deer, elk, and coyotes and shelter and nesting sites for numerous species of birds and small mammals. Juniper has been used for fenceposts and poles, railroad ties, charcoal, and mine timbers.

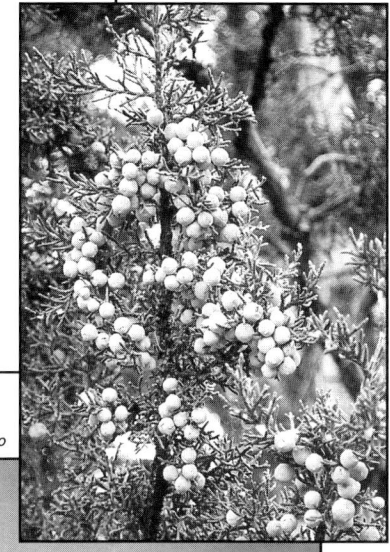

Photos: Kris Pizarro

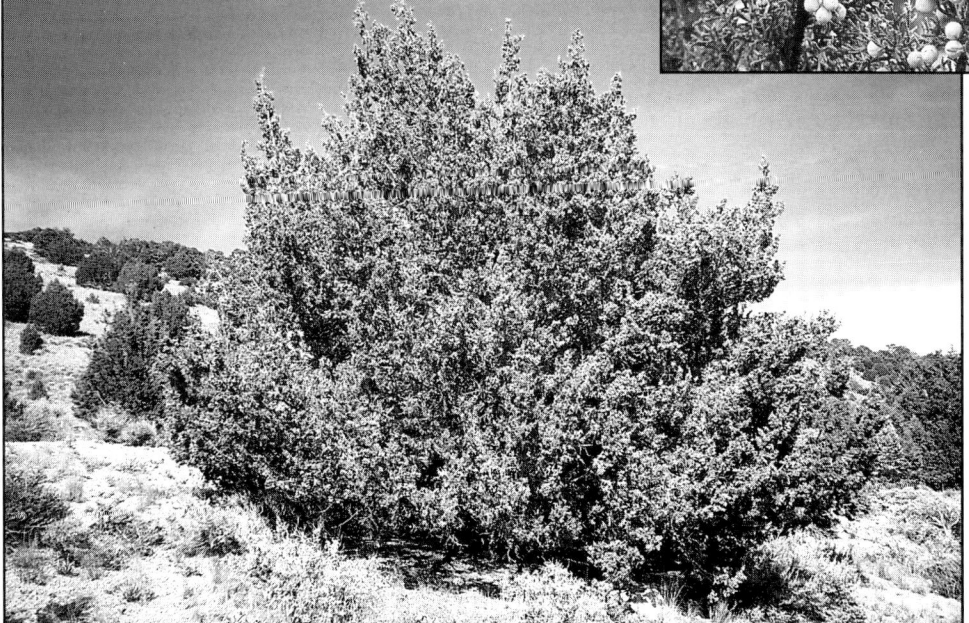

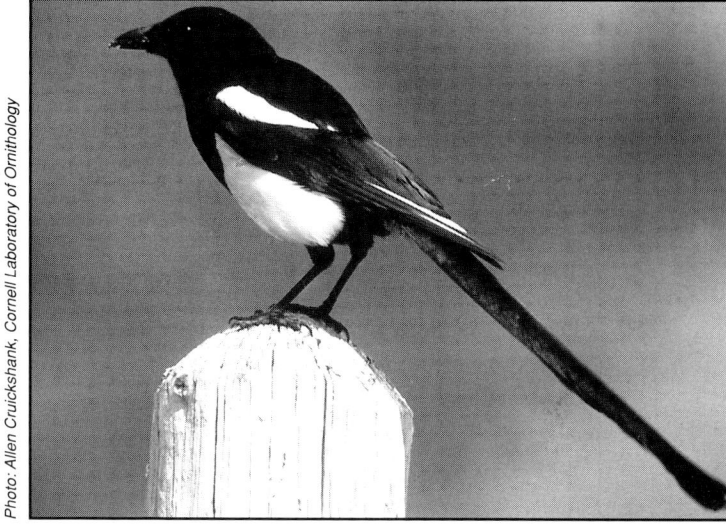

Photo: Allen Cruickshank, Cornell Laboratory of Ornithology

Black-billed magpie (*Pica pica*) Black, except for white belly and patches on wings; long, streaming, greenish-black tail; length 17½–22", 9½–12" of which is tail; wing spread 24"; large vocabulary of sounds, most often hear cack-cack-cack.

Members of the Lewis and Clark expedition were the first to report on the black-billed magpie in 1804. They were impressed with its quickness to take advantage of food supplied by man. The birds invaded their tents and filched meat from their dishes or from where hunters were dressing game. Black-billed magpies are frequently encountered, permanent residents of shrublands and woodlands in the Great Basin. They belong to the crow family and are known for their curiosity and intelligence. They are omnivorous but prefer animal food. They dine on a variety of insects, particularly grasshoppers when available, as well as roadside carcasses, coyote kills, ticks from the backs of deer and elk, grain, and fruit.

Nests are engineering marvels usually built within 25 feet of the ground. They are domed structures that may contain as many as 1,500 sticks and twigs cemented together with mud. A new nest is built each year over a period of six weeks or so, though an old one may then be repaired and used instead. Nests are often located in protective association with large hawks. The sturdy structures are often used in following years by other birds, particularly owls. Eggs (usually seven) hatch after 16–18 days.

interval	cumulative	milepost	
0.1	31.0	LA 31.0	Bob Scott campground. Pleasant Forest Service campground with piñon-shaded campsites, good water, and restrooms. A cool place to spend the night after traveling the stretch of desert highway to the west. If you pass through here in the early fall, after the first frosts and it happens to have been a good year for nut-cones the cones on the piñon pines will have opened and you can sample Nevada pine nuts. Remember, if you are going to get involved with piñon pine cones, be supplied with rubbing alcohol to clean your hands and clothing.
2.5	33.5		Blackbird Ranch. Some old ranch buildings and rock foundations on the right side of the highway. Sometimes occupied, sometimes not.
1.5	35.0		Paleozoic rocks are exposed in the cuts on either side of the highway. East of this point, both to the north and south of our route, the landforms are composed mainly of Paleozoic rock. We have just entered the Roberts Mountains thrust belt, a region characterized by sheet-like masses of sedimentary rock that have been moved, or thrust, over other more or less stationary rock masses. The rocks we see here in the road cuts are Ordovician Valmy Formation rocks, fine-grained, reddish-brown and green quartzite commonly found in the upper plate of the Roberts Mountains thrust.
1.0	36.0	LA 36.0	Intersection with State Route 376, on right. This is a good spot to pull off the road for a few minutes and enjoy the view to the south, down Big Smoky Valley.

BIG SMOKY VALLEY, HOT SPRINGS, AND MORE CALDERAS

Big Smoky Valley, given its name because of the natural haze usually seen over the valley, lies between the Toiyabe Range, which we have just crossed, and the Toquima Range, ahead to the right. Round Mountain, with its huge open-pit gold mine, is about 50 miles down the road on the west flank of the Toquima Range. When the valley isn't too hazy, the waste dumps can be seen from here. The active ghost town of Manhattan is a few miles beyond Round Mountain, and Tonopah is another 50 or so miles beyond that. The gravel road heading across the valley at about 1:00 crosses the northern Toquima Range at Petes Summit, and then descends into Monitor Valley to the east. The whitish patch to the left of the Petes Summit road, about three-quarters of the way across the valley, is Spencer Hot Springs. The water is clean and there are several hot pools and stock tanks at the site that can be used for a hot soak. Being on BLM-administered land, the area is open to anyone wishing to drive the short length of dirt road to get there.

The highest points in the Toquima Range, Wildcat Peak, at about 1:30, and Mount Jefferson (11,949 feet), at about 2:00 at the south end of the range, are composed of thick ash-flow tuffs filling three major calderas. The Northumberland caldera is to the south of Wildcat Peak; Mount Jefferson, the higher peak to the south, is the resurgent central mass of the Mount Jefferson caldera, and the Moores Creek caldera is in the lower part of the range between the other two. Across the valley west of Mount Jefferson (about the limit of our view to the south from here), the eastward bulge of the Toiyabe Range is the eastern margin of another large caldera (Darrough) comprising the southern high part of the range. The area including the highest peak, Arc Dome (11,773 feet), is within a U.S. Forest Service wilderness area. Extending for several miles north, the more rugged portion of the Toiyabe Range is composed of a series of thrust sheets of Paleozoic rock.

If you have time, a trip down Big Smoky Valley is well worthwhile. Almost every canyon in the Toiyabe Range between here and Arc Dome once supported its own mining camp, and there are ghostly remnants to be seen everywhere. Most canyons have running streams and a few, notably Kingston Canyon, offer rewards to the fisherman. Some of these streams contain native Lahontan Cutthroat Trout, Nevada's state fish. There are Forest Service campgrounds at Kingston Canyon about 12 miles south (we mentioned this campground as we approached the Toiyabe Range, west of Austin), and at Peavine Creek, about 60 miles to the south. A word of caution: except for the roads to the campgrounds, most of the roads leading into the mountains from State Route 376 are unmaintained dirt tracks and should only be traveled by four-wheel-drive vehicles with good clearance. In addition to ghost camps and campgrounds, two great hot springs, Spencer (described above) and Darroughs (at the south end of the valley, north of Carvers) provide quiet places where the traveler can soak away most cares of the world along with any accumulated alkali dust.

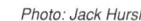
Photo: Jack Hursh

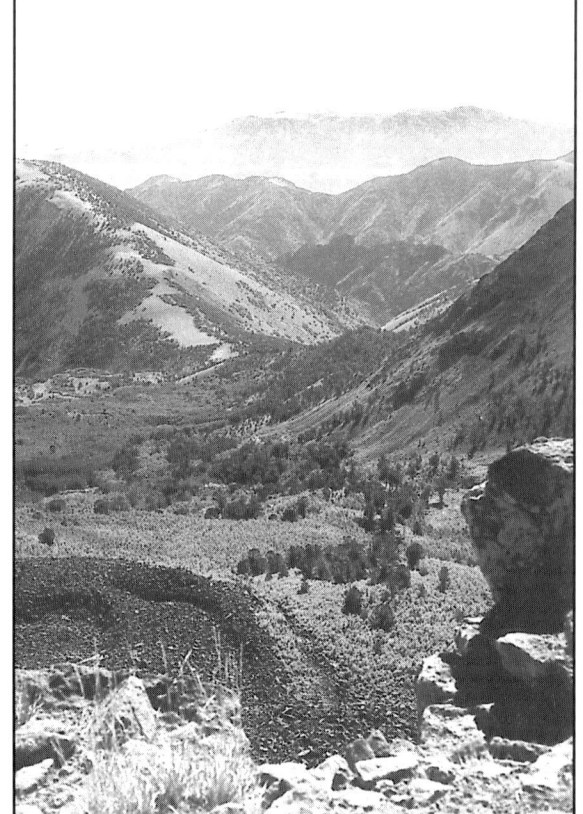

In the heart of a caldera. View from the top of Arc Dome looking east along the drainages of the North Twin River and South Twin River toward Mt. Jefferson in the hazy distance.

interval	cumulative	milepost
5.0	41.0	LA 41.0
7.0	48.0	LA 48.0
0.6	48.6	

Road to Linka tungsten mine to the right. The mine, located about 4 miles south of the highway, produced about $1 million in tungsten prior to the mid-1950s from a contact metamorphic deposit formed at the contact of Paleozoic limestone and Cretaceous granodiorite. Tungsten occurs in the mineral scheelite, a calcium tungstate, and is associated with silicate minerals such as diopside, epidote, and garnet. Garnet can be found on the dumps at the mine site, but crystals are rare. The dirt road leading to the mine is usually in fair shape, but there are very dangerous mine workings here and mineral collecting should be restricted to the dumps only.

The rocks on the north side of the highway at this point, and rocks in the road cuts from here to the summit ahead, are ash-flow tuff units of the Miocene Bates Mountain Tuff. These tuffs came from an as yet unlocated caldera.

Hickison Summit. Bates Mountain air-fall tuff overlain by ash-flow tuff; both units are faulted.

Hickison Summit (elevation 6,594 feet) marks the division between the Toquima Range, to the south, and the Simpson Park Mountains, to the north.

Turnoff to Hickison Petroglyph Recreation Area is to the left. The well-graded, gravel road leads to a BLM parking area and campground close to petroglyph sites on nearby cliffs. There is no water here but there are campsites with picnic tables shaded by large piñon and juniper trees. The petroglyphs at this site are typical of the Great Basin Curvilinear and Great Basin Scratched styles dating from the late prehistoric period (from approximately 500 years ago to the 1860s). Petroglyphs of these styles are typically carved on soft stone such as the ash-flow tuffs found in this area, and may represent hunting or fertility symbols. A trail leads to several areas where petroglyphs are scribed on vertical faces of welded ash-flow tuff and an informational brochure, prepared by the BLM, is available in a box at the start of the trail. Most of the trail is wheelchair accessible. Wild horses can sometimes be seen from vantage points on the trail that overlook the valley to the west. (plates 7f and 7g)

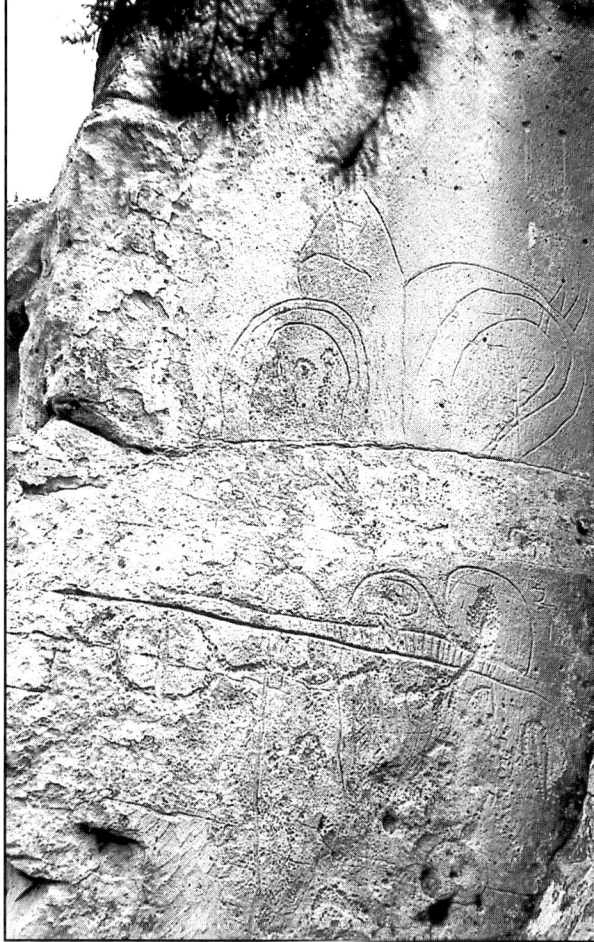

Photo: Kris Pizarro

Petroglyphs carved in ash-flow tuff seen along the trail in the Hickison Petroglyph Recreation Area.

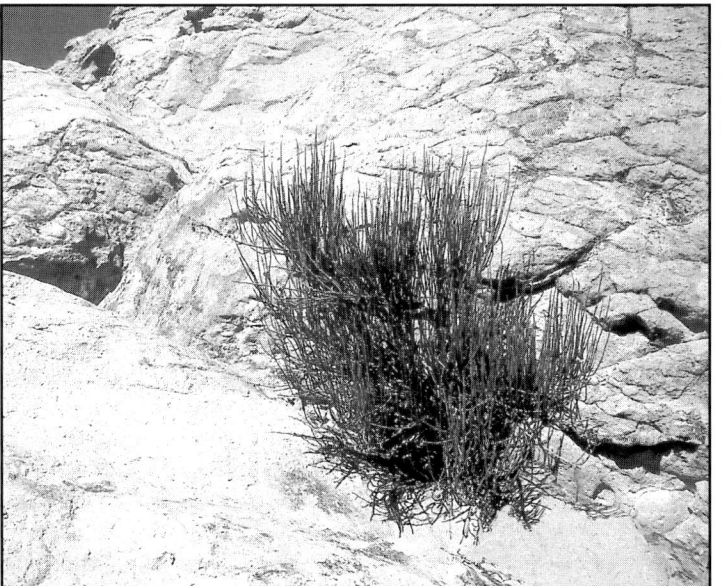

Photo: Jack Hursh

Green ephedra (*Ephedra viridis*) erect, rigidly branched, evergreen shrub; height 1½–4'.

Green ephedra (also called Mormon tea) is commonly found growing in sandy or rocky soils in association with big sagebrush, singleleaf piñon, and Utah juniper. This specimen found a home in a crack of some ash-flow tuff at Hickison Summit.

Numerous jointed, yellow-green branches give ephedra a broomlike appearance. Leaves are very small and scale-like and located opposite each other at nodes. Photosynthesis is carried out by the stems. Dense clusters of cone-like structures resembling small yellow flowers appear in early spring and produce seeds that ripen in July or August. Seeds are eaten by a variety of birds and mammals. Ephedra has been used for medicinal purposes both by indigenous peoples and early settlers.

THE ROBERTS MOUNTAINS THRUST BELT AND DISSEMINATED GOLD: NEVADA'S NEW GOLD BELT

The Roberts Mountains thrust fault, named for a structure first described in the Roberts Mountains of Eureka County, is a low-angle reverse fault of regional extent (throughout north-central Nevada) that places siliceous rock over the top of carbonate rock of the same general age. The siliceous rocks (quartzite, shale, and chert) are termed western assemblage rocks because they were originally deposited in a deep-water environment to the west of their present location. The carbonate rocks, the eastern assemblage, were deposited in shallow, near-shore waters and are in place at their site of deposition.

When the thrusting occurred, in Late Devonian to Early Mississippian time, the western assemblage rocks were moved to the east, overriding the carbonate assemblage. The total movement is estimated to have been as much as 90 miles. The siliceous rocks tend to be thin-bedded and weak, while the carbonate rocks formed in thick, stronger layers. When the thin-bedded rocks were shoved over the top of the stronger carbonate rocks, the effect was somewhat like sliding playing cards over the top of a stack of bricks. The siliceous upper plate rocks became contorted and crumpled, sometimes into tight, overturned folds while the more stalwart carbonate rocks in the lower plate occasionally broke into angular blocks but more or less stayed put. The fault surface is usually marked by crushed rock and clay and can range from a few feet up to several hundred feet thick.

The original surface of thrusting, although undulating, was probably nearly horizontal. Later block faulting cut the thrust sheet and tilted the various blocks so that now in places we can see the thrust contact in all kinds of attitudes from flat to vertical.

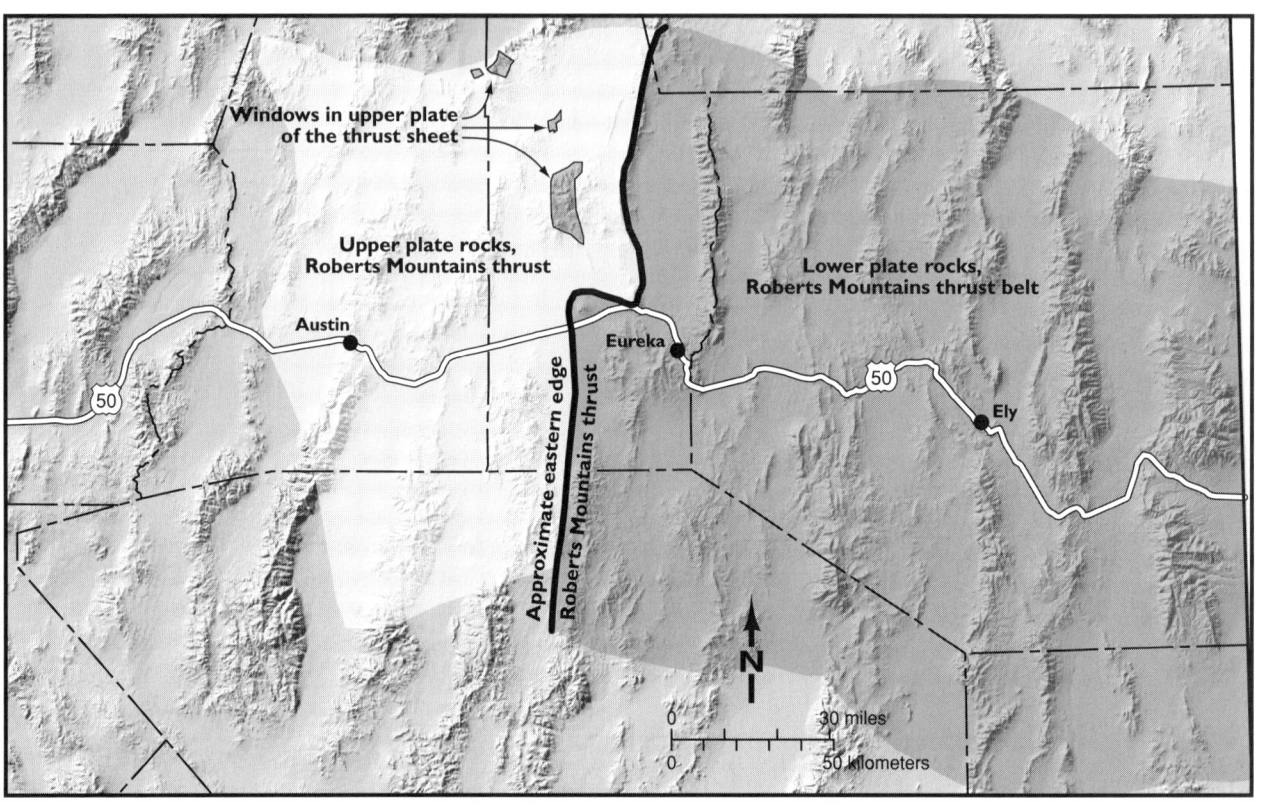

The Roberts Mountains thrust belt where it is crossed by U.S. 50. Upper plate rocks, shown in lightest shading, have been thrust from the west, overriding lower plate rocks, shown in darkest shading. The "windows" in the thrust sheet shown in the top center of the figure are in the Roberts Mountains, where the thrust fault was first studied by U.S. Geological Survey geologists in 1942.

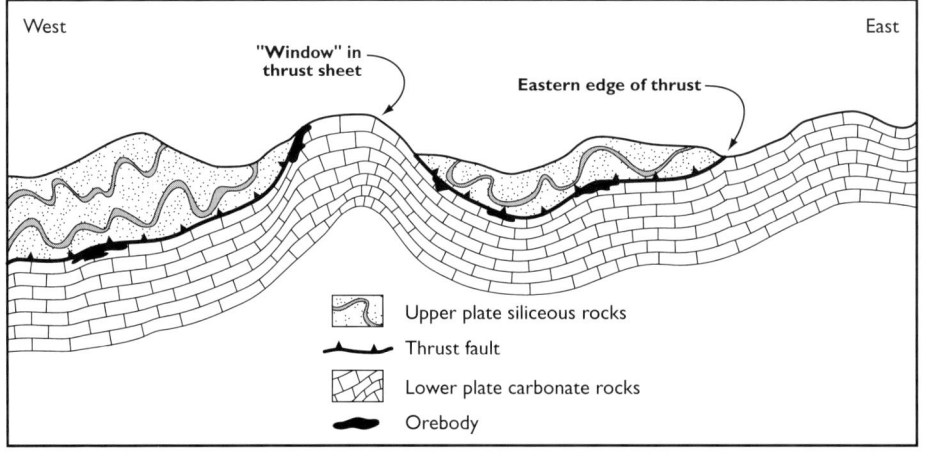

A typical cross section of the Roberts Mountains thrust belt showing favorable sites for gold deposits. This section could be along an east-west line through the "window" and thrust edge shown at top center in the figure above.

Beyond being fascinating geology, the Roberts Mountains thrust belt is important to Nevada's gold miners as the type locality of the Carlin-type gold deposit. The carbonate rocks below the thrust are good host rocks for gold, and the siliceous rocks in the upper plate, along with the clay and crushed rock along the thrust surface, sometimes formed a barrier to gold mineralization, causing gold to concentrate in the lower plate rocks. The first of these deposits to be recognized as a new type, the Carlin Mine located north of the town of Carlin in Eureka County, was discovered in 1961 in a "window" in the thrust sheet (a place where the overlying siliceous rocks had been eroded away exposing the underlying carbonate rocks). Prospectors then swarmed across the thrust belt, concentrating on the "windows," but staking almost every square foot of thrust exposure wherever they found it.

Like most generalizations, this prospecting clue did not always lead to bonanzas, but many of Nevada's major "Carlin-type" gold mines are located within or near the thrust belt and a lot are in "windows" in the thrust sheet. What is a "Carlin-type" gold mine, and what makes it different from an old-fashioned gold mine like those found along the Mother Lode in California? Among the main features of Carlin-type deposits that set them apart are their form and size, and the size of the gold particles within them. The deposits are large, usually containing from about one million to several tens of millions of tons of ore. They take on the shape of the bedded carbonate rocks within which they usually occur, forming thick, wide, but irregular masses. The gold occurs mainly as submicroscopic particles that are disseminated throughout the rock, hence the other terms for these occurrences-disseminated gold and micron gold deposits. Gold content of the ore is usually low; the original Carlin Mine averaged about 0.3 ounces of gold per ton of rock. Depending on the price of gold, rock containing even less than 0.01 ounces of gold per ton is sometimes counted as ore. Because of their size and because most are found near the surface, the Carlin-type deposits can be mined by low-cost, open-pit methods.

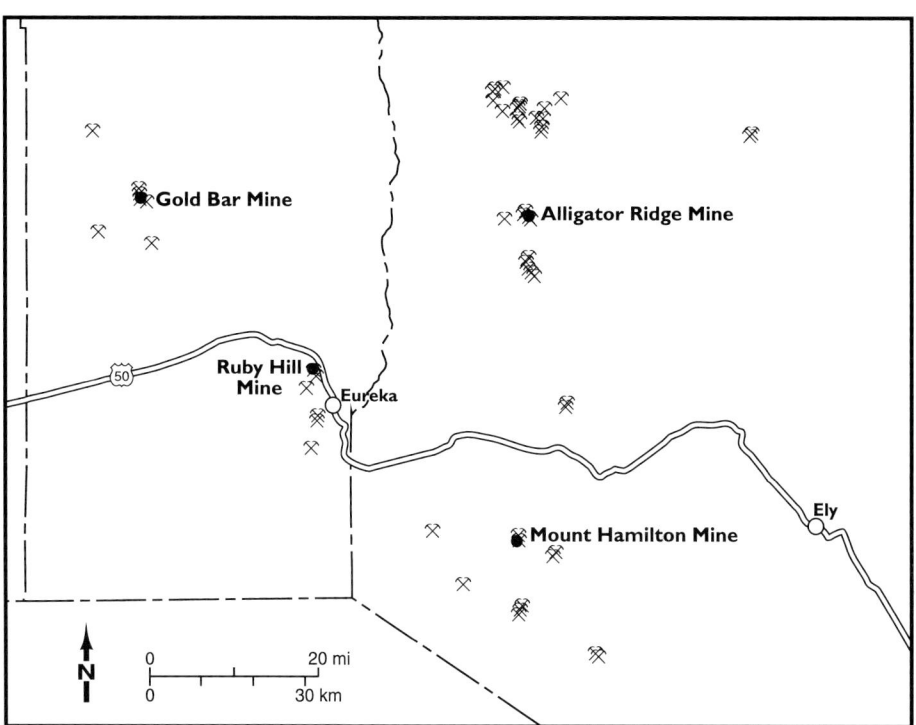

Carlin-type gold mines in the Roberts Mountains thrust belt, Eureka area. Every "⚒" is an individual mine, but most are not presently active.

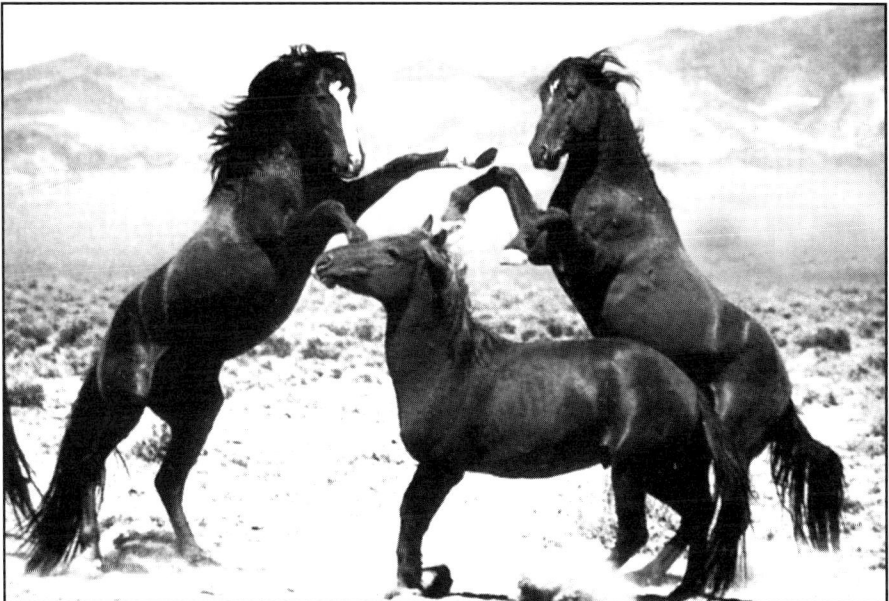

Horseplay in Nevada's wild horse country.
Photo: University of Nevada, Reno Equestrian Center

interval	cumulative	milepost
7.9	56.5	LA 56.53 EU 00
4.4	4.4	
0.3	4.7	
4.3	9.0	EU 9.0

Entering Eureka County, another zero point in the county mile marker column.

Northwest of here, stretching from Hickison Summit to about 10:00, are the Simpson Park Mountains. These mountains are underlain by Ordovician siliceous rocks (more upper plate rocks) to the north and by Miocene volcanic rocks to the south. To the northeast at 11:00 are the Roberts Mountains. The Roberts Mountains consist of complexly faulted Cambrian through Devonian carbonate rocks (these are the long-awaited lower plate rocks) that have been overridden by siliceous western assemblage rock moved from west to east along the low-angle Roberts Mountains thrust fault.

Lincoln Crested Wheatgrass Seeding. As part of a range-restoration program, 1,300 acres here were cleared of sagebrush and seeded with crested wheatgrass in 1955. As you can see, sagebrush is slowly but persistently reclaiming its space.

Bean Flat rest area, a place to pull over, rest or eat lunch, and enjoy the solitude.

The Monitor Range lies to the south of the highway here, Antelope Peak (elevation 10,220 feet) and Summit Mountain (elevation 10,476 feet) are the high points of this part of the range. The peaks are composed of early Oligocene ash-flow tuffs that fill the center of the Broken Back 2 caldera.

Twin Spring Hills, on right. These low hills are underlain by a block of Paleozoic limestone that is outside the northern margin of the Broken Back 2 caldera.

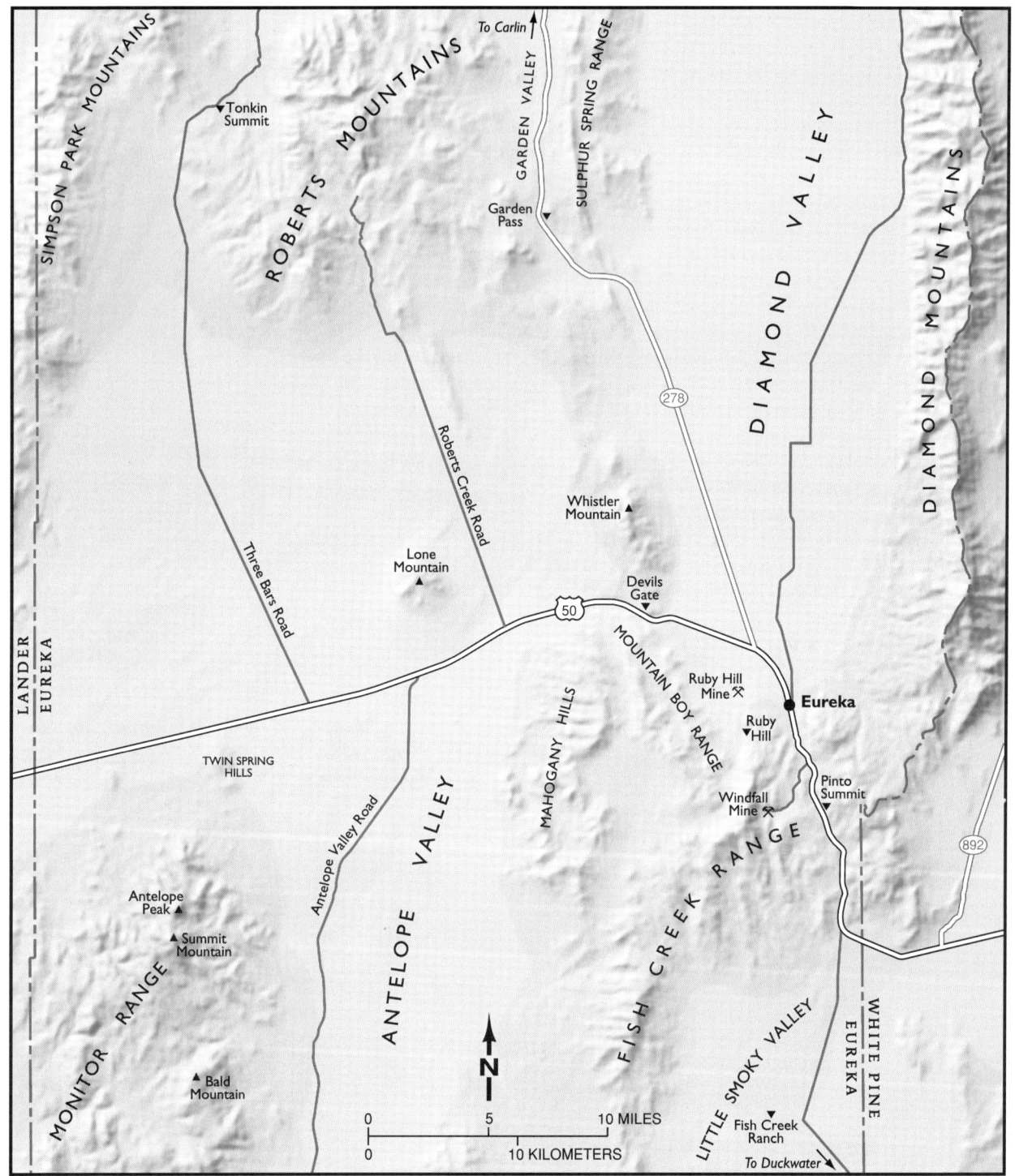

Route map, Eureka County.

Three Bars road, on left. Another gravel road leading to remote ranches and mines to the north. The Simpson Park Mountains are to the west at 8:00. The Roberts Mountains are in the distance at about 10:00. Three Bars road crosses the low divide between the two ranges at Tonkin Summit about 22 miles to the north. We are now within the heart of Nevada's gold belt, and two of several large gold deposits in the southern Roberts Mountains are visible on the south flank of the mountains at about 10:30 (look for a reddish-colored flash in the trees). These are the Goldstone and Gold Pick pits of the Gold Bar Mine. This mine is not in operation at present, but it produced over 500,000 ounces of gold between 1986 and 1994.

Antelope Road, to the right.

For the next few miles, we have a good view of Lone Mountain, on the left (north) of the highway.

Roberts Creek road, to the left, leads north into the Roberts Mountains.

The Mahogany Hills are on the right, south of the highway.

Paleozoic rocks exposed on the west flank of Lone Mountain. The prominent lower white band is Ordovician Eureka Quartzite. Light-colored Silurian Lone Mountain Formation dolomite caps the ridge in ▶ the top right side of the photo.

PALEOZOIC ROCKS ON LONE MOUNTAIN

Lone Mountain provides a good look at the layered nature of the Paleozoic rocks that lie below the Roberts Mountains thrust sheet. The oldest rocks, Ordovician Pogonip Group limestone, form the western edge of the mountain. These rocks are overlain by younger Ordovician quartzite and limestone, then by Silurian Roberts Mountains Formation limestone, and then by more massive, cliff-forming Devonian dolomite and limestone capping the mountain on the east. If you are a geologist new to Nevada, or just interested in rocks, this is a good place to stop, walk the section from west to east, and get a good introduction to the rocks of the area. A word of caution: this is not a casual undertaking. There is no handy way to get to the west base of Lone Mountain—it is too far to hike, and one needs maps and a four-wheel-drive vehicle to navigate around the mountain to find the described outcrops. Also, during the summer field season, you might encounter a traffic jam on the roads leading to Lone Mountain. On many summer days, vans filled with geology students from distant universities will be traveling to outcrops to study the complex rocks.

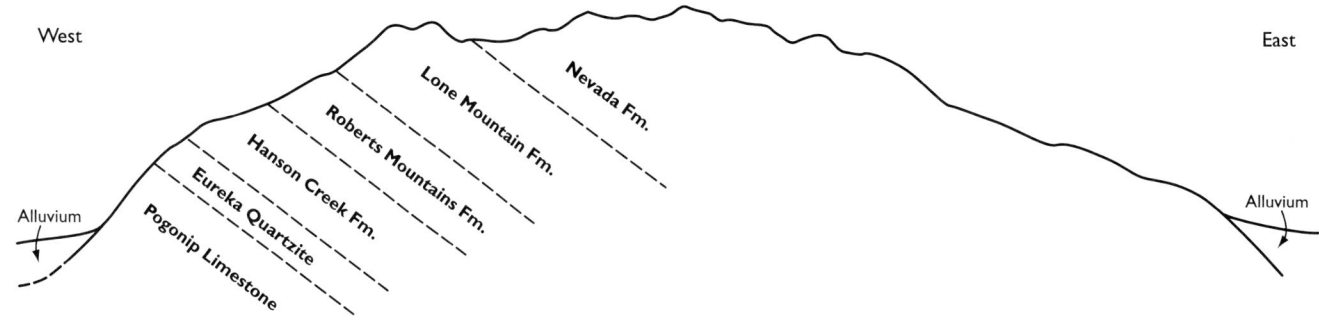

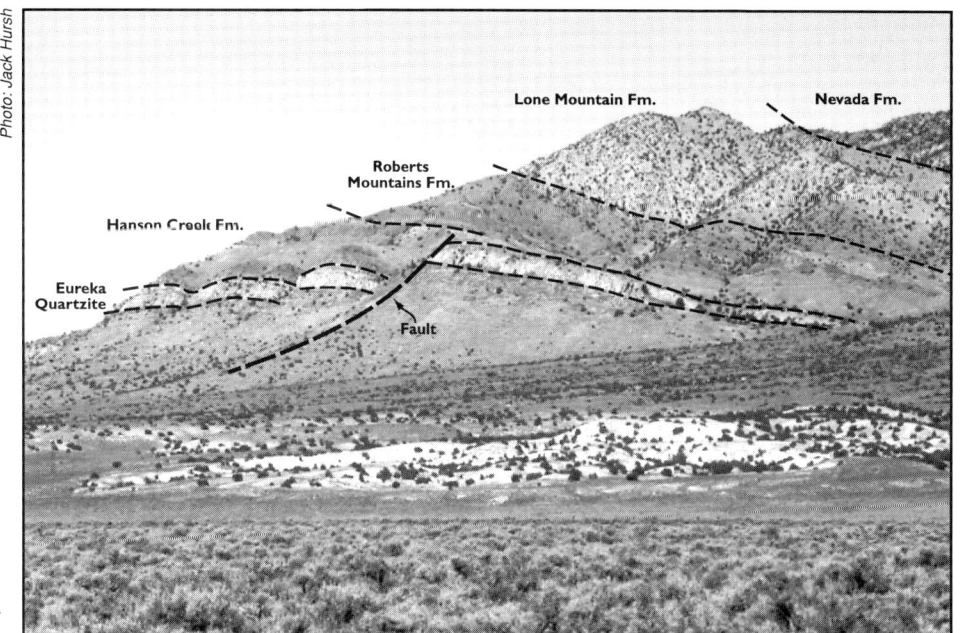

Photo: Jack Hursh

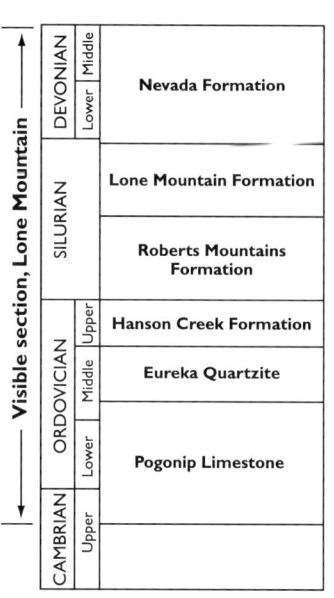

			Visible section, Lone Mountain
DEVONIAN	Middle		
	Lower		Nevada Formation
SILURIAN			Lone Mountain Formation
			Roberts Mountains Formation
ORDOVICIAN	Upper		Hanson Creek Formation
	Middle		Eureka Quartzite
	Lower		Pogonip Limestone
CAMBRIAN	Upper		

interval	cumulative	milepost	
5.8	28.0	EU 28.0	On the south (right) side of the highway as we enter the narrows of Devils Gate, the gently rounded, piñon-juniper covered hills are composed of thin-bedded shale and quartzite of the Ordovician Vinini Formation. These rocks are in the upper plate of the Roberts Mountains thrust sheet.
1.0	29.0	EU 29.0	Devils Gate. Good exposures of Devonian dolomite and limestone are on both sides of the canyon. These rocks are in the lower plate of the Roberts Mountains thrust sheet. Note the massive, cliff-forming nature of the rocks, the many small solution caves, and the narrow, white calcite veins that cut the carbonate rocks in the exposure on the left (plate 9b). This is a good place to note the contrast in outcrop style between the carbonate lower plate rocks, to the left, and the shaly upper plate rocks discussed back at Milepost 28.0.
0.5	29.5		Still on the north side of the canyon, the rock sheet displaying columnar jointing near the top of the ridge is a Jurassic alaskite sill that has intruded along bedding in the lower plate rocks.

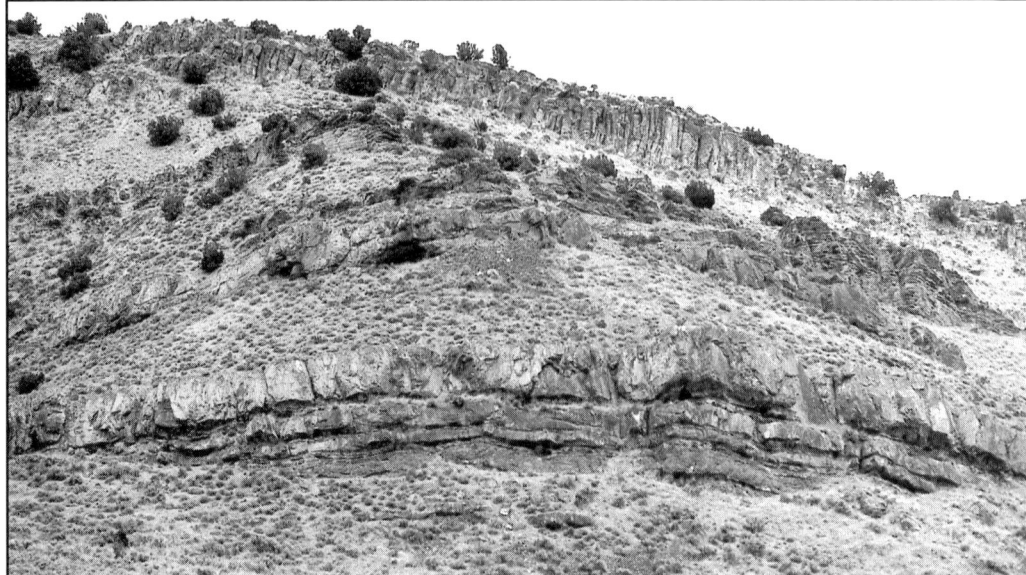

Photo: Kris Pizarro

Alaskite sill (unit on skyline with columnar jointing) intruded along bedding in Devonian dolomite at Devils Gate.

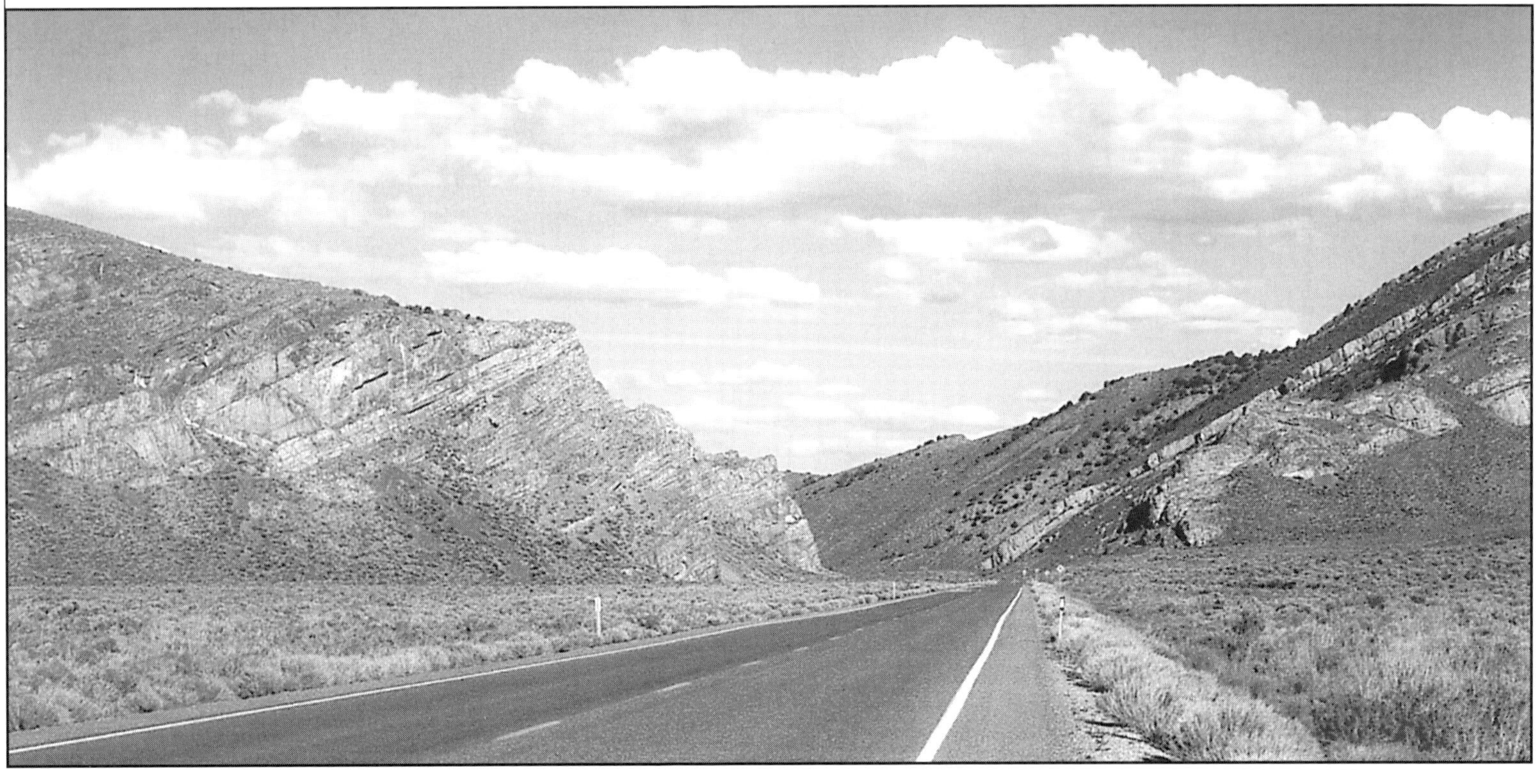

Lacy network of white calcite veins in dolomite at Devils Gate.
▼

Cliffs of Devonian dolomite and limestone at Devils Gate, west of Eureka.

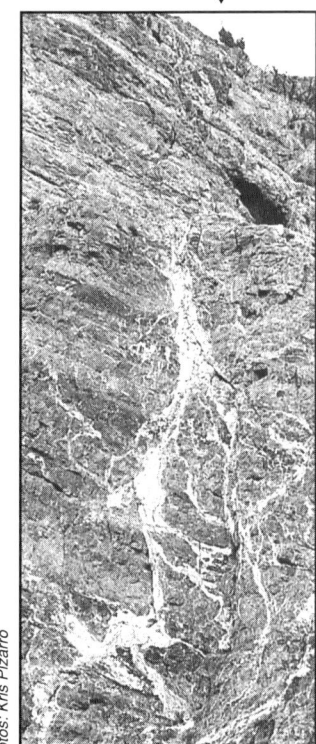

Photos: Kris Pizarro

80

interval	cumulative	milepost	
1.0	30.5		Diamond Valley, to the north. Ahead to the right, south of the highway, the buildings and waste dumps of Homestake Mining Co.'s new Ruby Hill Mine are visible. Discovered in 1993, the mine produced its first gold in 1998. In the background, on the north slopes of the Fish Creek Range, are the large headframe and buildings on Ruby Hill, the center of the historical Eureka mining district. The headframe is at the FAD Shaft, sunk to a depth of 2,500 feet during 1945–1949.
1.5	32.0	EU 32.0	The Diamond Mountains are on the eastern skyline, beyond Diamond Valley. The crest of the range, which forms the boundary between Eureka and White Pine Counties, is composed mainly of fine-grained shale, siltstone, and sandstone of the Mississippian Diamond Peak Formation. Note the general lack of trees on this range and contrast with the tree-covered lower range of hills to the west of the Diamond Mountains at about 9:30. These lower hills are composed of Silurian and Devonian carbonate rocks, apparently more friendly to vegetation.
1.8	33.8		Intersection, State Route 278 to the left. To the right is the entrance to the Ruby Hill Mine. The mine is not open to the public, but there is an informational sign in the turnout near the road intersection.

State Route 278 is the first paved road connecting with Interstate 80 to the north we have intersected since State Route 305 headed that way, west of Austin. This road heads north along the west side of Diamond Valley, crosses the south end of the Sulphur Spring Range at Garden Pass, then continues north through Garden Valley and Pine Valley to Carlin on the Humboldt River and Interstate 80, a long 85 mile trip. The road does, however, pass through some fine Nevada sagebrush and piñon-juniper lands. About 50 miles to the north, Pine Valley contains several small producing oil fields, and well sites can be seen along the road. Further north, the road passes large hay ranches in Pine Valley.

DIAMOND VALLEY: AGRICULTURAL STABILITY BETWEEN MINING BOOMS

Agriculture is Eureka County's second largest industry, providing an element of stability that helps equalize the sometimes boom-bust cycle of mining. Large farms in Diamond Valley produce alfalfa hay that is exported to California dairies. Some hay is even exported overseas to places like Japan. The alfalfa produced in Diamond Valley has high mineral content, and a substantial portion of the production goes to horse racetracks such as Santa Anita and Hollywood Park in California. ▼

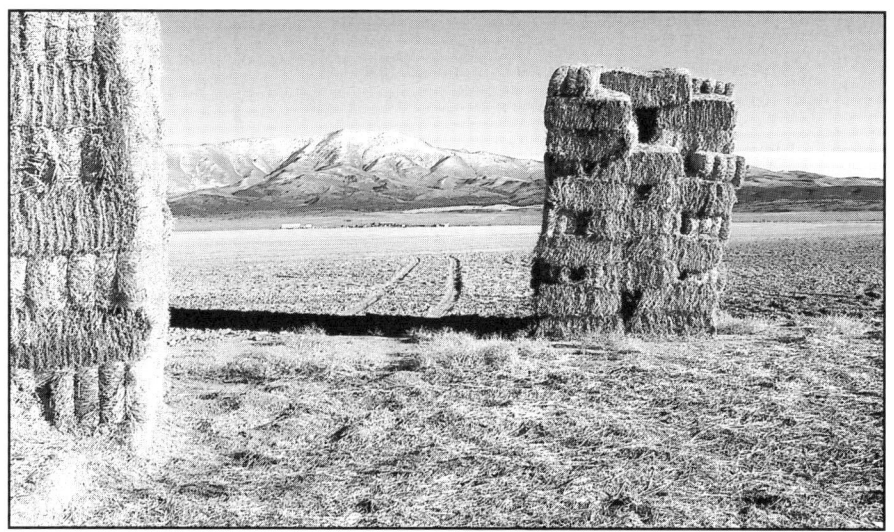

Photos: Jack Hursh

The Diamond Mountains with Diamond Peak (10,614 feet). ▼

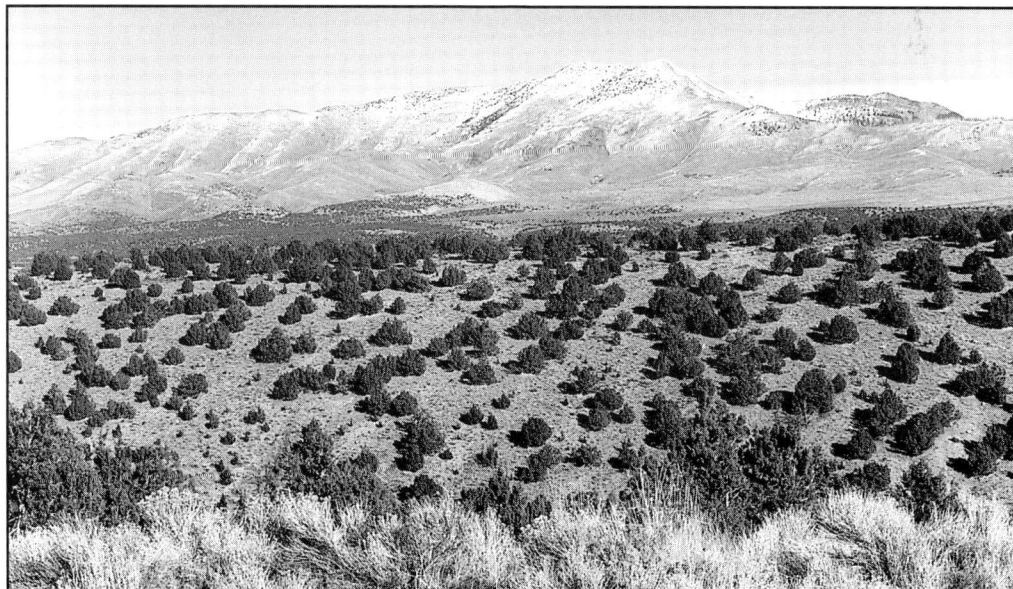

interval	cumulative	milepost	
2.6	36.4		Entering the town of Eureka, county seat of Eureka County.
0.1	36.5		Slag piles from 1870s-vintage lead smelters are to the left of the highway. The old smelter site is on the hill slope to the right, marked by an overgrown cut.
1.1	37.6		Ball field on the right, the craggy rock outcrops on the ridge behind the field are dolomite of the Ordovician Hanson Creek Formation capped by Eureka Quartzite, also Ordovician.
			Another slag pile, on the right
0.1	37.7		Tannehill Cabin on the right. This is one of Eureka's first houses, built in 1864. (NHM #222)
0.3	38.0		"Eureka!"—Nevada Historical Marker #11
0.1	38.1		Old county hospital on left side (brick building in the canyon)
1.0	39.1		Intersection, Windfall Canyon road to the right.

The Windfall Gold Mine is about 2½ miles up this road. Discovered in 1903, it was somewhat of an anomaly—a gold mine in a mainly silver district. After a long period of inactivity, the deposit was reopened and mined as an open-pit mine between 1975 and 1978.

EUREKA AND THE EUREKA MINING DISTRICT

Rich silver-lead ores were discovered about 2 miles west of the present town of Eureka in 1864. The deposit found on Ruby Hill, named for the large pockets of ruby silver (pyrargyrite) found there, was the first important lead-silver discovery made in America. The ores proved difficult to treat, however, and little was done in the area until 1869 when a successful smelting method was developed to treat the lead-rich ores. As many as 16 small smelters were soon in operation, and the resultant smoke and fumes earned Eureka the title of "Pittsburgh of the West." Eureka became the southern terminus of the Eureka and Palisade Railroad in 1875, connecting with the main Central Pacific rail line at Palisade about 80 miles to the north. The railroad allowed Eureka to become the center of wagon and stage transportation for most mining camps in central and eastern Nevada.

The orebodies in the Eureka district were massive replacement bodies of silver-bearing lead and zinc minerals formed in Cambrian limestone and dolomite. These are similar to what is called a "manto" in the famous silver camps of northern Mexico. The surface "discovery ores" were oxidized and were composed of ruby silver, lead carbonate (cerussite), lead sulfate (anglesite), and lots of colorful maroon, red, yellow, and brown iron-oxide minerals. In the deeper mines, the ores graded into masses of sulfide minerals (silver-bearing galena, sphalerite, and pyrite). Most of the production of the district was between 1871 and 1888 when, it has been said, Eureka controlled the lead markets of the world, producing about $95 million in silver, lead, and gold. Another $12 million was produced through the 1960s by small operations and lessees. Homestake will produce an estimated 100,000 to 110,000 ounces of gold per year from the New Ruby Hill Mine.

Although Eureka has a lot smaller population than it did in 1878 (with 9,000 residents then, it was Nevada's second largest city), it is far from being a ghost town. It is the county seat of Eureka County, currently one of the richest gold-producing counties in Nevada. The mines, mostly located north of Carlin in the northern tip of the county, generate tax revenues that continue to pump life into the old town. A new high school was built with mining taxes, and the town boasts a new enclosed swimming pool.

A stop in Eureka provides an urban rest-break from the sagebrush-covered wildlands that border the "loneliest road" to the west. The 1879 courthouse and several other buildings, including the Opera House, in downtown Eureka have been restored, and new buildings, including a modern motel on the main street, have been built in matching style. There is a small museum in the old Eureka Sentinel building, a short block west of the courthouse, where a brochure for a self-guided walking tour can be obtained.

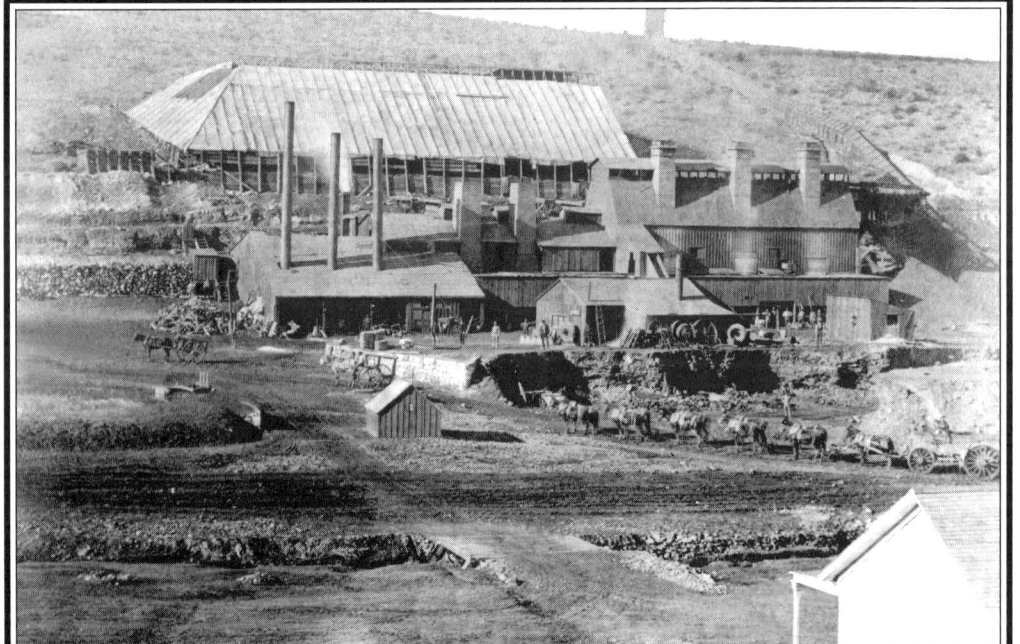

Eureka Consolidated lead smelter at the north end of Eureka, late 1870s.

Nevada Historical Society

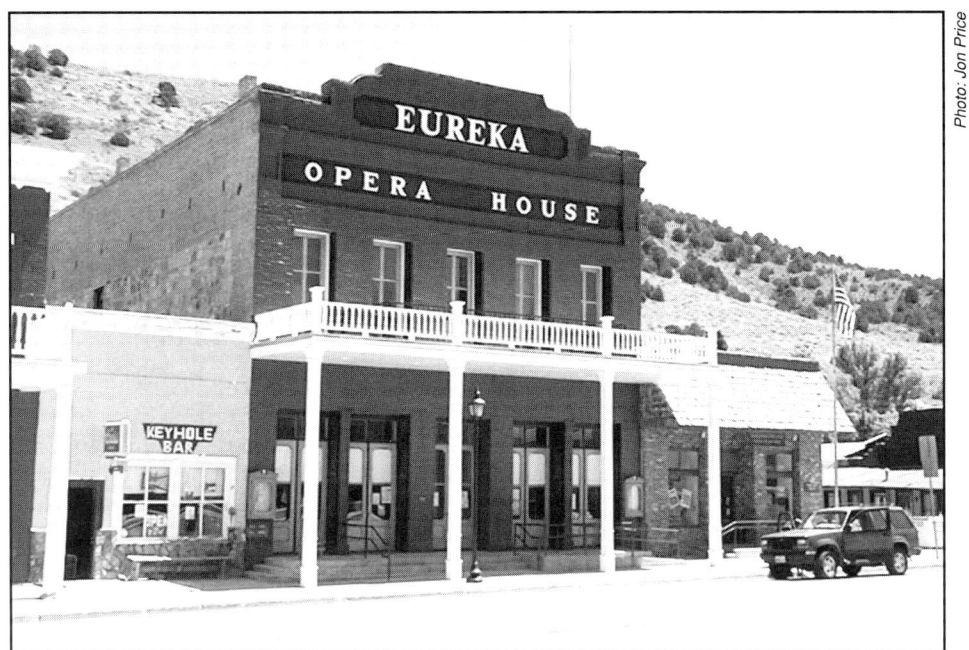

The restored Eureka Opera House, 1999.

Photo: Jon Price

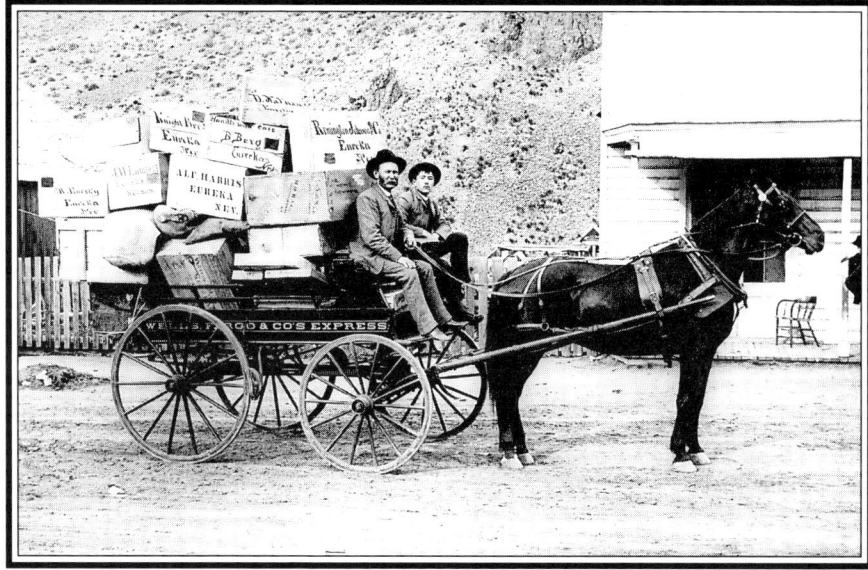

Wells Fargo Express, Eureka, 1905.

Nevada Historical Society

An 18-mule team hauling loaded freight wagons into Eureka, late 1870s.

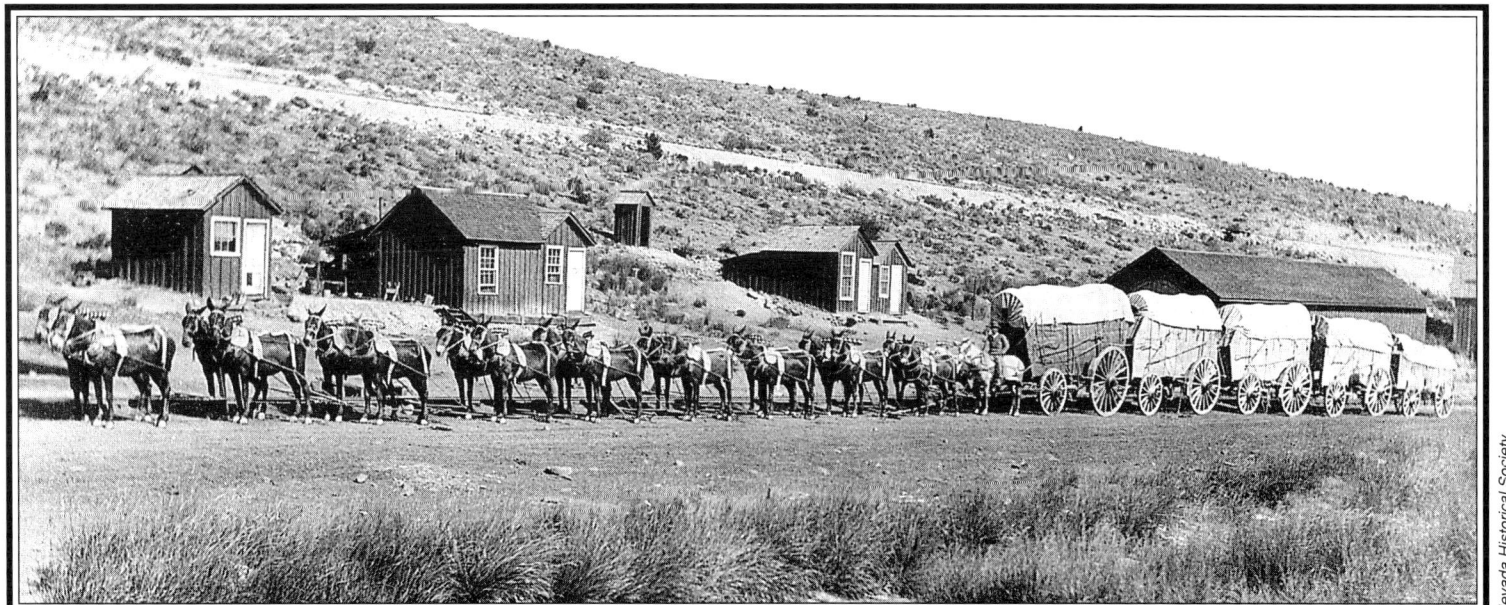

Nevada Historical Society

| 0.5 | 39.6 | |

Whitish outcrops straight ahead are bedded air-fall tuff. Tuff is also exposed in road cuts. The tuffs are Oligocene age, possibly originating from a volcanic center marked by the small basin ahead on the other (south) side of Pinto Summit.

| 1.5 | 41.1 | |

Picnic tables on left near a spring. There are currant bushes here, and wildflowers in the spring. A good place to stop and stretch your legs on a short walk to the crest of the hill beyond the spring. Who knows, you might see a deer or two.

| 0.5 | 41.6 | |

Pinto Summit (elevation 7,376 feet). Pinto Summit marks the division between the Fish Creek Range, to the right, and the Diamond Mountains, to the left.

Gravel deposits are exposed in the road cut to the left. These are poorly-sorted stream deposits consisting of pebbles and cobbles of Paleozoic rocks contained in a matrix of white tuff. They were deposited by small, fast-moving streams flowing from highlands that surrounded the basin in which the white tuffs were deposited.

| 1.3 | 42.9 | |

Cliff face of white air-fall tuff is exposed on the right at 2:00 to 3:00.

| 1.2 | 44.1 | |

Site of Pinto. Crumbling piles of brick and stone mark the site of the mill camp of Pinto. A 20-stamp mill was built here in 1871 to treat ore from mines in the Silverado (Pinto) district in White Pine County. The mill closed in 1884, and the camp was abandoned.

| 1.0 | 45.1 | |

Note the nifty outcrop of air-fall tuff in the road cut.

Coyote (*Canis latrans*) Weight 15–44 pounds; length 30–40" plus 12–16" tail; 15–20" tall at the shoulder; grizzled gray or reddish-gray with buff underparts; bushy tail with black tip; prominent ears; excellent runner with cruising speeds of 25–35 mph and short bursts of up to 40 mph.

Extremely intelligent and adaptable, the coyote is expanding its range despite loss of traditional habitat and human hunting pressures. Two hundred years ago most of the coyote population was concentrated in the northwestern U.S. Today the coyote can be found in desert, grassland, mountain, and suburban environments as far north as Alaska and as far south as Central America.

Photo: Terry Nelson

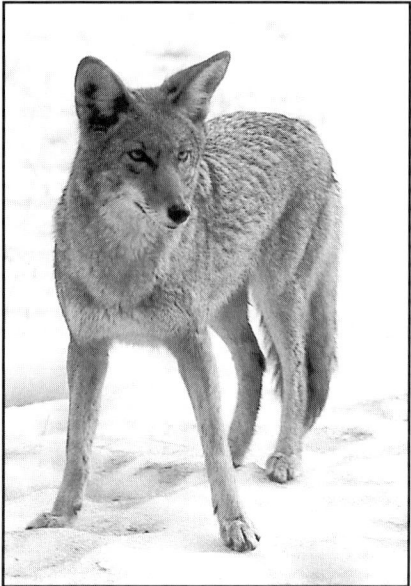

The coyote is an opportunistic hunter employing a variety of methods to obtain food. It patiently stalks and pounces on small mammals. Coyotes have tremendous endurance and can simply chase prey until it is worn out. Where the food supply is predominantly small animals, it hunts alone or in breeding pairs, while in the presence of large prey such as deer, coyotes will hunt in packs. Coyotes will also make do with insects, lizards, carrion, fruit, and even pine nuts.

Coyotes mate in late winter or early spring, and pups (usually six) are born about two months later. If the population is low, a pair may remain together for several years or for life. In the Great Basin, coyote numbers closely follow the cyclical rise and fall of the jackrabbit population.

Coyote vocalization is varied, but the familiar calls that evoke images of the wide open spaces of the American West are usually heard between dusk and dawn. Barks and yelps followed by a drawn-out howl serve to announce location, strengthen social bonds, and reunite separated members of a band.

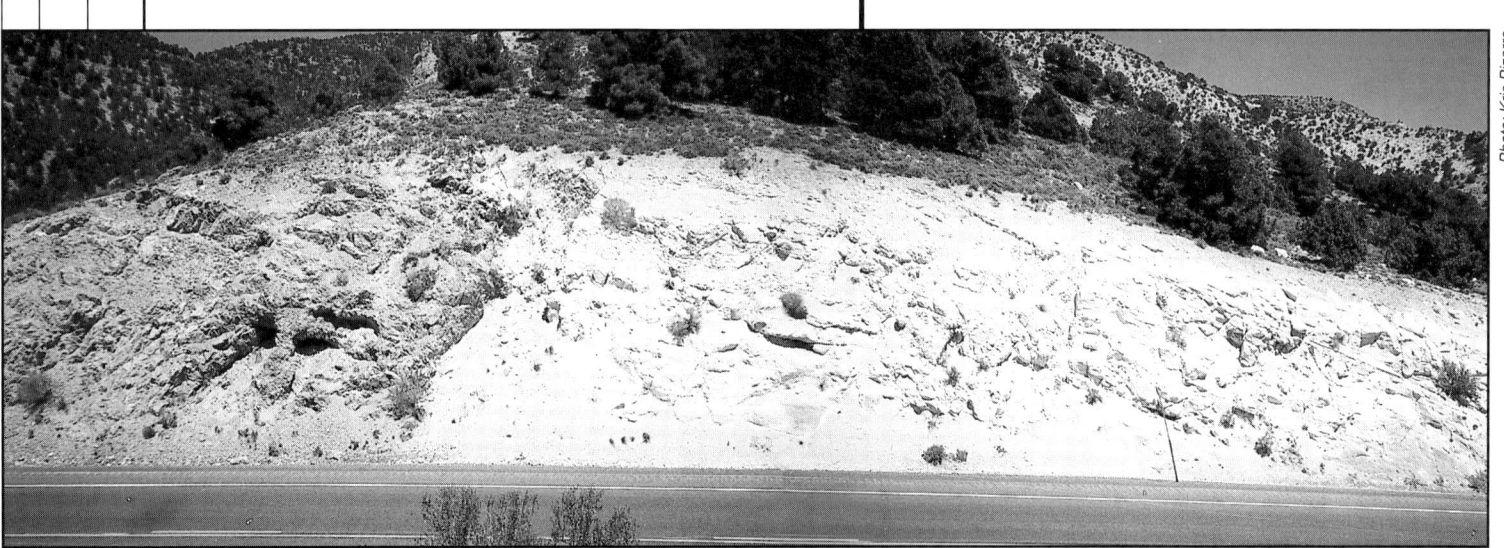

Photo: Kris Pizarro

White air-fall tuff, on the right, in contact with limestone, on the left. Near Milepost 45, east of Pinto Summit.

interval	cumulative	milepost	
2.1	47.2		The road to the right goes down Little Smoky Valley to Fish Creek Ranch (about 10 miles), then crosses the Pancake Range and leads to the Duckwater Indian Reservation, about another 20 miles to the south in Nye County.
			Fish Creek Ranch was the site of the major action of the "Fish Creek War of 1879."
0.2	47.4	EU 47.38 WP 00	Enter White Pine County. Mile markers begin at zero for the last time—White Pine County extends east to the Utah border and we cross no more county lines.
1.5	1.5		Silverado Mountain is on the left at about 9:00. Small mines along the west side of the mountain were the source of ores for the Pinto mill. The barren, gray and buff-colored rocks on Silverado Mountain are Devonian-age limestone. The normally gray limestone (calcium carbonate) has been locally altered to dolomite (magnesium carbonate) causing the mottled, buff coloration. The alteration is related to the silver-lead-zinc deposits exploited by the Pinto district mines.
			To the right, Little Smoky Valley extends south along the west side of the Pancake Range. Black Point, so named because of the dark volcanic rock (Oligocene andesite) forming it, is at about 2:00.
2.9	4.4		Note the bedding in the carbonate rocks on the hills north of the highway. The beds dip to the north at a low angle. The large gray dump at 9:00 came from an open-pit lead-zinc mine up on the mountain to the north.
0.1	4.5	WP 4.56	Intersection with State Route 892 on the left. This road travels north through Newark Valley, crosses a low divide into the valley of Huntington Creek, follows the western margin of the Ruby Mountains, and eventually reaches Elko on the Humboldt River. This is over 110 miles of "really lonely" road with no services or settlements in the first 78 miles. The first and last parts of the road are paved, but a 40-mile segment in the middle is gravel surfaced. A scenic route, but one to avoid unless you have lots of time, a well-equipped vehicle, and plenty of gas.
1.5	6.0	WP 6.0	The Pancake Range is ahead on the right. This range is composed mostly of Mississippian to Pennsylvanian-Permian limestone and shale, with patches of andesite cropping out along its lower west flank.
			Several small to moderate-sized disseminated gold deposits have been mined in the Pancake Range, evidence that we are still in Nevada gold country.
			In the far distance, at about 9:00, are Big Bald Mountain and Little Bald Mountain in the southern Ruby Mountains. Buck Mountain is at 10:00.

THE CARBONARI AND THE FISH CREEK WAR OF 1879

Also called the Charcoal Burners War, this was a confrontation between Italian-Swiss charcoal burners (the carbonari) and the smelter owners in Eureka. The fuel used to fire Eureka's lead smelters was made from piñon and juniper that was cut, made into charcoal, and hauled into Eureka by crews of carbonari who worked from camps scattered throughout the surrounding hills. The dispute was supposedly over the difference between 30 cents a bushel (carbonari price) and 27½ cents a bushel (smelter-owner price) for charcoal delivered to Eureka, but there may also have been some anti-immigrant feelings directed against the Italian-Swiss community.

If this seems like a small matter today, consider that in 1879 manufacture of charcoal was one of the most important industries in Eureka County. From the early 1870s until 1887, thousands of acres of piñon and juniper were clear-cut for miles around Eureka, and several thousand men were employed producing charcoal that was consumed at a rate of over 1,200,000 bushels per year by the hungry smelters. At the height of the dispute, the charcoal burners refused to deliver charcoal until their demands were met.

On August 11, 1879, about 2,000 members of the Charcoal Burners Association stormed Eureka, resisted all civil authority, and took control of the town for a short period. The Governor was urged to call out the state militia to quell the "insurrection." Things cooled down for about a week, but on August 18, a sheriff's posse attacked a "coal ranch" at Fish Creek, opened fire on about a hundred charcoal burners and killed five. This ended the Fish Creek War.

Who won? Well, this is not clear, but charcoal was reported to be selling for 22 cents a bushel in Eureka in 1880.

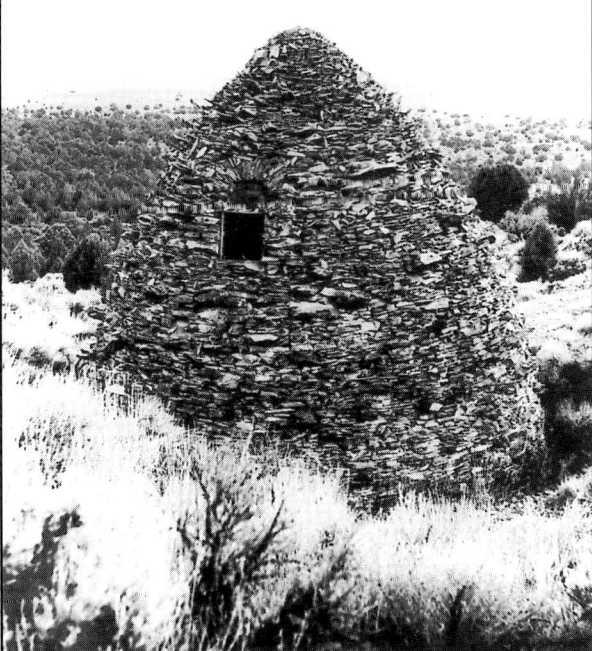

Nevada Historical Society

Beehive charcoal oven, overlooking Diamond Valley northwest of Eureka.

PLATE 9

9a Summer storm clouds. *Photo: Jack Hursh*

9b Cliffs of **Devonian limestone** at **Devils Gate**, Eureka County. *Photo: Kris Pizarro*

9c **Fritillary butterflies** are very fast flying and are commonly seen as an orange blur, darting about meadows and canyons. *Photo: Roy W. Cazier*

9d **Arrow-leaf balsam root** (*Balsamorhiza sagittata*) grows in clumps 1 to 2 feet tall. Flower stems rise from large, arrowhead-shaped basal leaves covered with silvery, feltlike hairs. Balsam root is found in dry, open areas from the sagebrush steppe to open coniferous woodlands. *Photo: Kris Pizarro*

9e **Golden eagle (*Aquila chrysaetos*) Adults dark brown; golden wash over head and neck difficult to see at a distance. Body 30–41" long; wingspan 76–92"; weight between 8 and 13 pounds; female is larger than male.**

The golden eagle is a common permanent resident in the Great Basin. It is most often seen perched on utility poles along the highway or flapping and gliding along in its search for prey. Occasionally one or two can be seen on the highway dining on carrion.

Golden eagles are thought to mate for life. A pair usually has several nests in an area and may alternate from year to year.

They favor ledges on cliff sides or canyon walls, as well as power transmission towers. They often share a cliff with nesting prairie falcons or ravens, and small birds may form protective nesting associations with them. Eggs, usually two, are laid at 3–4 day intervals in late February or March, and eaglets emerge 43–45 days later. A young eagle makes its first flight 65–70 days after hatching, though it remains dependent upon its parents for a month or more after that.

In Nevada the jackrabbit is the eagle's main food source (bird in photo is holding one), along with carrion and large rodents. When chasing birds, it can surprise them from above by diving at speeds of 150–200 miles per hour. Snakes, including rattlesnakes (which are decapitated) are often fed to the young. *Photo: Roy W. Cazier*

9f Gravel deposits exposed at **Pinto Summit**. Pebbles and cobbles are deposited in a matrix of white tuff. Note the steep fault cutting the beds to the left of center. *Photo: Kris Pizarro*

PLATE 10

10a **Winterfat** (*Ceratoides lanata*) is a small (1–3 feet tall) shrub whose flowers resemble little cotton balls. It is an important winter browse for deer, elk, and sheep. *Photo: Jack Hursh*

10b Green **copper carbonate** mineral coatings on a boulder in the display at the **Robinson Mine** viewing area. *Photo: Kris Pizarro*

10c **Liberty Pit** at the **Robinson Mine**, from the viewing area. Iron-stained rocks of the leached capping of the copper orebody are exposed in the upper benches in the background. The dark areas are the remains of the enriched ore (see figure on page 100). The large gray rock slide cutting across the mine benches on the left is an area of pit wall failure. *Photo: Kris Pizarro*

10d **Blue elderberry** (*Sambucus cerulea*) is a many-branched shrub that can reach 25 feet in height. It grows in moist soils along streams, roadsides, and in clearings. Berries are used in pies and preserves. *Photo: Kris Pizarro*

10e The **Rocky Mountain bee plant** (*Cleome serrulata*) grows to three feet tall and is often found along roadsides and in disturbed areas. It is rich in nectar and extremely attractive to bees. It blooms from May through August, depending on elevation. *Photo: Kris Pizarro*

10f **Bedded tuffaceous sediments**, on the left, separated by a fault from gravel deposits, on the right, located near the Belmont Mill road intersection west of Little Antelope Summit. The colorful iron-oxide staining occurs along the fairly wide fault zone. *Photo: Kris Pizarro*

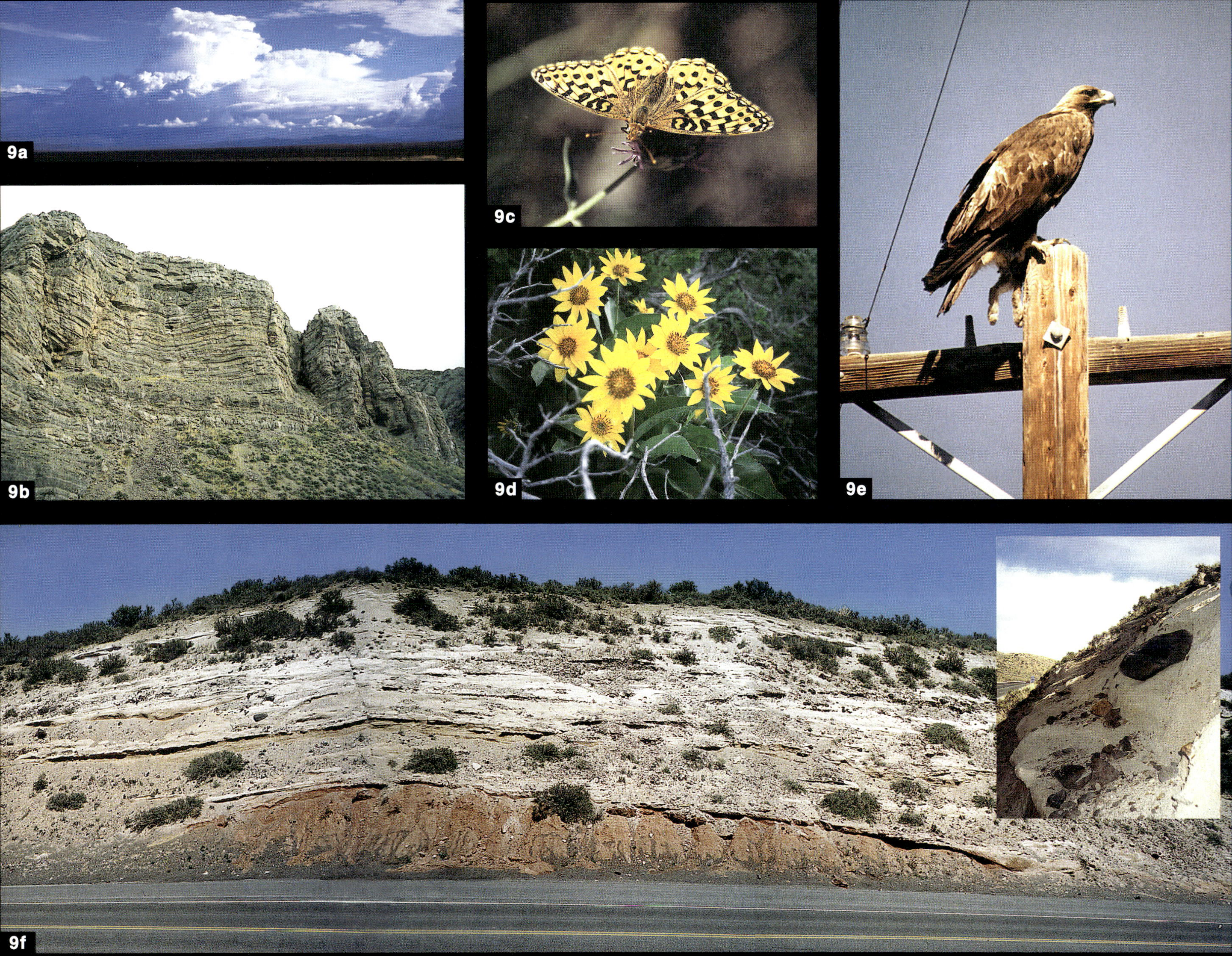

9a

9b

9c

9d

9e

9f

PLATE 9

87

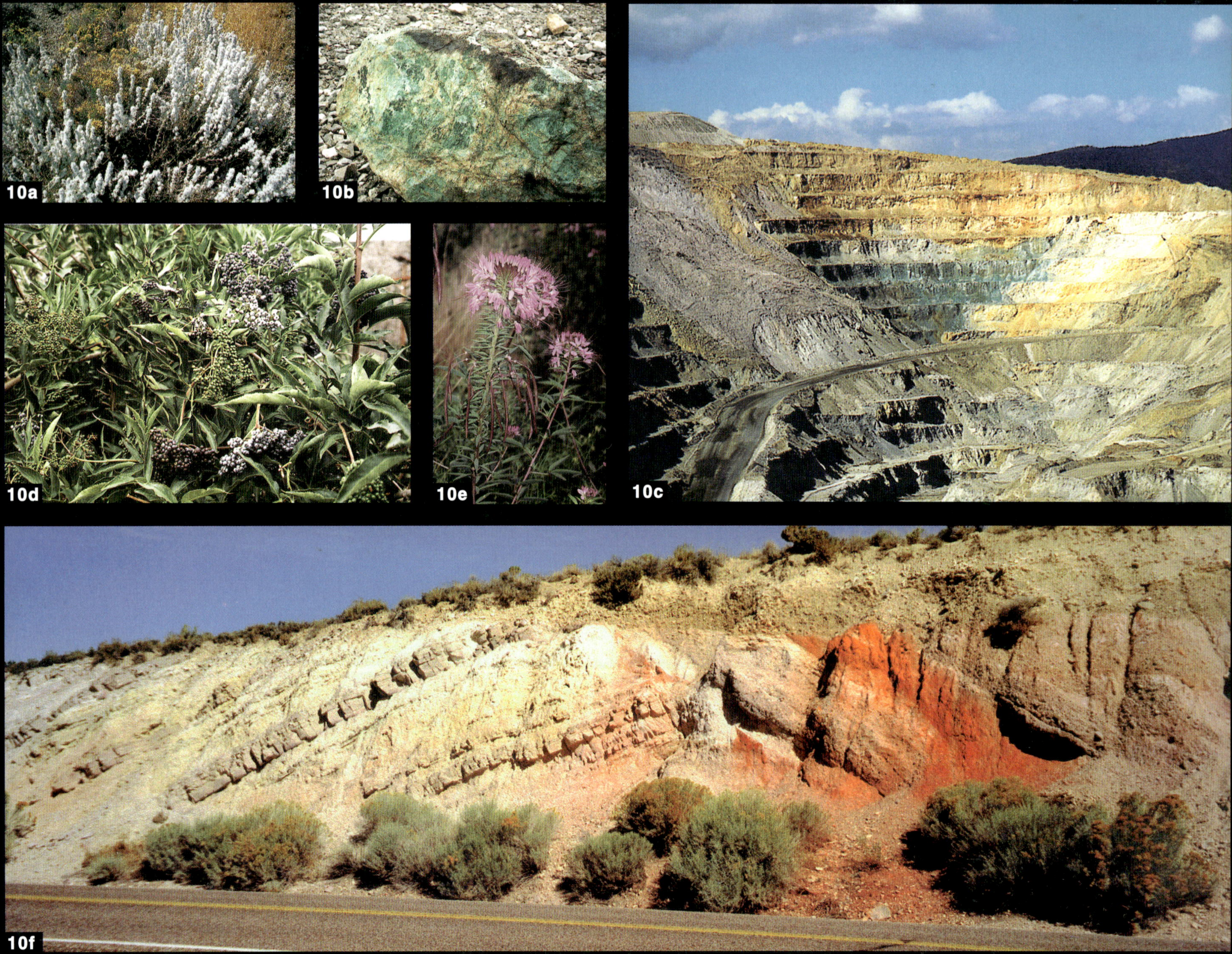

10a

10b

10c

10d

10e

10f

PLATE 10

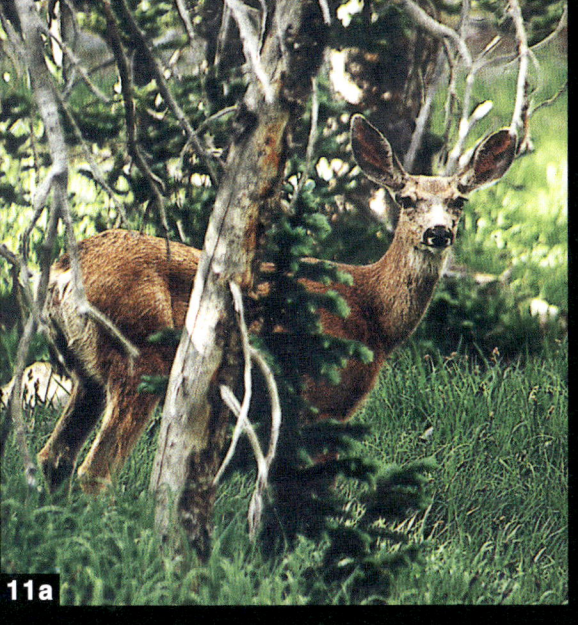

11a

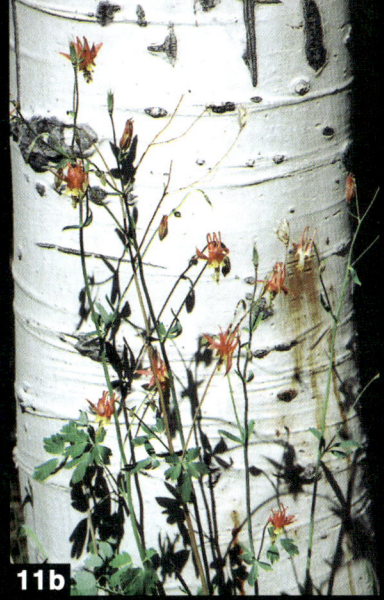

11b

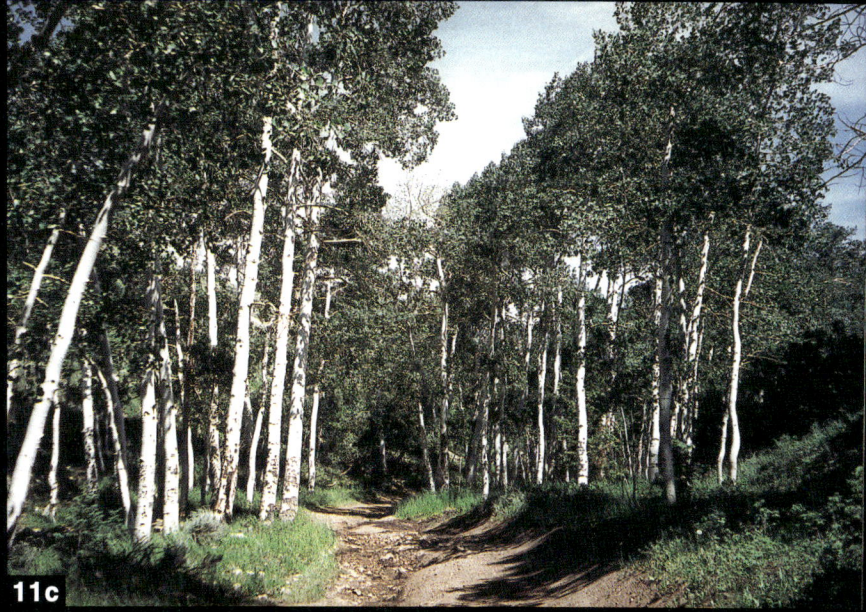

11c

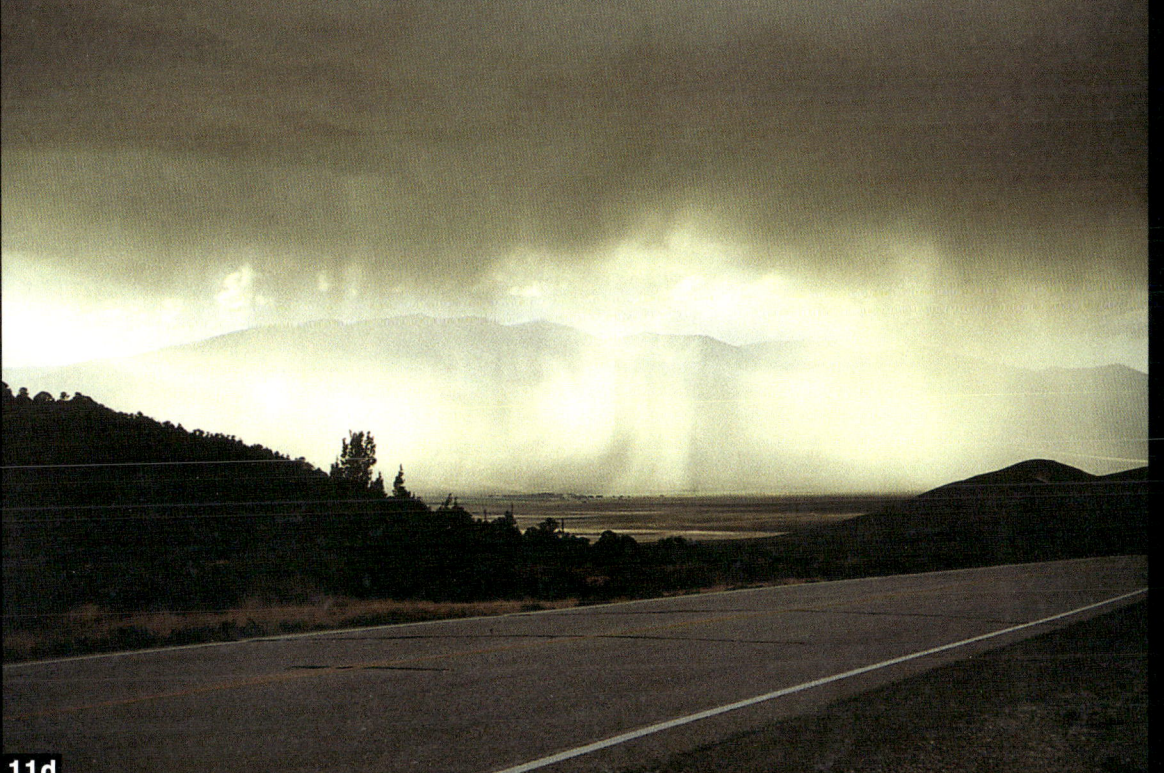

11d

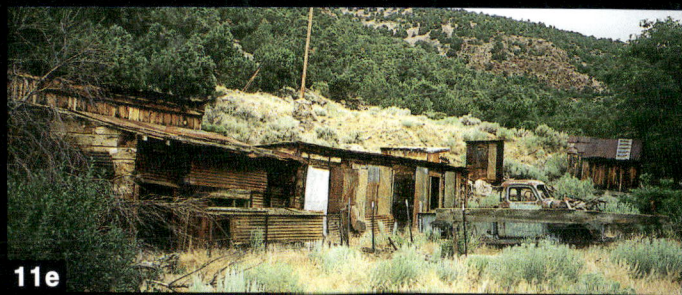

11e

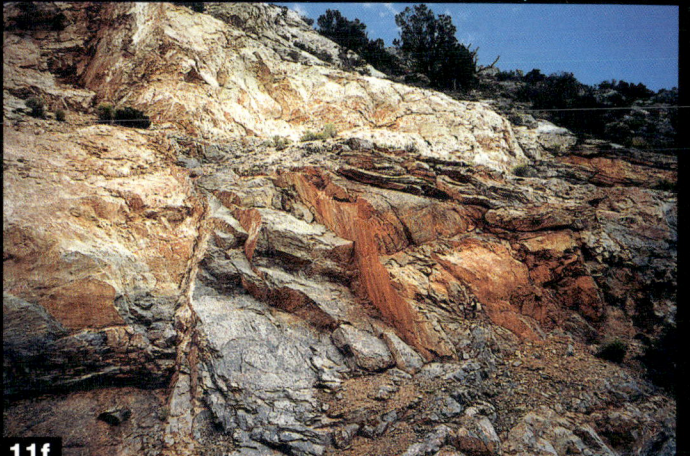

11f

PLATE 11

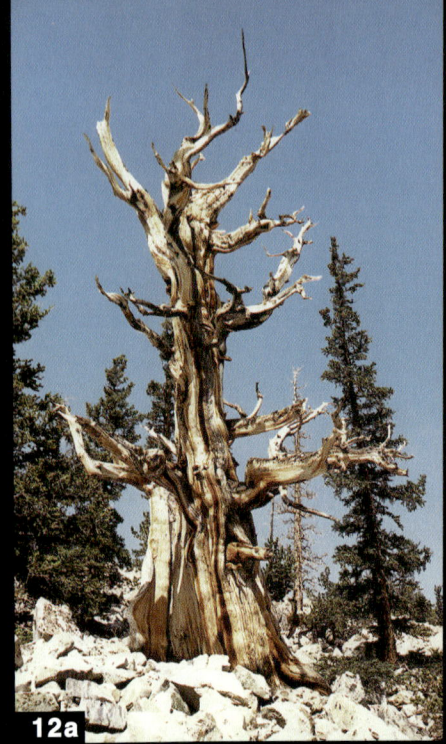

12a

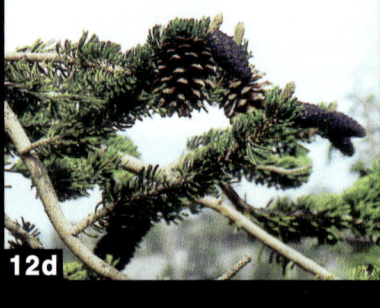

12d

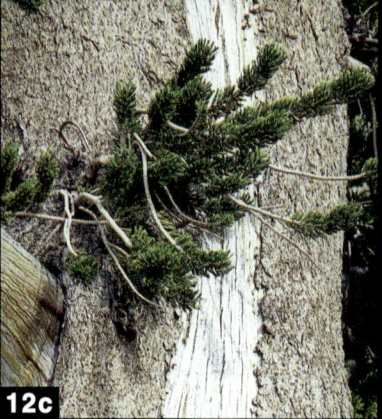

12c

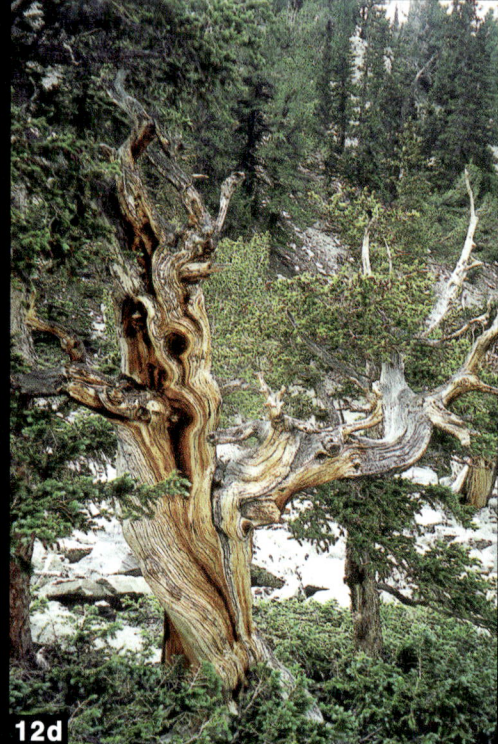

12d

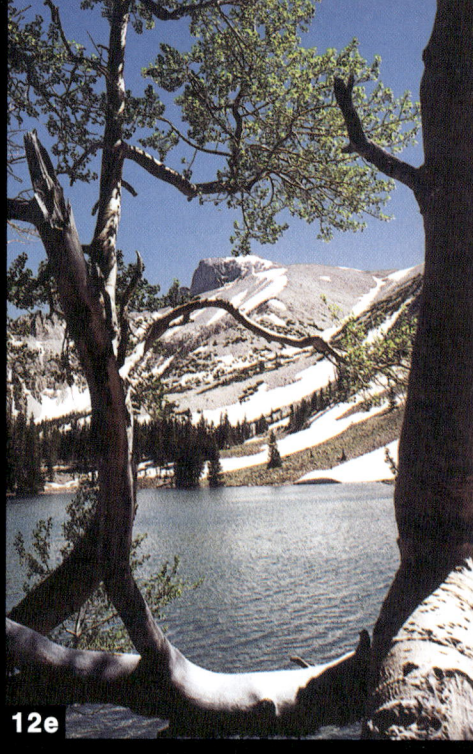

12e

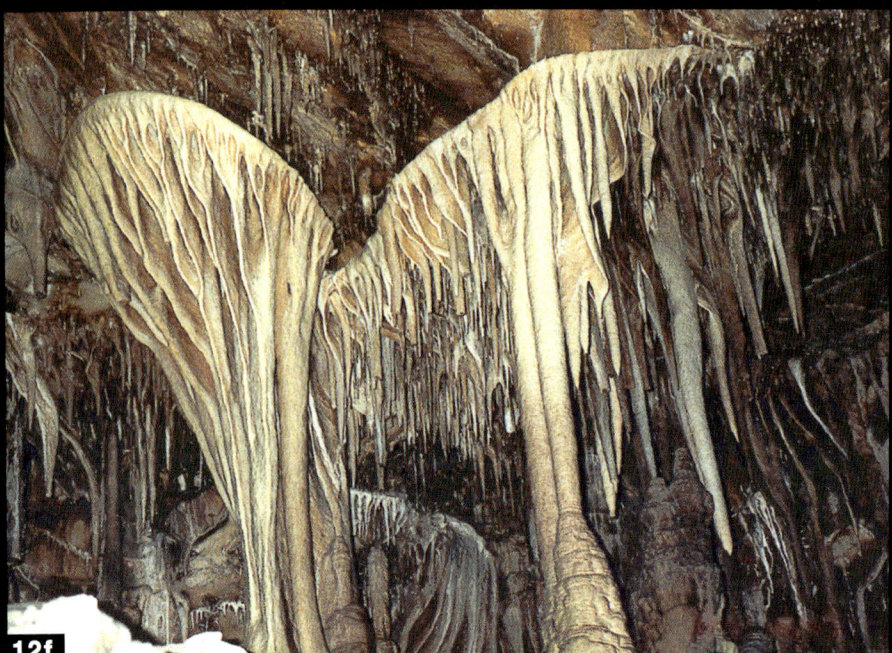

12f

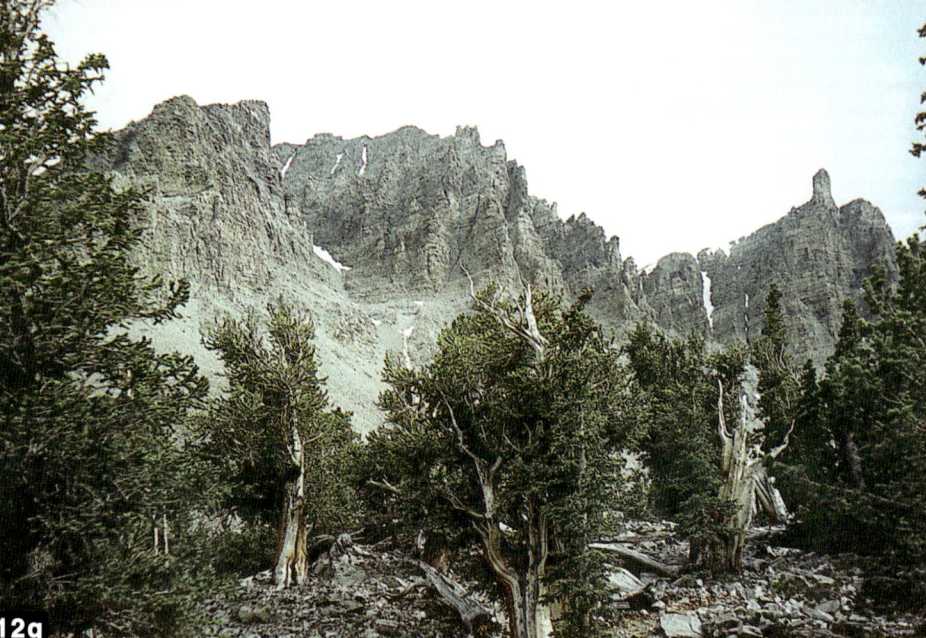

12g

PLATE 12

COLOR PHOTO CAPTIONS

PLATE 11

11a Doe spotted near Lehman Creek, **Great Basin National Park**. *Photo: Roy W. Cazier*

11b **Red columbine** (*Aquilegia formosa*) is found in open woods and banks from 4,000 to 10,000 feet elevation. It is particularly favored by hummingbirds as a nectar source. *Photo: Kris Pizarro*

11c **Quaking aspen (*Populus tremuloides*) Deciduous; slender in form, up to 70' tall; branchlets droop at end; bark smooth and whitish, on large trunks becoming dark gray and furrowed; leaves round and finely toothed, about silver dollar size with short point, turning brilliant gold in fall; male and female flowers (catkins) on separate trees; found in relatively moist areas.**

The quaking aspen is widespread in moist areas of Nevada's mountains. Its name refers to its leaves, which flutter with the slightest air movement. It is often seen in stands with straight, vertical trunks, but where it is exposed to high winds and heavy snows it becomes distorted or nearly prostrate. Deer and elk browse on its twigs and foliage.

Aspen in the Great Basin seldom reproduces from seed. Instead it sends up root sprouts, or suckers, which produce new trees that are essentially clones of the parent. The cloned groups bloom, leaf out, turn color, and drop their leaves in unison. Aspen depends on fire or other disturbance for survival. The destruction of the tree above ground stimulates production of suckers and provides for the renewal of a grove. *Photo: Kris Pizarro*

11d Late afternoon shower near **Sacramento Pass**. *Photo: Kris Pizarro*

11e Old homestead along the **Osceola Road**. *Photo: Kris Pizarro*

11f Variably bleached and iron-oxide-stained **limestone** exposed on the east side of **Connors Pass**. The triangular wedge of gray rock in the center foreground is caught between two steep faults that merge near the upper left corner of the photo. Note the offset of the thick light-colored limestone band across this fault. *Photo: Kris Pizarro*

PLATE 12

12a–d **Great Basin bristlecone pine (*Pinus longaeva*) height 20–40'; irregular crown of spreading branches; shrub size at timberline; evergreen; five needles in a bundle, very short and crowded in a long dense mass similar to a foxtail.**

Great Basin bristlecone pines are among the oldest trees in the world. Some have been dated at nearly 5,000 years. The oldest trees are found living at high elevations near the tree line, where conditions are most harsh. They tend to be under 30 feet tall, and many are multi-stemmed with few branches and only a portion of the trunk alive. They are remarkably energy efficient and adapt quickly to a changing environment. Most other conifers retain needles for a just a few years, while the bristlecone can retain its needles for as many as 40 years. In favorable years the bristlecone grows well, but when conditions deteriorate its foliage dies back until what is left can be supplied with nutrients provided by the root system. Its extremely dense, resinous wood doesn't rot in the dry, cool air, so it will remain standing for centuries after most of it has died. *Photos: 12a-Jack Hursh, 12b–d-Kris Pizarro*

12e **Stella Lake** and Wheeler Peak. *Photo: Aleta Hursh*

12f Unique **shield formations** in **Lehman Caves**, Great Basin National Park. The shield on the left is called "The Prospector." *Photo: Terri Garside*

12g View to the southeast from the trail leading to the **bristlecone forest** and the **icefield**, Great Basin National Park. The cliffs on the right form the north face of **Jeff Davis Peak**. All of the rugged cliffs and peaks in the photo are composed of Cambrian **quartzite**. *Photo: Kris Pizarro*

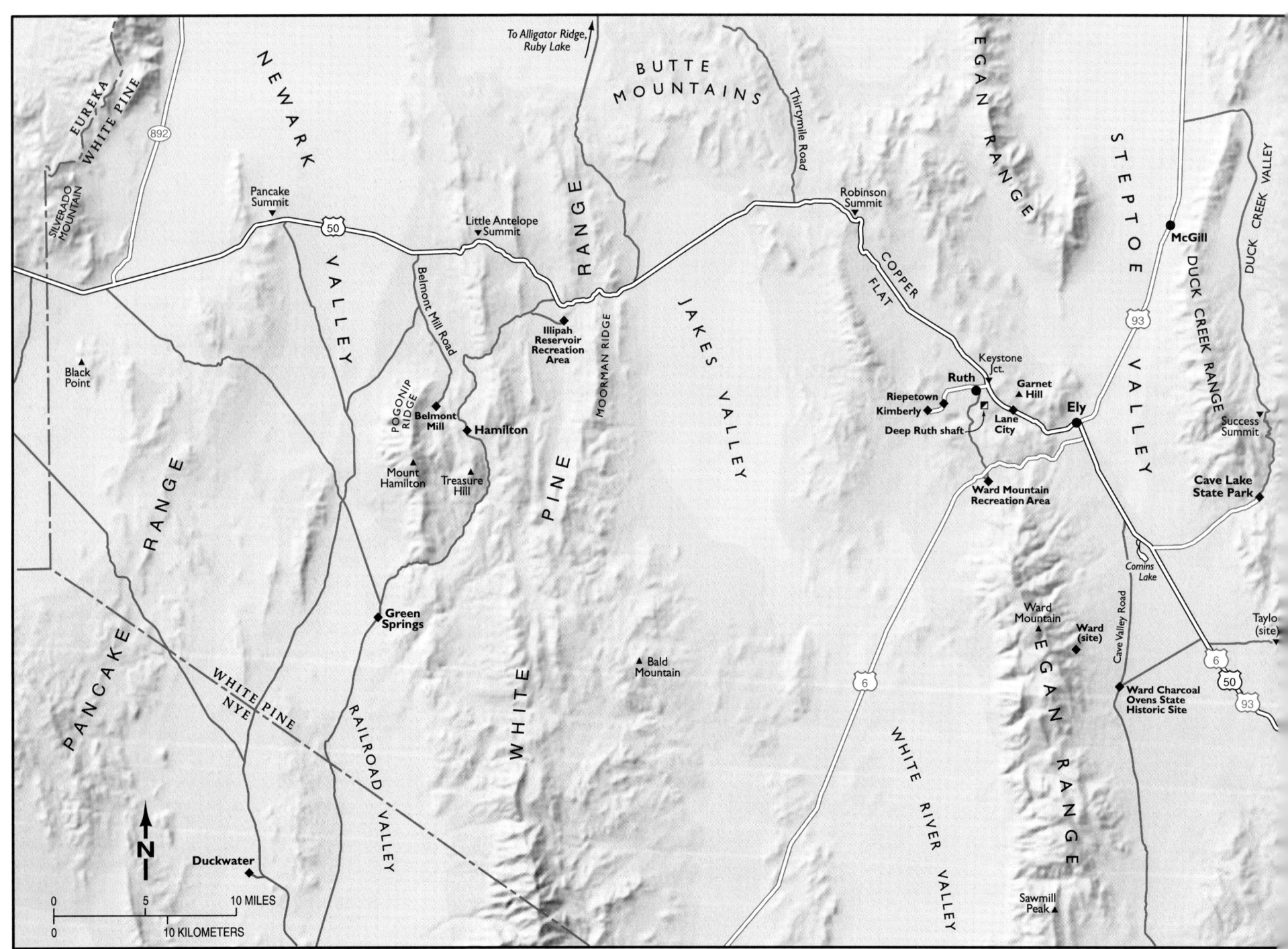

Route map, White Pine County.

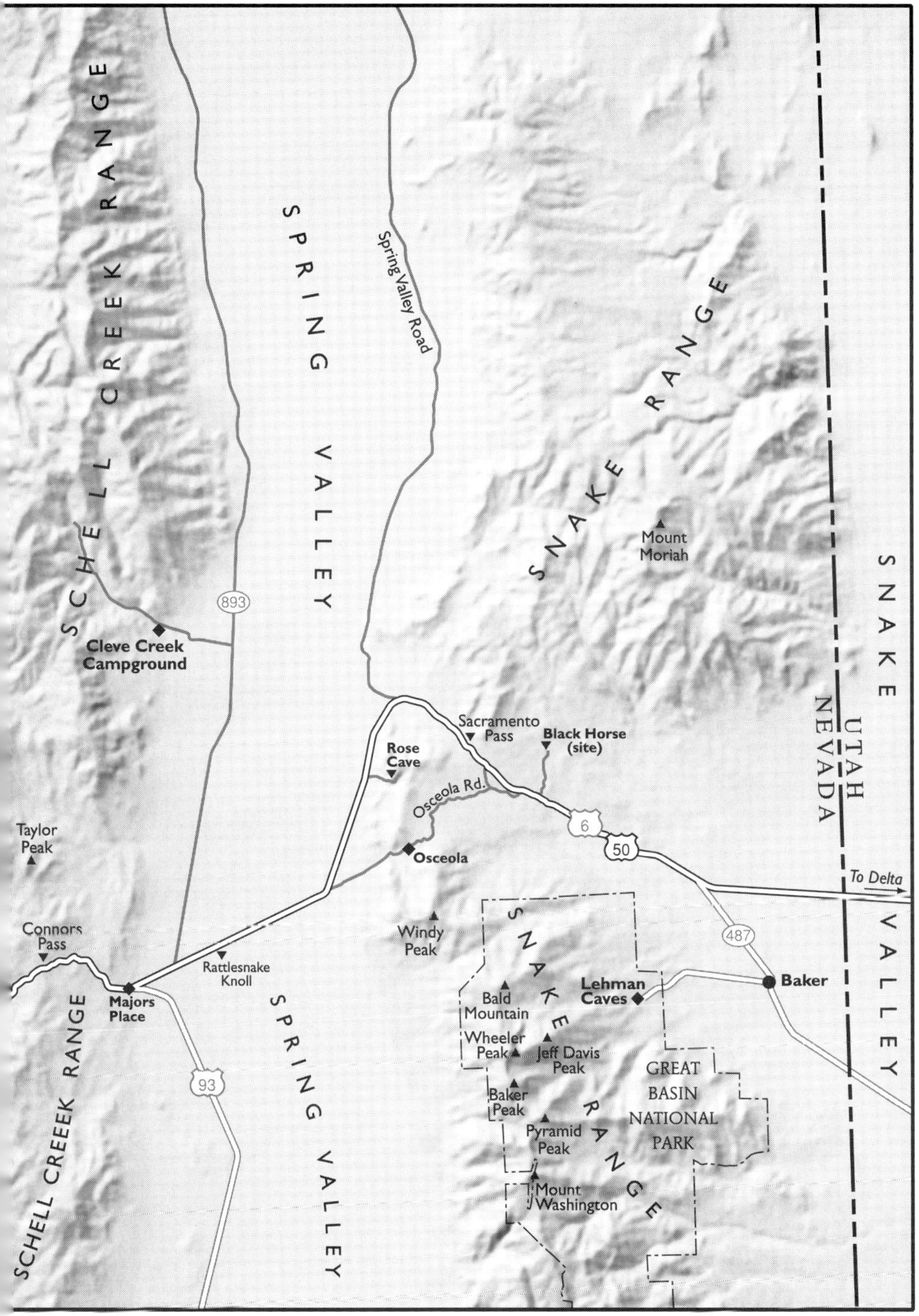

The map shows the Schell Creek Range, Spring Valley, Snake Range, and Snake Valley areas. Labeled features include: Cleve Creek Campground, Taylor Peak, Connors Pass, Majors Place, Rattlesnake Knoll, Rose Cave, Osceola, Windy Peak, Bald Mountain, Wheeler Peak, Jeff Davis Peak, Baker Peak, Pyramid Peak, Mount Washington, Sacramento Pass, Black Horse (site), Mount Moriah, Lehman Caves, Baker, Great Basin National Park. Roads: Spring Valley Road, Osceola Rd., 893, 93, 6, 50, 487. State line: UTAH / NEVADA. To Delta.

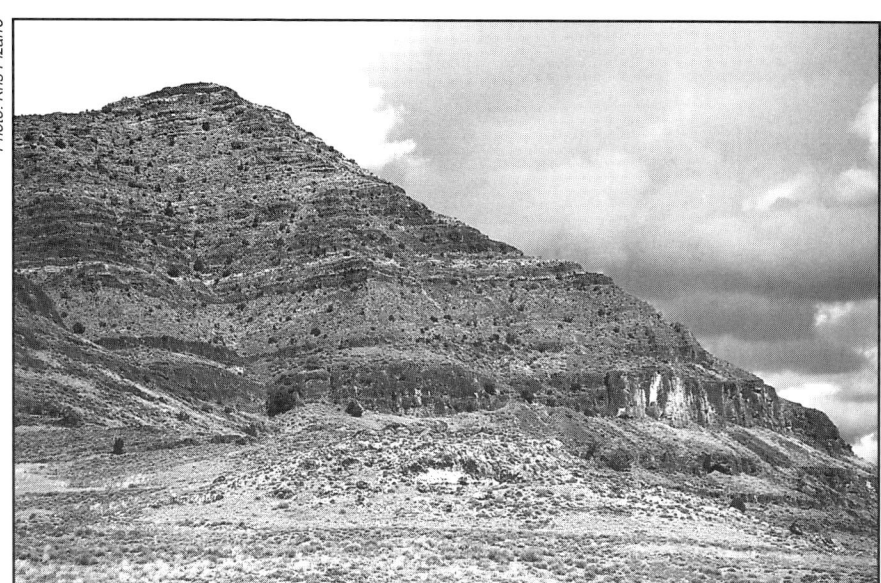

Photo: Kris Pizarro

Prominent bedding in Devonian carbonate rocks exposed in the hills east of Silverado Mountain.

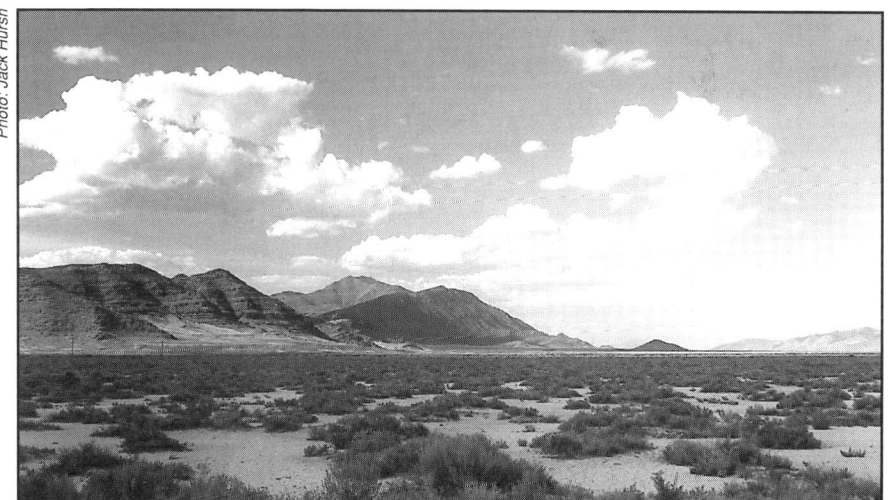

Photo: Jack Hursh

View north into Newark Valley near the intersection of State Route 892 and U.S. Highway 50. The Diamond Mountains are to the left. The west flank of Buck Mountain is in the distance to the right.

interval	cumulative	milepost	
4.4	10.4		Volcanic rocks crop out to the north of the road.
0.8	11.2		East-dipping conglomerate and sandstone beds of the Mississippian Diamond Peak Formation are exposed in the road cut.
1.0	12.2		Pancake Summit (elevation 6,517 feet).
1.2	13.4		The volcanic rock in the road cut on the right (south) side of the highway contains a layer of perlite, volcanic glass that, when processed commercially, expands, or "pops" like popcorn to form a material used for insulation. This particular deposit is fairly low quality and has never been mined. South of here about a mile, the Pancake Coal Mine is located on the middle of the eastern slope of the mountain. Nevada is not renowned for its coal deposits, and to call this one a mine is a real exaggeration. Three thin seams of coal occur in a clay-shale unit in the upper part of the Mississippian Diamond Peak Formation. This deposit was never mined, and it has not been looked at with any interest since 1892.
1.6	15.0	WP 15.0	Newark Valley. The Pancake Range forms a prow-like mass that extends from the south into Newark Valley. We crossed the western fork of the valley a few miles back, west of Pancake Summit. Now we are dropping into the center of the eastern, main part of Newark Valley. To the left, at about 9:00, Buck Mountain can be seen in the distance. Ahead, the highway climbs into the White Pine Range. At 2:00 the highest part of this range, standing to the west of the lower ridges in the background, is Pogonip Ridge. Mount Hamilton (elevation 10,745 feet), at the south end of Pogonip Ridge, is the highest peak.

The White Pine Range is composed mainly of thick sections of Paleozoic carbonate rocks. The rocks are folded, with the axes of the folds trending generally north-south. Cambrian rocks form the west side and a lot of the top of Pogonip Ridge. To the east, the long, lower ridges of the range are underlain by Mississippian and Pennsylvanian-Permian limestone, sandstone, and shale.

Several large open-pit gold mines have recently been active along the west side of the White Pine Range. The large cut visible at about 2:30, below and to the northwest of the high peak, is the Mount Hamilton Mine.

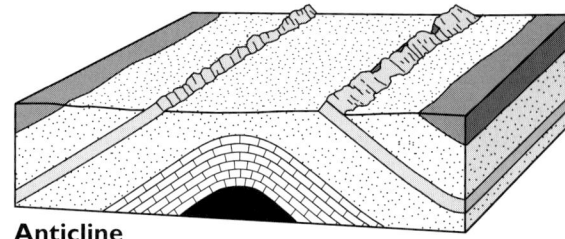

Anticline

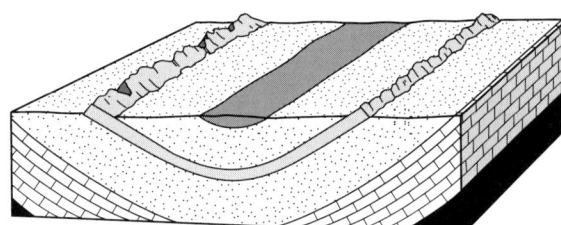

Syncline

Generalized anticline (beds dip away from the axis) and syncline (beds dip toward the axis).

East-dipping conglomerate and sandstone beds of the Mississippian Diamond Peak Formation exposed in a road cut west of Pancake Summit.

Photo: Kris Pizarro

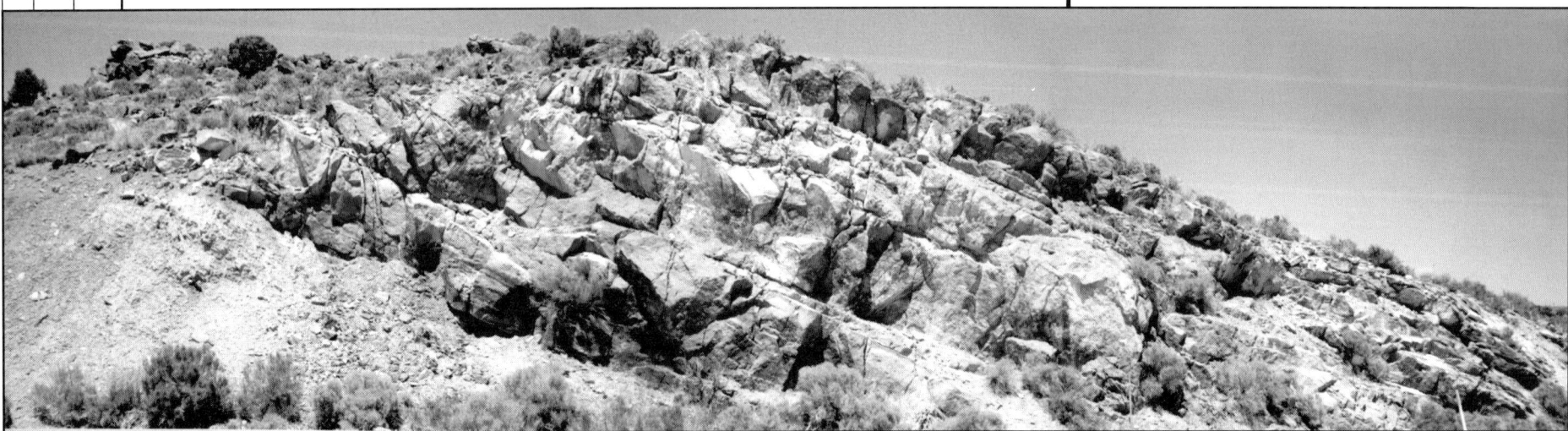

interval	cumulative	milepost	

3.8	18.5	Note the low, stubby juniper trees along the road here. Cold, windy conditions result in everything growing close to the ground.
2.7	21.2	Road cut on the left exposes varicolored Tertiary sedimentary rocks. Note the fault contact between bedded tuffaceous sediments to the west and gravel units to the east. (plate 10f)
0.6	21.8	Intersection with Belmont Mill road, to the right. This is the access road to the Mount Hamilton Mine, other sites on the east side of Pogonip Ridge, and Green Springs. By connecting with branching roads of varying quality one can travel south from here and connect with U.S. 6 at Currant in Nye County. Not recommended for most vehicles.
		Looking again toward Pogonip Ridge, Treasure Hill is visible on the skyline at about 2:00 (the peak with the slightly inclined flat top located east of Mount Hamilton, the highest peak on the west). The ghost town of Hamilton lies on the north flank of Treasure Hill, about 3 miles east of Mount Hamilton. Hamilton and Treasure Hill are discussed in a following section.
0.4	22.2	For the next 11 miles, we pass through a section of gently folded Mississippian Chainman Shale and overlying Pennsylvanian-Permian Ely Limestone. Beds will first be seen dipping to the west, then to the east, then to the west again as we move from one fold into another.
1.9	24.1	Good turnout here. It's only a short way to the summit, but you can stop and rest for the last push to the top.
1.2	25.3	Little Antelope Summit (elevation 7,433 feet). U.S. 50 crosses the White Pine Range at this point.
0.5	25.8	Illipah Mine road, on the left. A small open-pit gold mine, located about 3 miles to the north, operated between 1987 and 1989.

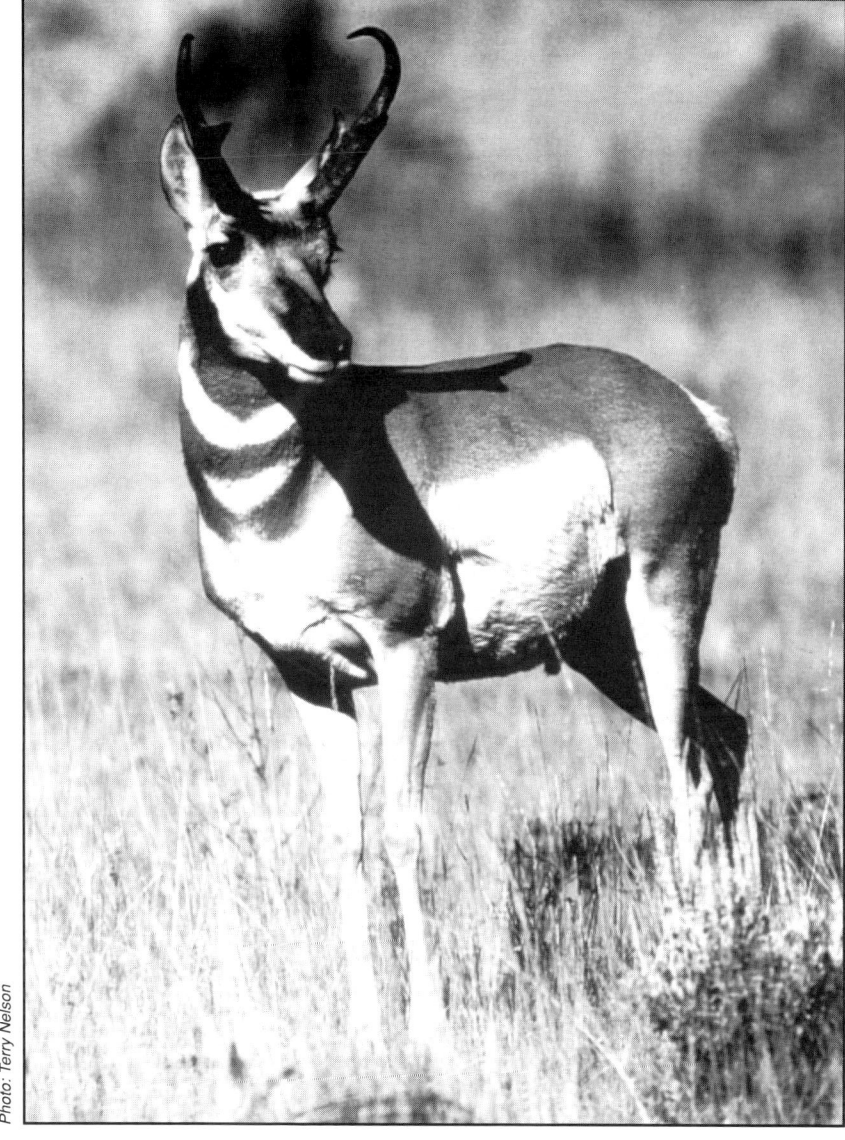

Photo: Terry Nelson

Pronghorn antelope (*Antilocapra americana*) 49–57" long; deerlike; tan or reddish tan above; chest, belly, inner legs, and rump patch white; buck has black mask on face; horns black (12–20" long on buck, 3–4" long on doe), curving back and in at tips, with prong about two-thirds of the way up.

The pronghorn antelope is the fastest animal in the Western Hemisphere. It covers great distances at cruising speeds of 25–30 miles an hour and has been clocked at up to 70 miles an hour for short periods of time. It inhabits grasslands and open shrub-grasslands where it can rely on its superior vision and speed to elude danger.

Pronghorn mate from late summer to fall, and fawns are born the following May or June. A fawn can walk just hours after birth and can run by its fifth day. It spends most of its first weeks lying in seclusion, is grazing by the third week and weaned by fall.

Directly ahead are great exposures of bedded Mississippian limestone.

Gravel road to the right (west) goes to the ghost town of Hamilton, center of the White Pine mining district. The years and some recent mining operations have taken their toll on Hamilton, and there are only a few partial walls standing from the historical buildings. The 11-mile gravel road into Hamilton is usually kept in good repair, but it is a winding mountain road, narrow in spots, and should not be attempted in bad weather.

Just 0.1 mile off the highway on this road, a road to the left leads to Illipah Reservoir Recreation Area (follow the signs another 1.3 miles). A good spot to rest, camp, fish, or just enjoy the view of more outcrops of bedded rocks, which here are the Permian Rib Hill Sandstone and Arcturus Formations. The reservoir overlook area is equipped with campsites, tables, shelters, and restrooms.

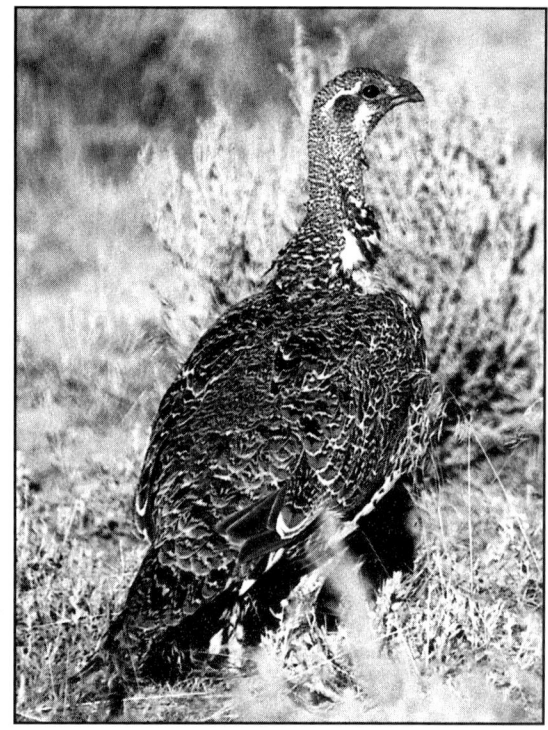

Sage grouse (*Centrocercus urophasianus*) Sagebrush provides a major source of food, as well as cover and nesting sites for this grouse.

Photo: Roy W. Cazier

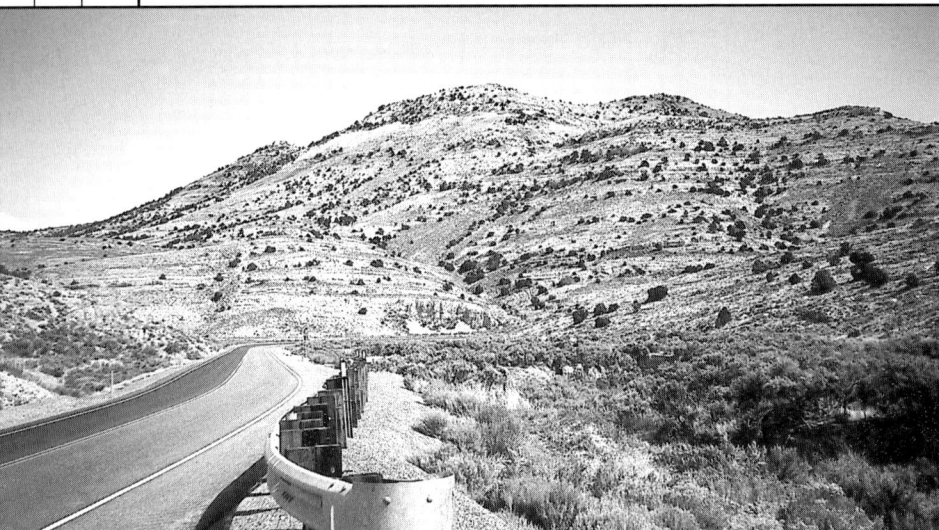

Photo: Kris Pizarro

◄ **A hillside of bedded Mississippian limestone near the Illipah Reservoir-Hamilton turnoff. Notice small juniper trees aligned along limestone bedding in the right side of the photo.**

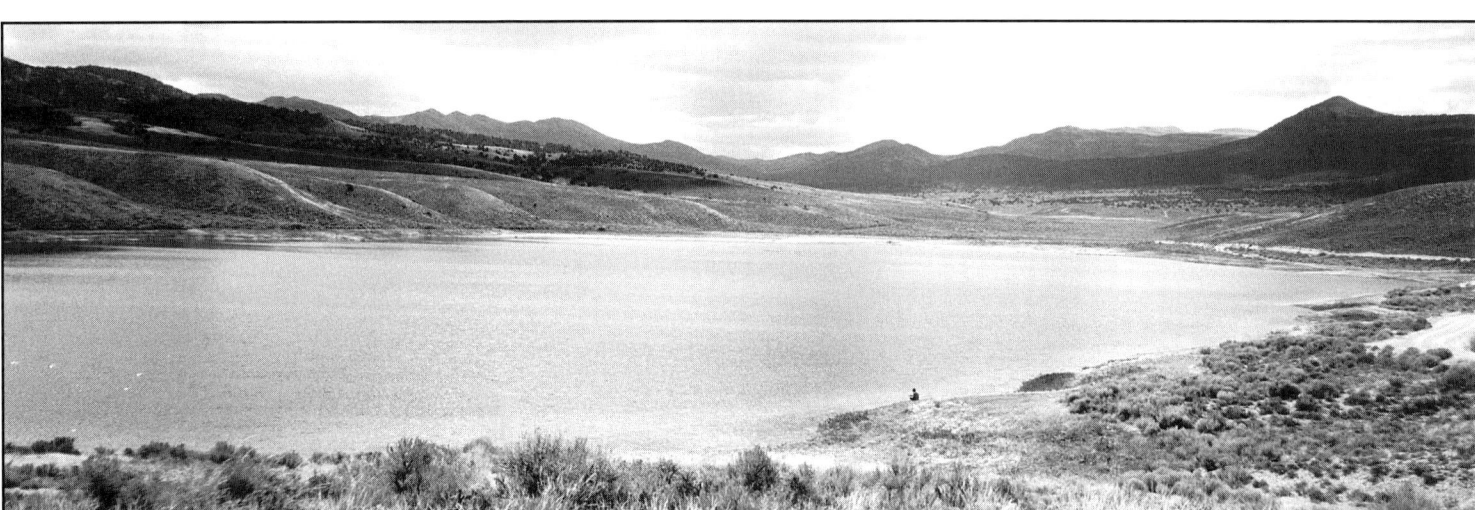

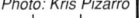
Photo: Kris Pizarro

Illipah Reservoir.

HAMILTON AND THE RUSH TO WHITE PINE

In 1867, rich deposits of horn silver, the soft chloride of silver, were discovered on Treasure Hill, high in the White Pine Range about 11 miles to the south of here. The silver ores found on top of Treasure Hill were described by U.S. Geological Survey geologist Arnold Hague, who visited the area soon after its discovery, as "...probably the most remarkable occurrence of horn silver on record," and the resulting "Rush to White Pine" was perhaps the most sensational mining stampede in the history of the West. Within a few months, it was reported that more than 25,000 adventurers, more than had flocked to the Comstock 10 years earlier, swarmed over Treasure Hill. Miners bragged of digging on a "mount of solid silver," causing extravagant speculation in world silver markets.

At the height of the boom, Hamilton became the county seat of White Pine County. The rich ores did not persist with depth, however, and the spectacular bonanzas were virtually exhausted within two short seasons. The population drifted away over the next 20 years and Hamilton became a ghost town. In 1885, a disastrous fire finished off the remains, and the county seat was moved to Ely.

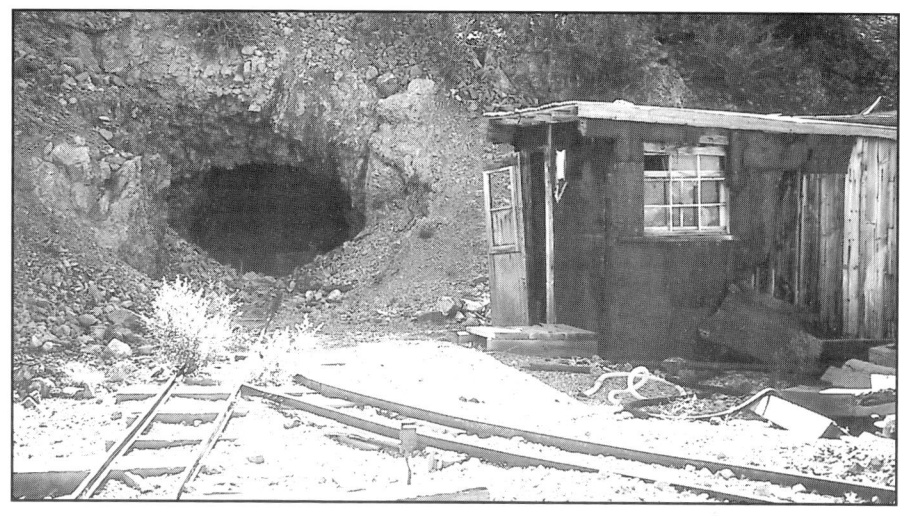

Portal of the Onetha Mine, a more recent "abandoned mine" on the west side of Treasure Hill in the White Pine mining district. Remember, for all old mines, "Stay Out and Stay Alive." *Photo: Richard Jones*

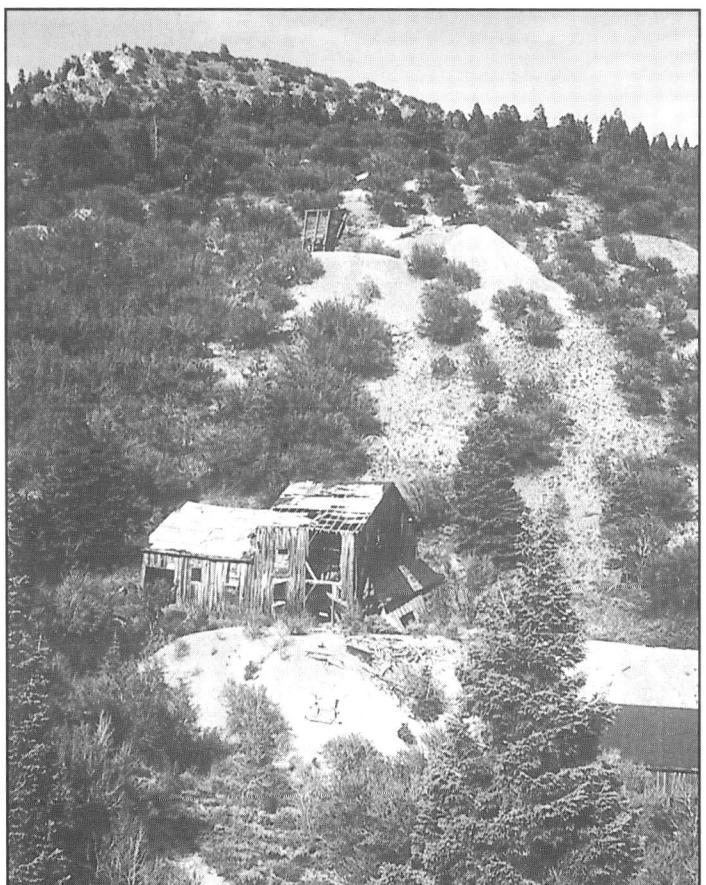

Old buildings and dumps at the Ne Plus Ultra Mine, White Pine mining district.

Photos: Joseph V. Tingley

Stone ruins remaining at the site of an 1869 stamp mill in Shermantown Canyon, White Pine mining district.

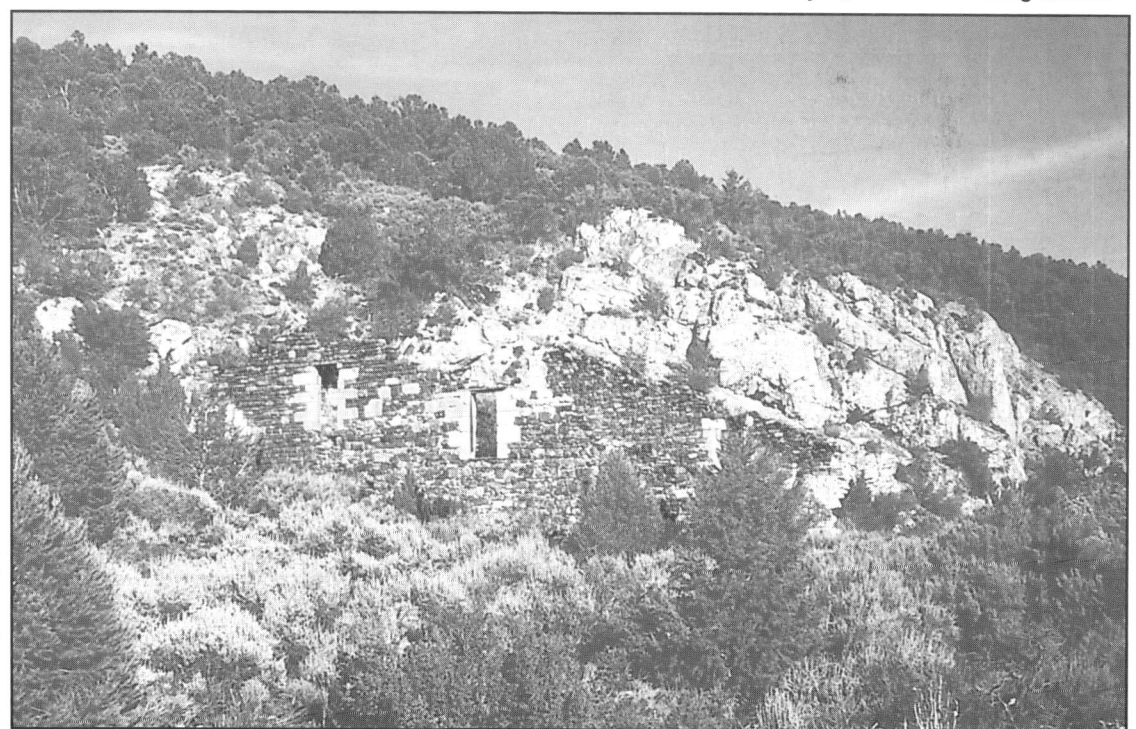

interval	cumulative	milepost	
0.8	31.3		For the next few miles, we will be driving through upper Paleozoic formations, mostly medium- to thick-bedded limestones. The rocks become progressively younger from west to east. Here, on the left side of the highway, is a contact between the Pennsylvanian-Permian Ely Limestone and the overlying Lower Permian Reipe Spring Limestone. There are no good (safe) places to pull off here, so save any urge to examine rocks for other spots.
0.8	32.1		Abandoned homestead on the left. Rocks exposed on both sides of the highway are massive limestones and platy siltstones of the Permian Arcturus Formation.
1.2	33.3		Straight ahead is a good view of a bedded limestone outcrop. Notice that the trees grow lined up along joints (cracks) in the limestone.
1.3	34.6		Moorman Ranch on the right (south) side of the highway. This ranch was established soon after the Hamilton boom and has been an active ranch since that time.
1.6	36.2		Road to Alligator Ridge Mine and Ruby Lake, on left.
			The Alligator Ridge Mine is 40 miles to the north. Discovered by a modern-day "lonely prospector" in 1976, Alligator Ridge was the first large disseminated gold deposit to be found in White Pine County. Its discovery led to other similar deposits, and this area is an important contributor to Nevada's gold production.
			Some 20 miles beyond the mine, this road reaches the Ruby Lake National Wildlife Refuge, a great place for birding, fishing, and just plain enjoyment of some of Nevada's most spectacular scenery. The Ruby Mountains, just west of the lake, are one of the highest ranges in Nevada with peaks rising to over 11,000 feet. It is 87 miles to pavement, another 35 to Interstate 80, and 20 more miles to Elko and gas, so don't attempt to visit the Rubys from here without first carefully planning the trip.
9.8	46.0	WP 46.0	Volcanic rocks reclaim the landscape for awhile, beginning here.
1.2	47.2		Thirtymile road, to the left (north). We are now climbing into the Egan Range. Oligocene volcanic rocks make up the landscape on both sides of the highway from here to Robinson Summit, ahead.
0.2	47.4		Good exposure of columnar jointing in the volcanic rocks to the left.
2.3	49.7		Robinson Summit (elevation 7,607 feet).
			Outcrops at the summit are still Oligocene volcanic rocks, but as the road drops into Copper Flat to the east, it passes through the volcanic rocks and back into the underlying Permian limestone.

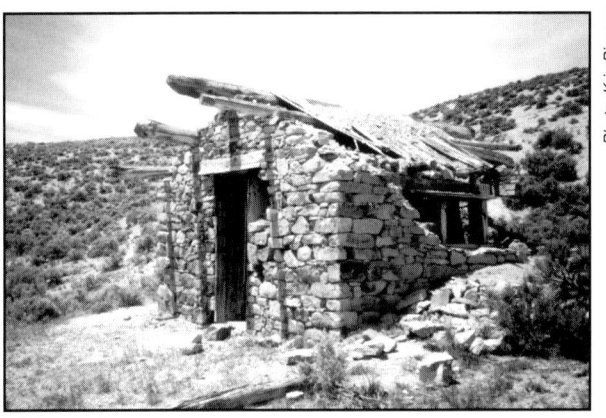

Photo: Kris Pizarro

Basque Sheepherders

White Pine County has a strong history of sheep ranching. The first large domestic bands of sheep to cross Nevada came through Deep Creek Valley in northeastern White Pine County in 1852. Resident bands of sheep were raised in Carson Valley below Genoa starting about the same time. Large sheep outfits were active in what is now White Pine County starting in the early 1860s, and sheep ranching on a somewhat smaller scale still continues in the area. The herders who tended these sheep were traditionally from the Basque provinces of Spain and France. Nevada, along with other sheep grazing areas of the mountain west, have been enriched by the heritage of the Basque sheepherder.

The Highway 50 traveler has a chance of encountering sheep anywhere along the route from Austin east to the Utah state line. Large bands are not commonly seen close to the highway, however, and you are more likely to find them in the high valleys and adjacent mountains farther removed from traffic and towns. The valley east of Robinson Summit in White Pine County is a possible exception, as the "watch for sheep" sign seen along the road here warns us. If you encounter a herd of sheep on U.S. Highway 50, or some dirt road in the backcountry, stop and let the sheep pass around you. You won't miss the time lost, and will probably earn a friendly wave from the sheepherder.

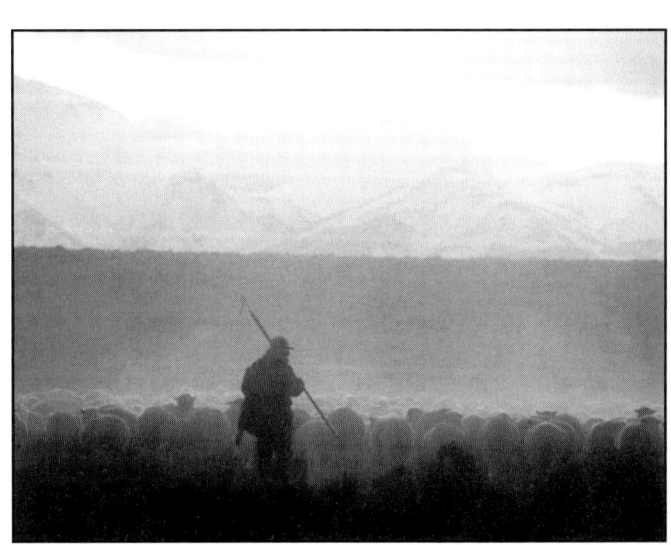

Photo: Linda Dufurrena

interval	cumulative	milepost	

| 4.0 | 53.7 | Entering Copper Flat, a small basin bounded by north-trending prongs of the Egan Range. Ahead on the right are the large waste-rock dumps from copper mining operations at Ruth in the Robinson mining district. |

Ward Mountain, at 1:00, elevation 10,936 feet.

| 1.5 | 55.2 | BLM Copper Flat-Gleeson Creek seeding area. Experimental seeding of crested wheatgrass along the right side of the highway. |

| 6.2 | 61.4 | To the left, BLM sign for Garnet Hill. This is another worthwhile side trip/adventure for mineral collectors. Access is good, but over a gravel road best traveled when dry. Garnet Hill, the prominent hill about 3 miles east of the junction, is composed of a Tertiary rhyolitic intrusive body which contains vesicles lined with quartz and near gem-quality garnets. The area is open to collecting, and the BLM has provided limited camping and picnic facilities in the area. |

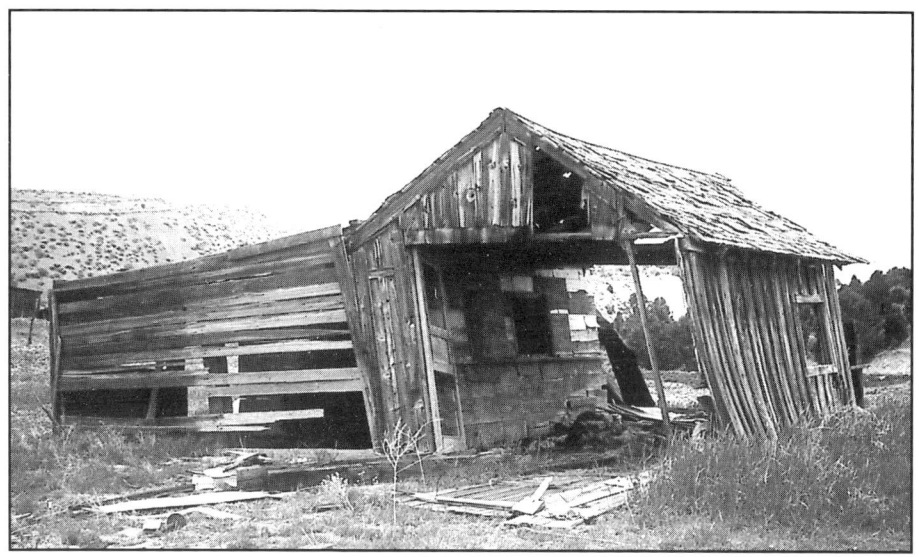

Photo: Kris Pizarro

Old buildings in Tonopah Canyon south of the Robinson Mine.

The lower photo is a framed view of Ward Mountain.

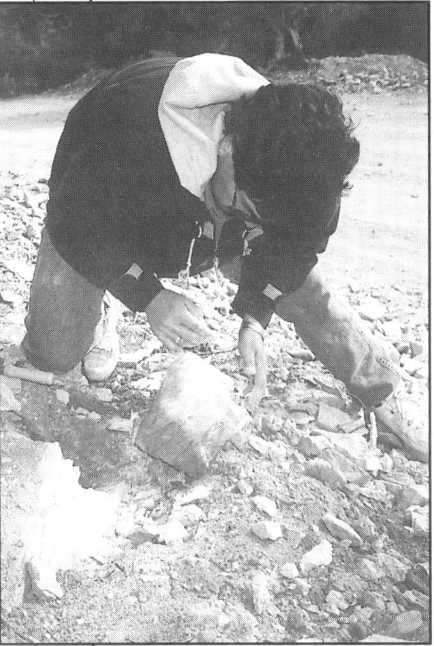

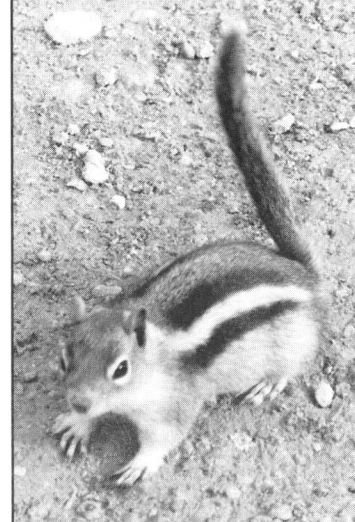

Photos: Kris Pizarro

Checking a rhyolite boulder for garnets in the collecting area at Garnet Hill.

Golden-mantled ground squirrel (*Spermophilus lateralis*) inhabits mixed woodlands, coniferous forests, and alpine habitats. It feeds mainly on seeds, fruits, and nuts, and is often seen in campgrounds and picnic areas prospecting for snacks.

| 0.2 | 61.6 | Keystone Junction. State Route 44, on the right, leads to the town of Ruth and to the Robinson Mine, last operated by BHP Copper North America. The company has provided a viewing area overlooking the large open pit. Signs describe the operation and the geology of the deposit, and there are ore specimens for examination and souvenirs. The overlook is a short mile from the intersection. (plates 10b and 10c) |

Copper in the Robinson Mining District

Robinson, the largest copper mining area in Nevada, began with discovery of small gold-silver deposits in 1867. Only about $1 million was produced from these deposits, which proved to be peripheral to a large porphyry copper orebody discovered in 1902. The Nevada Consolidated Copper Co. was incorporated in 1904, and a railroad was constructed between Copper Flat and the Southern Pacific main line 150 miles to the north at Cobre. Copper production began in 1908 and by 1978, when operations closed, the district had produced more than $1 billion. A revival of the copper market brought a new company to the district in 1992 with plans to produce another 2.5 billion pounds of copper and over 1 million ounces of gold over a 16-year period. Low copper and gold prices brought this new operation to a halt in mid-1999, however, and the future of copper in Ely is again uncertain. During the first period of operation, ore was mined from the pits ahead at Ruth, shipped by rail through Ely and north 12 miles to a large mill in McGill. There the ore was concentrated and fed to a smelter to produce copper metal. Metal ingots were shipped north on the company railroad, the Nevada Northern, to a connection with the Southern Pacific Railroad and eastern markets. The concentrator and smelter at McGill were removed after 1978 and, for the latest short period of mining, a new mill was constructed on site at the mine. Copper-gold concentrates produced by the new mill were shipped to company-owned smelting facilities at San Manuel, Arizona.

The Robinson district is one of four districts in Nevada (the others are Yerington in Lyon County, Battle Mountain in Lander County, and San Antone in Nye County) at which "porphyry-type" copper deposits have been mined. In these deposits, low concentrations of copper sulfide (chalcopyrite), along with iron pyrites and various other sulfide minerals, formed in a granitic intrusion and the older metamorphic rocks. As the sulfide-bearing rock was exposed to surface weathering, the copper minerals were oxidized. Copper was leached from the rock and moved down to the water table where it was again deposited. As this process continued over a long period of time, copper concentration built up in what is called a supergene orebody. The primary grade of a porphyry copper orebody (the original, non-enriched material) is generally between 0.4 percent and 1 percent copper, while the enriched orebody can contain ore up to 5 percent copper. The oxidized, leached rock left behind after most of the copper has moved out forms what is called a "leached capping" above the supergene orebody. Although not usually commercially valuable, this capping is highly colored by maroon, red, and yellow iron oxides splashed with varying amounts of green and blue-green from trace amounts of oxide copper minerals left behind. Prospectors quickly determined the relationship between leached capping and copper ore and still use the type and intensity of certain mineral stains to search for new copper orebodies.

In the Robinson district, the copper deposits occur in an east-west-trending zone defined by outcrops of bleached limestone, leached porphyry, and iron-oxide staining (the leached capping). Most of the copper ore occurs in the altered porphyry, although about 20 percent of the production has been derived from metamorphosed sedimentary rocks adjacent to the porphyry bodies.

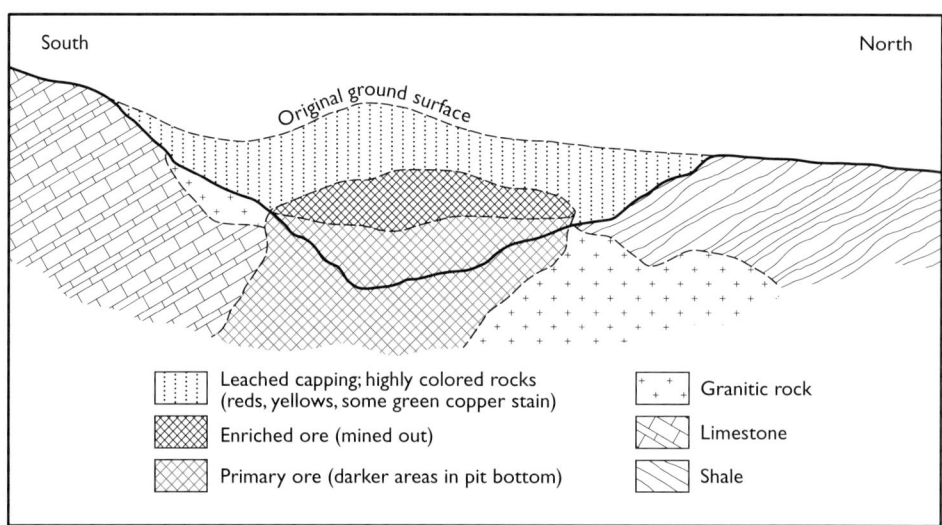

South   North

Original ground surface

	Leached capping; highly colored rocks (reds, yellows, some green copper stain)
	Enriched ore (mined out)
	Primary ore (darker areas in pit bottom)
	Granitic rock
	Limestone
	Shale

Generalized cross section of the Liberty Pit, showing location of copper orebody and approximating what can be seen from the viewing area on the east side of the pit (*see photo below*).

Photo: Roy W. Cazier

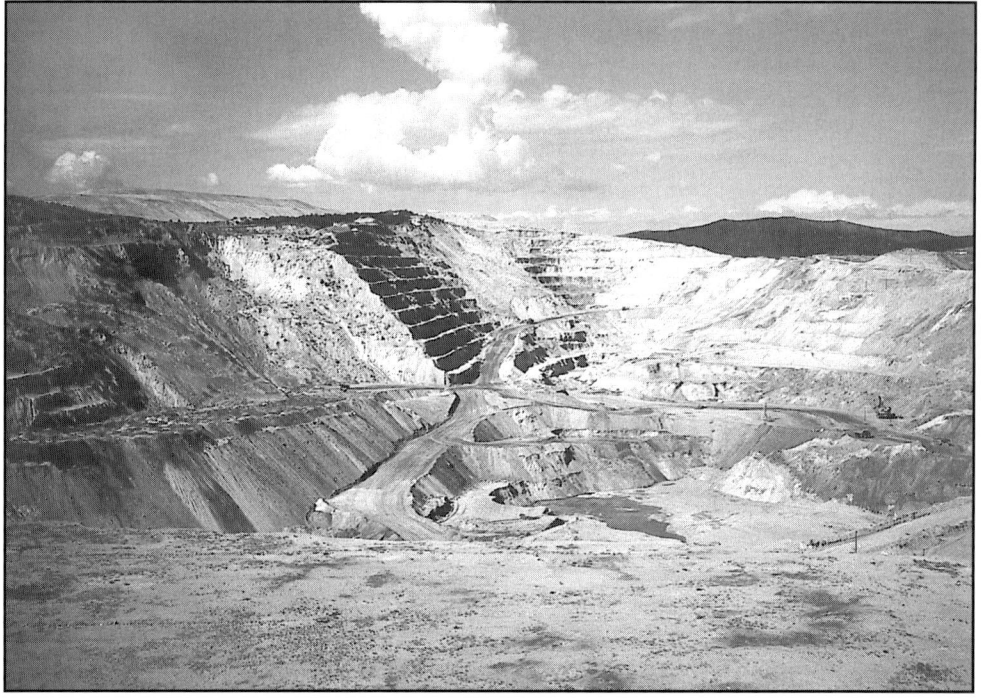

Liberty Pit at the Robinson Mine, from the viewing area on the east side of the pit.

interval	cumulative	milepost
0.6	62.2	
0.8	63.0	
1.0	64.0	WP 46.0
0.5	64.5	
0.2	64.7	

Mississippian-age Joana Limestone forms the outcrops on the left. Note the tight folds in the limestone.

The large red steel headframe to the right (south) at about 3:00 is at the Deep Ruth shaft, which was sunk in the 1970s to develop and mine the Deep Ruth orebody. The underground project was abandoned and much of the ore was later mined from the nearby Ruth pit. The mine waste dumps on the right have been contoured as part of a reclamation program.

On the left, old buildings mark the site of the ghost town of Lane City. Known as Mineral City when it was founded in 1870, it was the milling center for the Robinson district following discovery of silver ores nearby in 1867. Activity peaked for Mineral City in 1872–1873 when the population reached about 600.

The Chainman and Joana Mines, on the right, produced modest amounts of gold and silver between 1889 and 1901 from oxidized orebodies in limestone (Joana Limestone) near shale contacts (Chainman Shale). The two Mississippian rock formations were named for these mines.

See if you can spot the "gossan" in the road cut on the left. The gossan, looking like a big gob of rust in the surrounding rock, is what the surface expression of many orebodies looks like. The rusty, red-brown material is composed of iron and some copper, lead, and zinc oxide minerals formed by weathering of minerals such as pyrite, chalcopyrite, galena, and sphalerite.

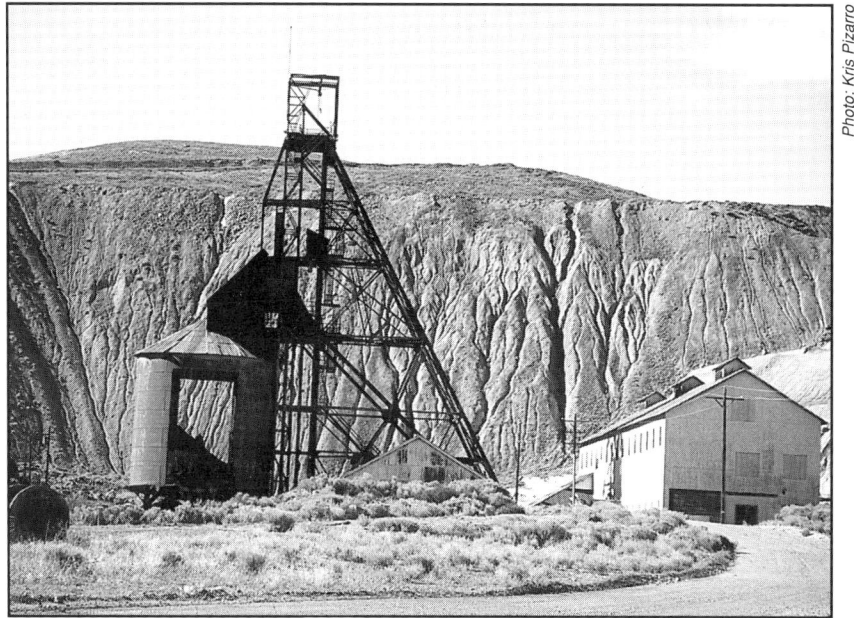

Photo: Kris Pizarro

Headframe at the Deep Ruth shaft.

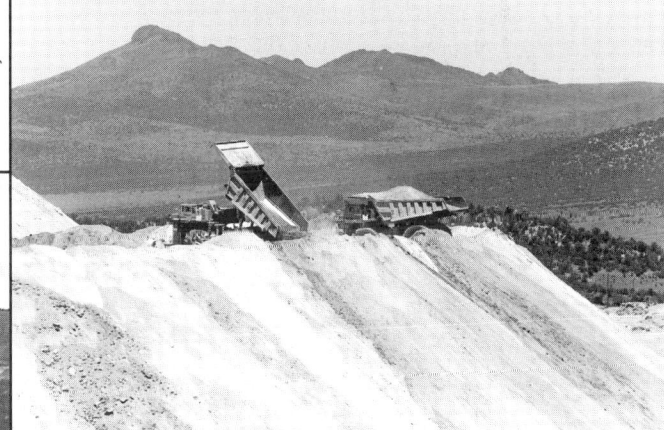

Photo: Roy W. Cazier

Haul trucks dumping waste at the Robinson Mine.

Photo: Kris Pizarro

Panorama of Ward Mountain and the Robinson Mine.

ELY

Ely was founded in 1870, and its prosperity depended entirely on mining activity around Mineral City (Lane City) and other small camps to the west. Ely became the county seat of White Pine County in 1887, but the town did not grow much until the arrival of the Nevada Northern Railroad in 1906 and the completion of the smelter at McGill (12 miles to the north) in 1908. With these developments, the town became the shipping and distribution center for the mines and ranches of east-central Nevada.

Today Ely is still the business and population center for all of this part of Nevada. Ely suffered when the Kennecott copper mine and smelter closed in 1978; businesses closed and the population dropped. Several large gold mines opened in the county in the mid-1980s and 1990s, and copper mining resumed for a short period between 1996 and 1999. These events have given new life to Ely, although its future is still somewhat dependent on the ups and downs of metal prices.

Ely has numerous good motels and restaurants, shopping centers, a hospital, and an airport. The White Pine County courthouse is in the center of town, and there is a shady park where you can stop, rest, and decide what to explore in Ely. There is a museum east of downtown, near the cemetery (museum on the left, cemetery on the right), and the railroad is still in town, but it is now a first-rate tourist attraction rather than a freight and passenger line. Billed as the "Ghost Train of Ely," it makes short runs on the old tracks east and west of town. Although the rails are still in place, it no longer has connecting service with the Union Pacific main line to the north. The "Ghost Train" runs scheduled weekend excursions between May and September from its restored East Ely depot. There is a railroad museum in the old depot, and tours are given of the restored railroad shops and other facilities. This was the last operating short line railroad in Nevada and is said to be the best preserved short line in North America.

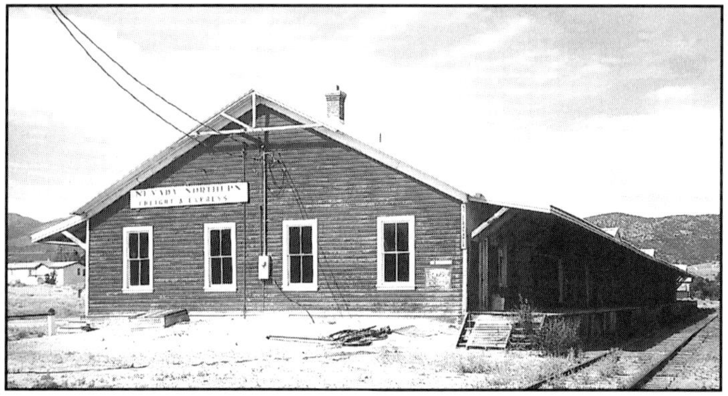

The "Ghost Train" emerging from a tunnel just west of Ely.

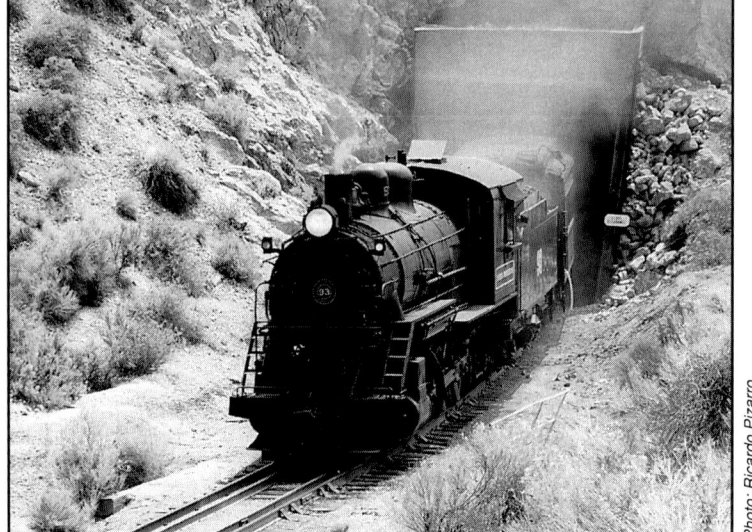

Photo: Ricardo Pizarro

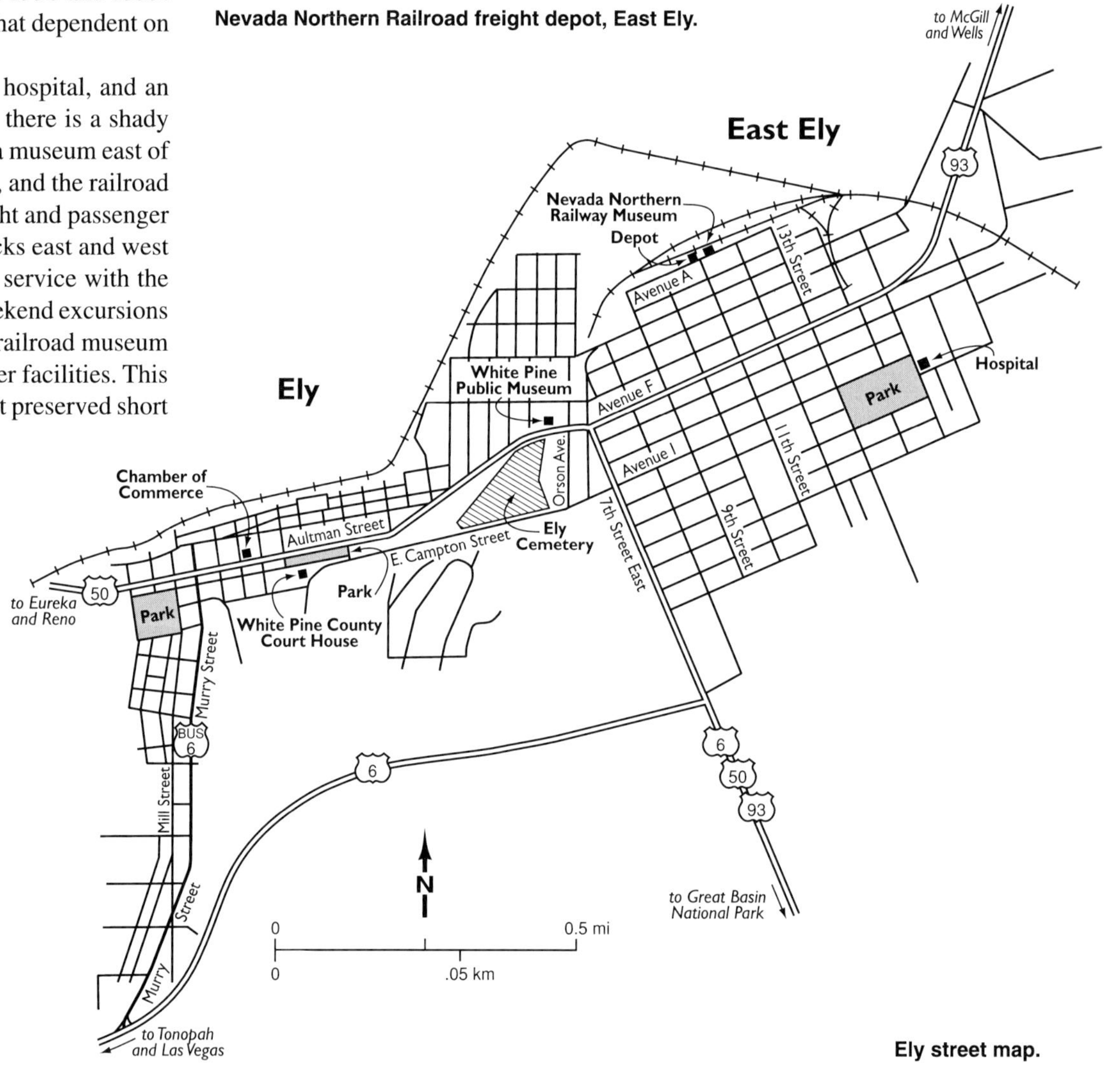

Photo: Kris Pizarro

Nevada Northern Railroad freight depot, East Ely.

East Ely

to McGill and Wells

93

Nevada Northern Railway Museum Depot

13th Street

Avenue A

White Pine Public Museum

Avenue F

Ely

Hospital

Park

11th Street

Avenue I

Orson Ave.

Chamber of Commerce

7th Street East

9th Street

Aultman Street

E. Campton Street

Ely Cemetery

to Eureka and Reno

50

Park

Park

White Pine County Court House

Murry Street

Mill Street

BUS 6

6

6

50

93

to Great Basin National Park

N

0   0.5 mi

0   .05 km

Murry Street

to Tonopah and Las Vegas

Ely street map.

interval	cumulative	milepost
1.1	65.8	

Railroad tunnel on the left.

| 0.3 | 66.1 | |

Entering Ely, county seat of White Pine County. This is the last town on U.S. 50 in Nevada. There are accommodations, food, and gas at Baker, 5 miles south of U.S. 50 at the Utah state line some 60 miles ahead. For bright lights and most other things, however, this is it until Delta, Utah, about 145 miles away.

Intersection with shortcut to U.S. 6, on the right, a connection with Tonopah, Las Vegas, and other points to the south and west.

| 0.3 | 66.4 | |

About 5 miles from this intersection, a road to the right leads to Ward Mountain Recreation Area and Ward Mountain Ski Hill. There is a campground with both picnic and overnight facilities, and access to a 20-mile system of signed hiking-cycling-ski trails that meander through high elevation piñon and juniper woodland as well as sagebrush and grassland.

Intersection at stoplight. The road straight ahead is U.S. 93 (north), which enters East Ely, and continues to intersect with Interstate 80 at Wells, 137 miles to the north.

For a side excursion, turn left at this intersection on 11th Street, and go five blocks to the Nevada Northern "Ghost Train" depot and museum.

| 1.2 | 67.6 | |

Our route, U.S. 50, is to the right. For the next 28 miles, U.S. 50 and U.S. 93 (south) are combined.

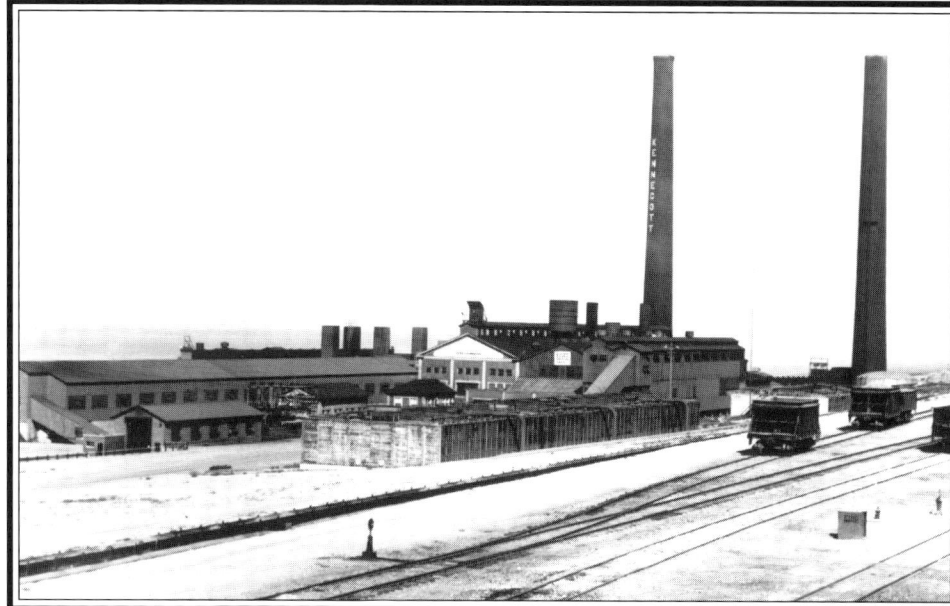

Kennecott copper smelter at McGill, circa 1959.

Nevada Historical Society

Nevada Northern Railroad coaling station, East Ely. ▶

Hydrant (*below*) still on duty at the Nevada Northern Railroad passenger depot (*below left*), East Ely.

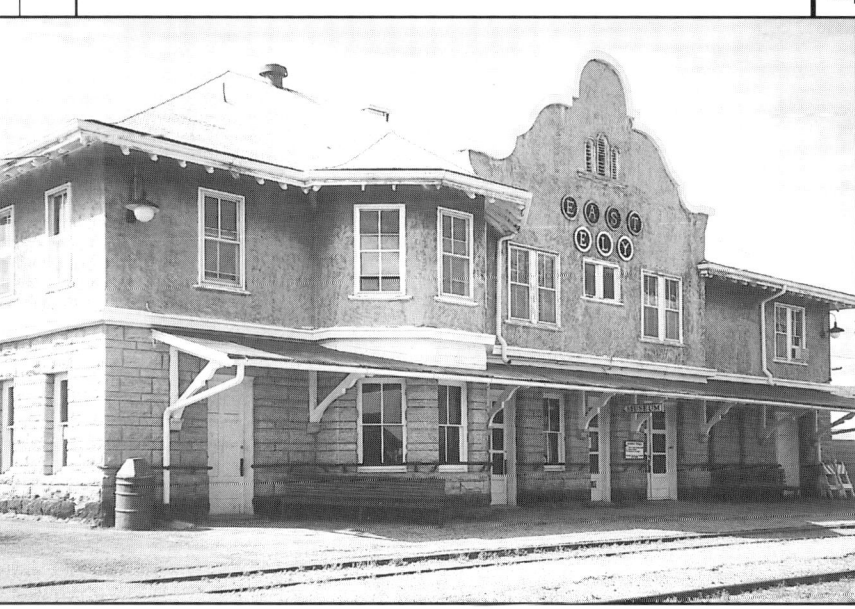

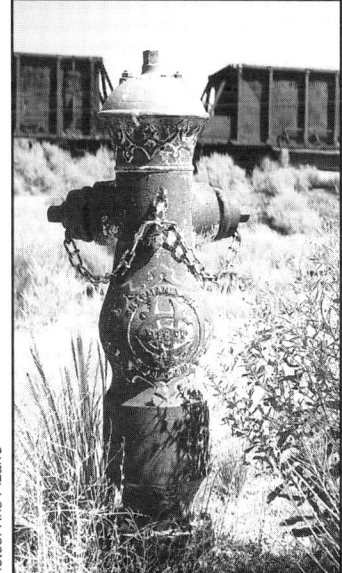

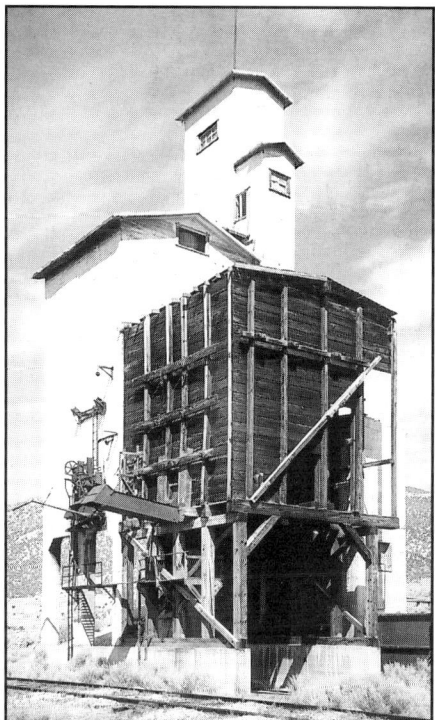

Photos: Kris Pizarro

Photo: Kris Pizarro

interval	cumulative	milepost	
1.9	69.5		Intersection with U.S. 6, on the right. Note that from this point to the Nevada-Utah state line, the highway markers are labeled for U.S. 6 rather than U.S. 50. The highways are combined here, and the highway department chose to use U.S. 6 mileage on the mile markers rather than U.S. 50 mileage. To be able to use the mile markers, we begin counting anew at the next intersection, milepost WP 40.00.
2.7	72.2 (00.0)	WP 40.0	Cedar Park West turnoff on the right, this is milepost WP 40.00 on combined U.S. 50, U.S. 6 and U.S. 93. (*Note cumulative mileage starts at zero here*)

Continuing southeast, U.S. 50 begins its crossing of Steptoe Valley. The Egan Range is to the right; the Schell Creek Range is to the left. Both are block-faulted ranges composed almost entirely of Paleozoic limestone with some dolomite and shale. |
3.0	43.0	WP 43.0	There is a good view of dissected alluvial fans along the lower east front of the Egan Range at about 3:00. The alluvial material washed from the range was deposited in the fan-shaped area at the mouth of each canyon. The fan deposits are now being eroded by the very streams that originally deposited them. On the left, Steptoe Creek, flowing north more or less parallel to the highway, is eroding the toes of alluvial fans formed along the west flank of the Duck Creek Range.
0.8	43.8	WP 44.0	Intersection with Cave Valley Road. The wide gravel road that angles to the right goes south to the ghost mining camp of Ward and to Ward Charcoal Ovens State Historic Site. The site is open all year, but the road is best traveled by passenger vehicles only between May and October. There are no facilities at the site.
1.4	45.2		Comins Lake. The old Argus and Monitor stamp mill sites are on the southeast shore of this lake, just west of the highway. These mills were built in the early 1880s to treat silver ores from both Ward and Taylor (to the east in the Schell Creek Range). Comins Lake usually contains water, but in years of drought, it can completely dry up. There's fishing here (when the lake is not dry) and good birding.
0.8	46.0	WP 46.0	Intersection, to the left, with the Success road.

Picnic tables and shelters at a site within the Ward Mountain Recreation Area on U.S. Highway 6 about six miles west of Ely. The view is to the south with Ward Mountain in the background.

Photo: Kris Pizarro

Alluvial fans deposited at the mouths of streams draining the Duck Creek Range. Each fan is now being cut through and eroded by the stream that formed it, and Steptoe Creek is washing away the coalesced toes of the fans.

WARD AND THE WARD MINING DISTRICT

The Ward deposits, discovered in 1869, produced lead, silver, and copper until 1879 and again between 1906 and 1920. Recent attempts to begin large-scale mining have not been successful, but the district did see some production in the 1980s. Ore was trucked to the mill at Taylor for treatment. The camp is now inactive, but the cemetery can still be seen.

The road forks about 4 miles past the turnoff to Ward (a total of 9 miles south from U.S. 50). The right hand fork goes to Ward Charcoal Ovens State Historic Site on Willow Creek. The monument protects six large charcoal ovens built in 1876 to furnish fuel for furnaces at the Ward mines. These ovens were built by Italian-Swiss charcoal workers, the "carbonari" whom we mentioned earlier when we talked about the Fish Creek War (page 85). These ovens are essentially intact and are well worth the short side trip to visit them. On the return trip, just 1 mile north of the charcoal oven turnoff, a straight road to the right leads directly back to U.S. 50. This saves a return trip north, but will miss Comins Lake and the Success road turnoff.

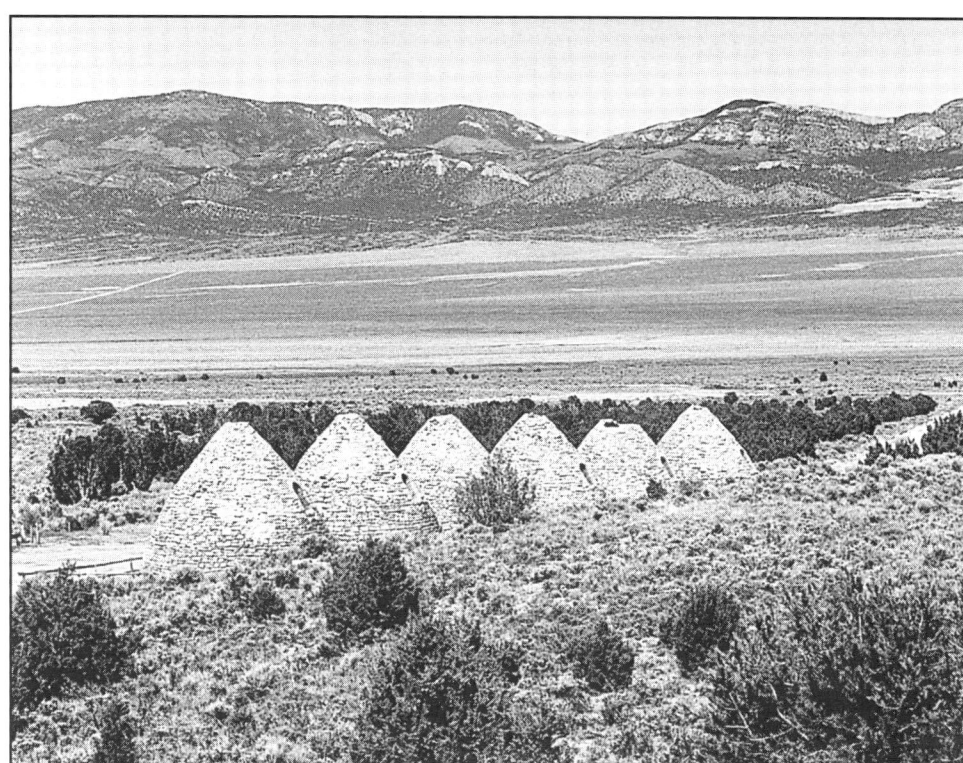

Beehive ovens at Ward Charcoal Ovens State Historic Site. The view is to the east across Steptoe Valley toward the Schell Creek Range in the distant background. ▶

Ward Charcoal Ovens in the foreground, Charcoal Ovens Tuff in the background (*lower right photo*).

Photos: Roy W. Cazier

Outcropping of the Oligocene Charcoal Ovens Tuff, showing columnar jointing, located just south of the Ward Charcoal Ovens State Historic Site. This rock, a welded ash-flow tuff, was quarried from a site to the southwest and used to construct the ovens in 1876. ▼

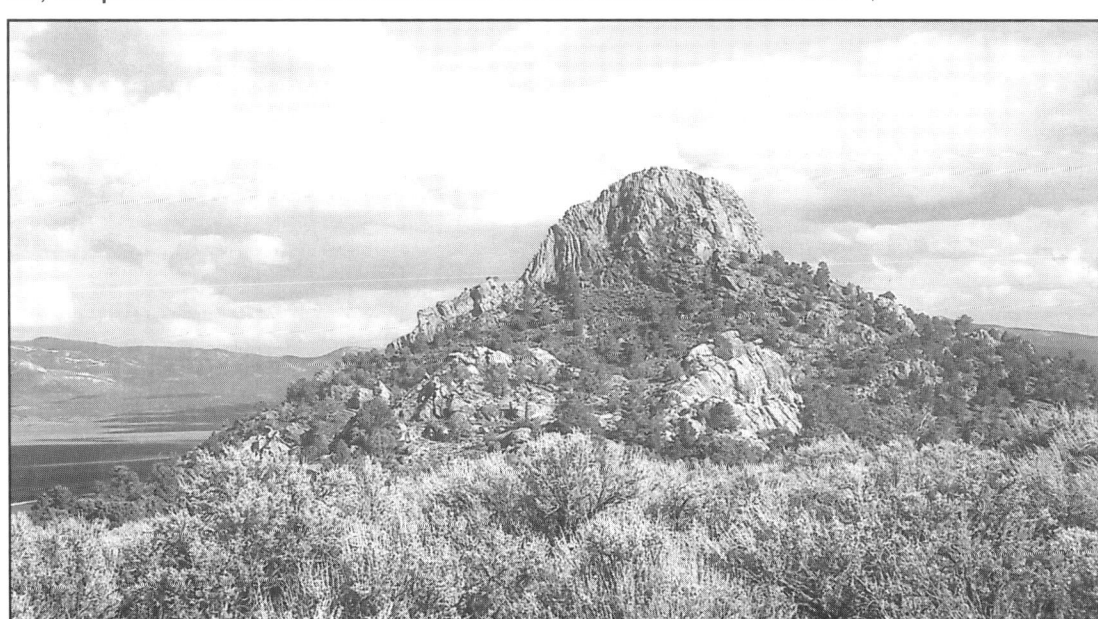

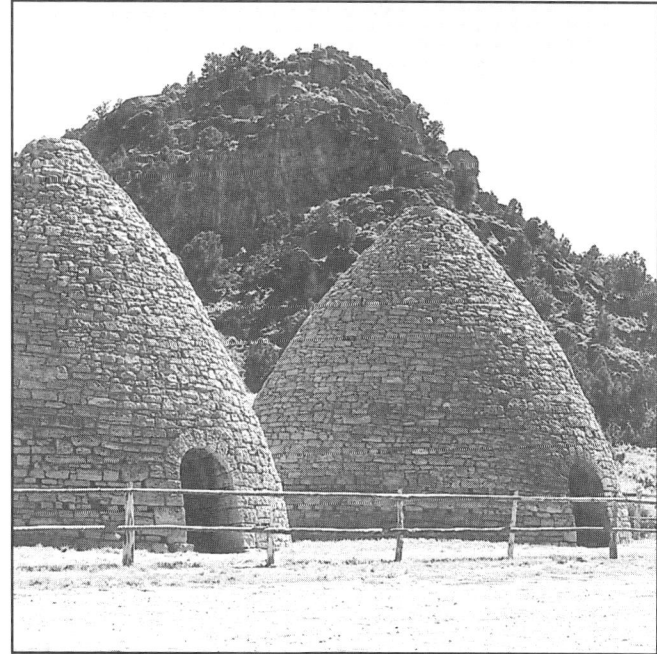

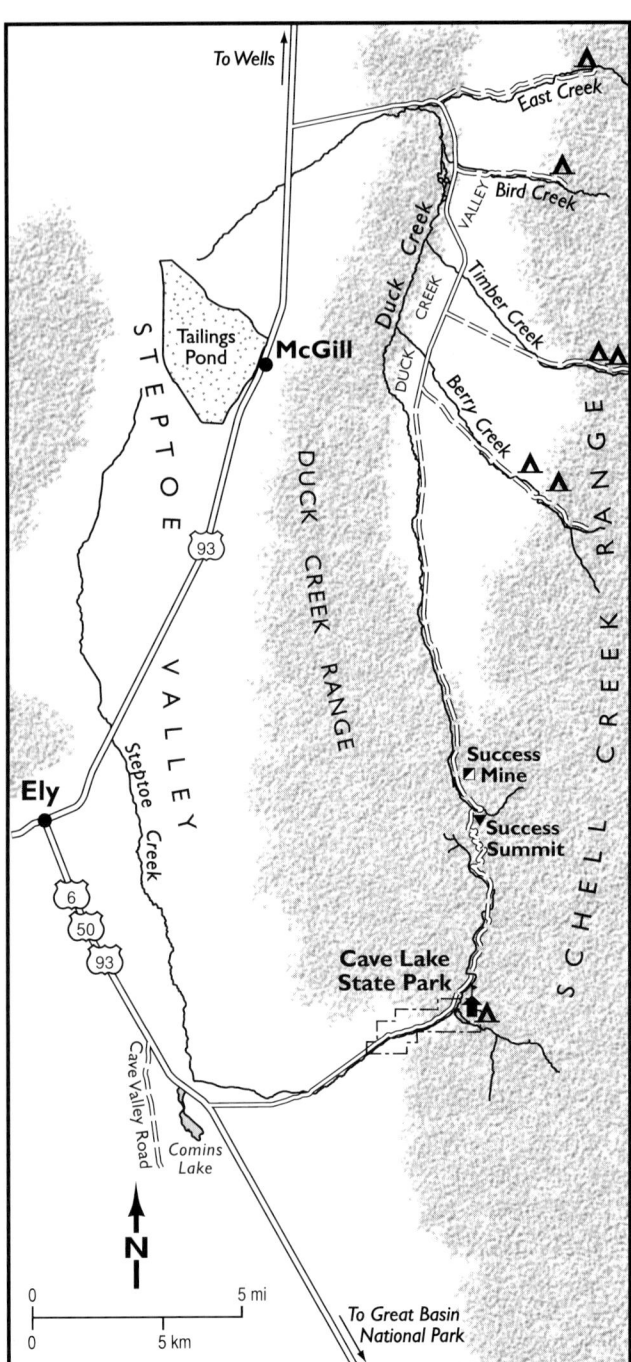

Side trip 7 map, Cave Lake and the Success Loop.

SIDE TRIP 7, CAVE LAKE AND THE SUCCESS LOOP

This 33-mile long scenic road travels up Steptoe Creek to Cave Lake State Park. Cave Lake has camping and picnicking spots, lake fishing, and mountain scenery. The shoreline of the 32-acre reservoir offers good birding, and you might see mule deer, elk, and the ever-present coyote in the surrounding meadows and mountains. There is a 5-mile interpretive trail above the lake campground for both hikers and mountain bikers, and there is a 3-mile trail along Steptoe Creek. The Nevada Division of Wildlife keeps the lake stocked with rainbow and German brown trout, and claims that there are more fish caught in Cave Lake than at Lake Tahoe. Beyond the state park, the road continues to climb north along Steptoe Creek to Success Summit, and then drops down Duck Creek into Duck Creek Valley. The road is not paved beyond Cave Lake, but is usually passable in standard passenger vehicles unless it's been too long since the last grading. To be safe, check with the staff at the State Park before you set out to drive this stretch of road. There are several campgrounds and picnic areas up various creeks that flow from the North Schell Creek Range, east of the road. Roads up these smaller creeks are even more primitive, however, and most should not be attempted without a four-wheel-drive vehicle.

The road continues through the Duck Creek basin. During the earlier period of copper mining, water from this drainage was collected and piped to McGill, on the west side of the Duck Creek Range, for use in the copper mill and smelter. Kennecott Mining Co., the mine operator, maintained the large ranch at the mouth of the valley for use by visitors and company officials. The Success road connects with U.S. 93 just north of McGill, 12 miles north of Ely and U.S. 50.

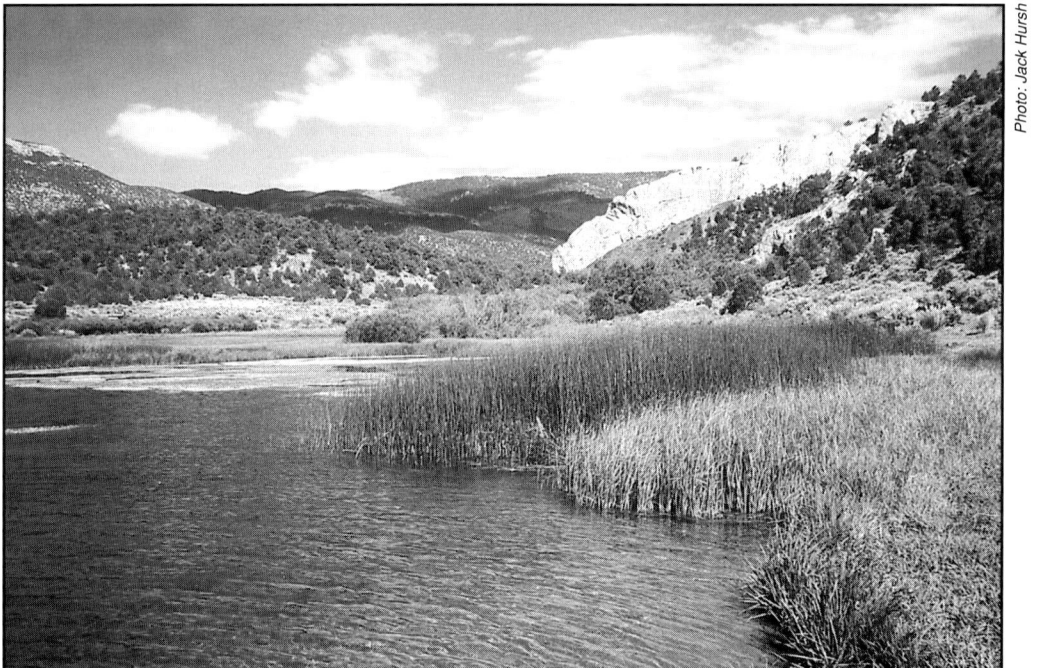

A view to the southeast across Cave Lake. Cave Lake derived its name from a cave found at the base of the prominent limestone cliff in the right background. This cave is not open to the public.

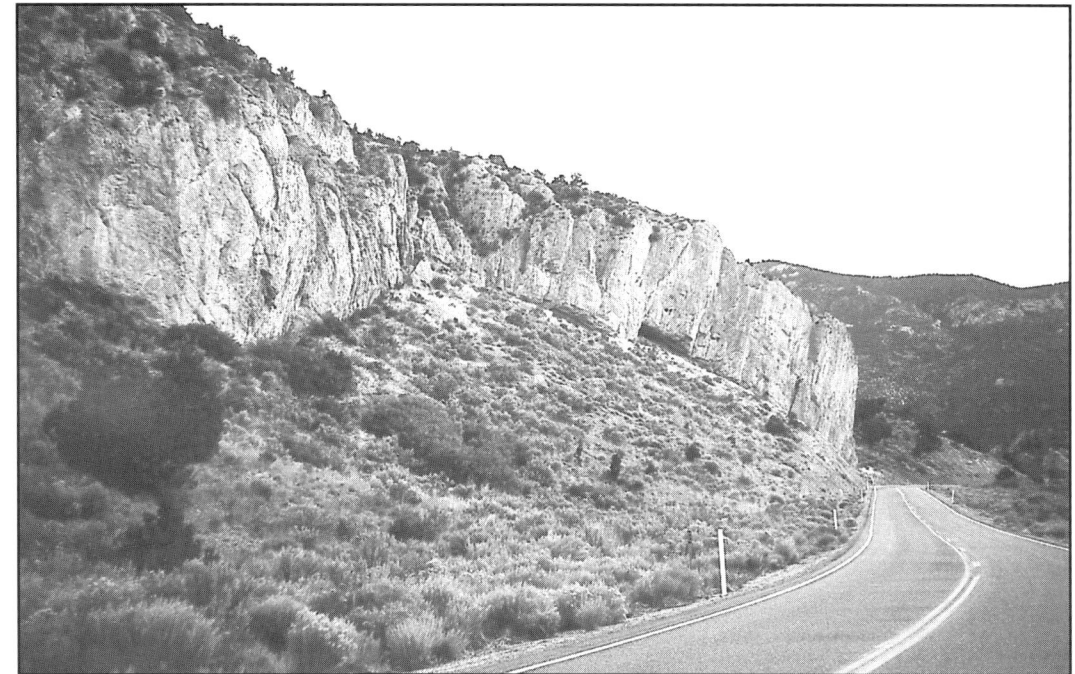

Photo: Kris Pizarro

On the Cave Lake Road, near the turnoff to Cave Lake. Cliff-forming limestone overlies a less-resistant shale that forms the sagebrush-covered slope to the left of the road. Notice the rock overhangs and caves formed along the limestone-shale contact.

Campsite in an aspen grove beside Berry Creek, Success Loop Road north of Success Summit.

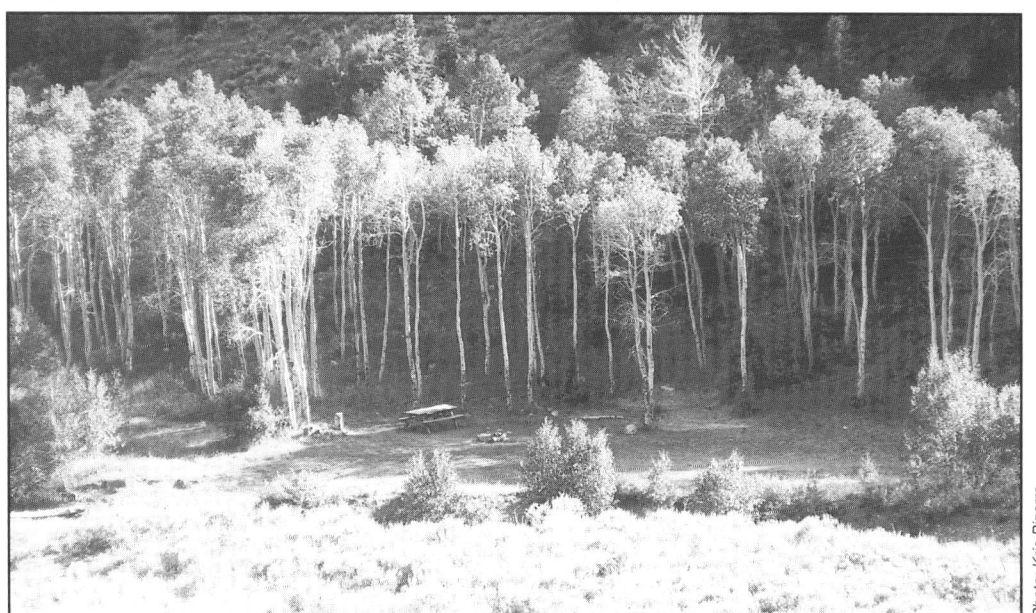

Photo: Kris Pizarro

Photos: Roy W. Cazier

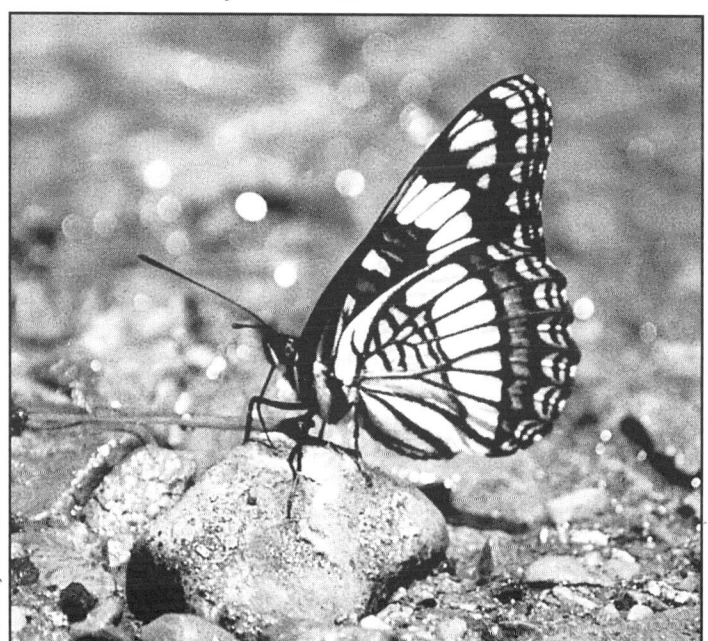

Top view of Weidemeyer's Admiral (*Limenitis weidemeyerii*), commonly found near willow, cottonwood, and aspen.

Many insect larvae feed on specific plant species. Nevada, with its extreme variation in topography, has a wide diversity of plant species, thus supporting the presence of many insect species. Some mountain ranges in Nevada host some of the greatest concentrations of different butterfly species in all of North America.

Side view of Weidemeyer's Admiral.

Photo: Roy W. Cazier

107

interval	cumulative	milepost
4.6	50.6	

Elk Viewing Area on the left, small parking lot with picnic tables and shelters, grills, and an interpretive display on the local Rocky Mountain elk herd. Elk spend the summers in the high country of the Schell Creek Range but spend the rest of the year in the lower foothills east of here. Even if you don't see elk here, you will probably see some antelope. The viewing area also provides good views of the ranches of the southern Steptoe Valley and, to the southeast, the upper benches of the inactive Taylor open-pit silver mine.

interval	cumulative	milepost
1.0	51.6	

Road to Ward Charcoal Ovens to the right, intersects the Cave Valley Road (see comments at mile point 43.8).

interval	cumulative	milepost
0.5	52.1	

Taylor Mine road to the left. The large metal mill building and tanks visible in the foothills of the range to the east mark the site of the Taylor open-pit silver mine, active in the 1980s and early 1990s. Silver was discovered at Taylor in 1872, and several small underground mines operated there for the following 20 years. The district was then only intermittently active until 1962, when work began that led to the opening of the large open-pit mine in 1981.

The Schell Creek Range at this point is composed of faulted blocks of Devonian, Mississippian, and Pennsylvanian limestone and shale. The lower foreground consists of Devonian dolomite and limestone; the slope break is underlain by Chainman Shale and Joana Limestone; and the blocky, ridge-forming limestone on the skyline is Ely Limestone.

Photo: Joseph V. Tingley

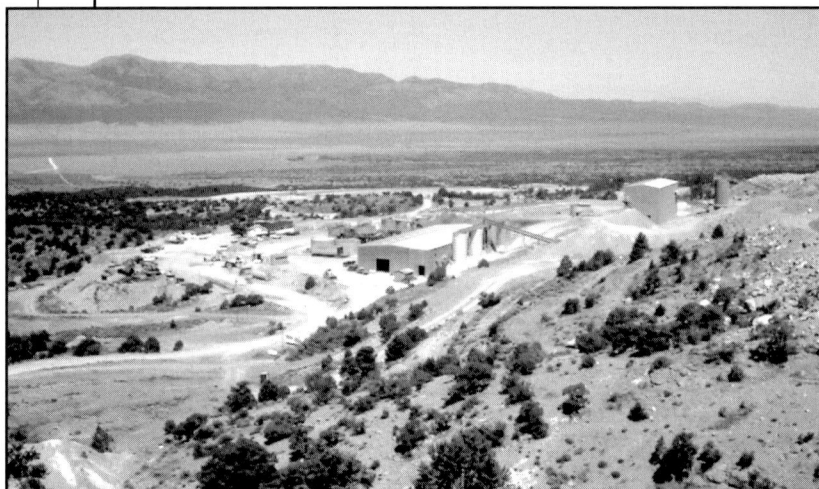

Silver King Mining Co. mill facility at the Taylor Mine, 1981. The tall building on the right houses the crusher, the silver recovery plant is in the center, and the tailings dam is in the trees beyond the mill building. Steptoe Valley and the Egan Range are in the background.

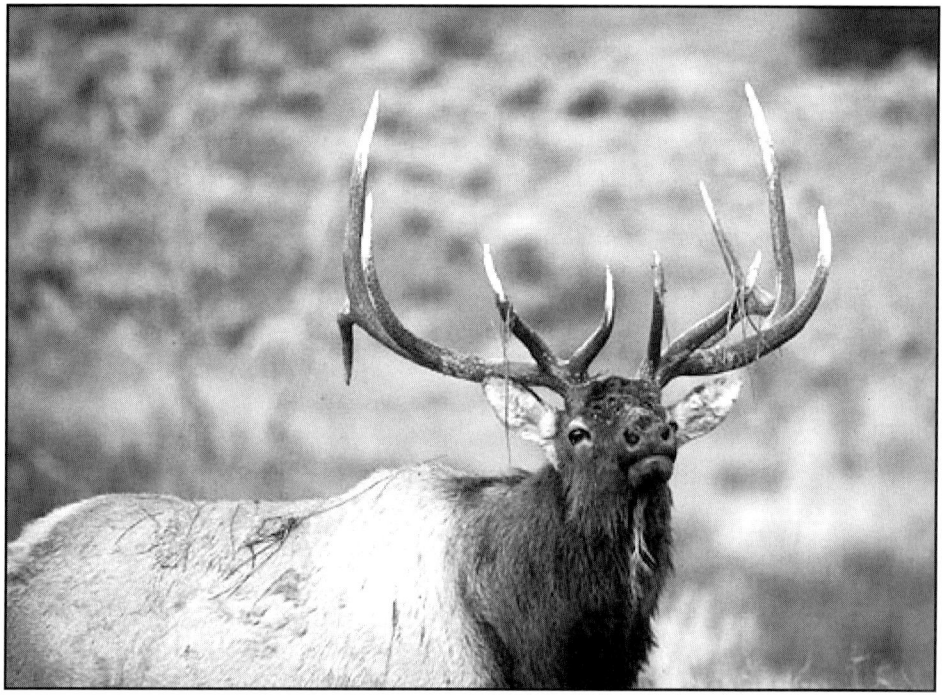

Photo: Eugene Hester, U.S. Geological Survey, Department of the Interior

Elk (*Cervus elaphus*) large deer 6¾–9¾' long; brown or tan with darker underparts and yellowish tan rump patch and tail; males have dark mane on throat.

Each year males grow antlers up to five feet long, with as many as six tines on each side. Bull elk are the most polygamous of all North American deer and during the fall rutting season may collect large harems of cows. In late spring cows give birth in seclusion, usually to one calf. About a week later, mother and offspring rejoin the herd. Except during mating season, bulls live apart from cows and calves.

Elk are nocturnal and are most likely to be seen at dawn or dusk. They bed down in sheltered spots during the day. Despite their large size, they move very quietly through vegetation. Elk consume a variety of grasses, forbs, and woody browse, depending upon availability.

interval	cumulative	milepost	
4.9	57.0	WP 57.0	We are now entering Connors Canyon and beginning our climb into the Schell Creek Range.

Rocks on either side of the highway at the mouth of the canyon are Pennsylvanian-Permian Ely Limestone, which forms the upper plate of a thrust sheet that has overridden Mississippian Chainman Shale. The thrust contact is to the left, just beyond the first curve into the canyon. The limestone forms more rugged outcrops on the ridge tops to the left, while the shale forms subdued, rounded hills on both sides of the highway for the next 2 miles. |
1.0	59.0	WP 59.0	Here we climb from Chainman Shale back into upper plate Ely Limestone. The thrust contact cannot be seen, but it is marked by the change from rounded, smooth hill slopes to the left and right, contrasted with the outcrop of cliff-forming limestone on top of the hills to the left. As we continue around the curve ahead, the road cuts will be in crushed, contorted upper plate limestone.
1.3	60.3		Connors Pass (elevation 7,723 feet).
0.7	61.0	WP 61.0	We are still in Ely Limestone. On the north (left) side of the highway the limestone is cut by numerous vertical faults. The fault zones are marked chalky, broken rock and rusty staining. There are no safe turnouts here, so look but don't stop.

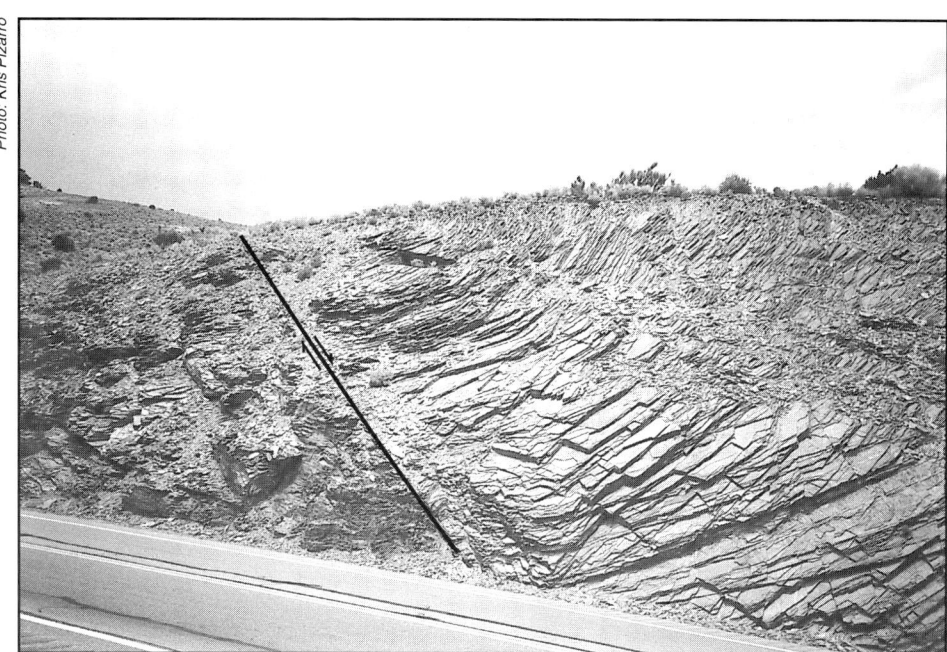

Thin-bedded shale and phyllite exposed east of Milepost 63, east of Connors Pass. The beds are folded and cut by faults; one fault is marked on the photo.

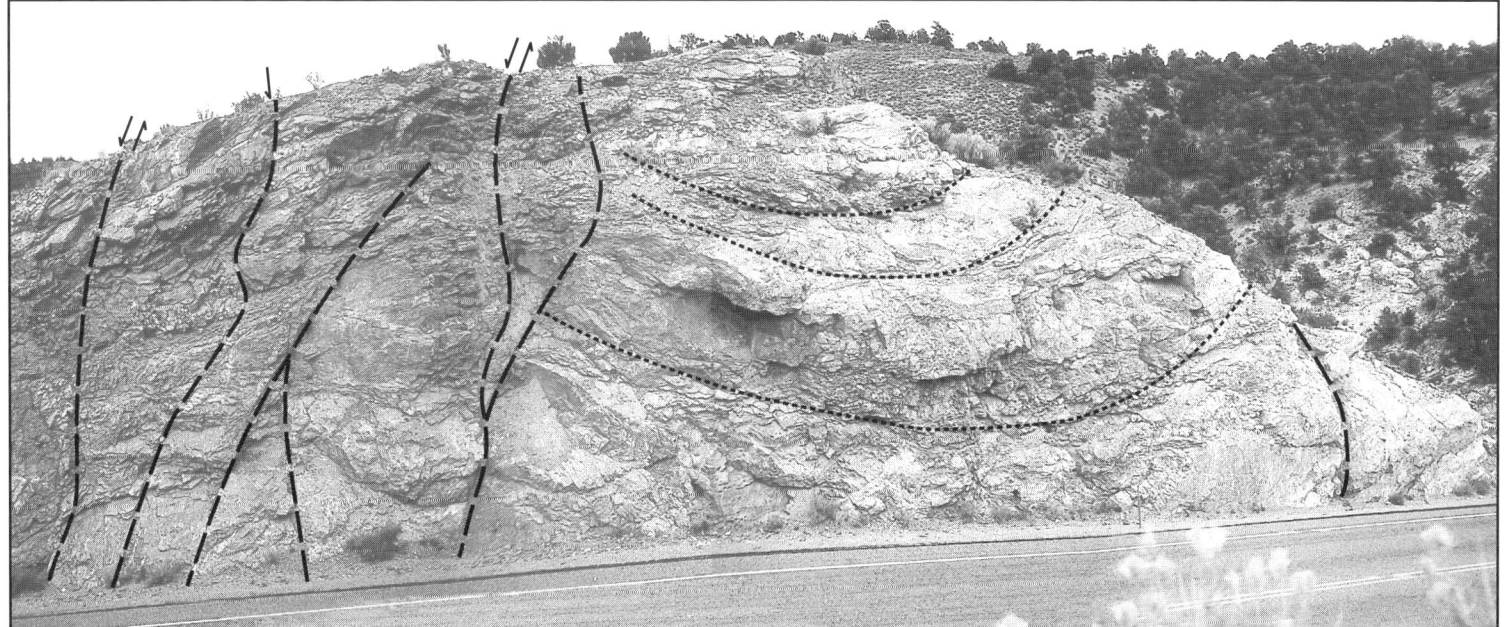

Several near-vertical faults (on the left) and a synclinal fold (on the right) in limestone exposed in a road cut near Milepost 63, east of Connors Pass. The arrows indicate relative movement on the faults.

interval	cumulative	milepost	
2.7	63.0	WP 63.0	Turnout on the west (right) side of the highway. Thrust fault contact between limestone (upper plate) and shale (lower plate). The largely gray limestone is folded and cut by faults. The outcrops are streaked with red iron-oxide that marks some of the faults and fractures. There are also thin, vuggy veins of calcite along fractures, and some large lenses of white, crystalline calcite that have formed in cavities in the limestone. (plate 11f)
0.7	63.7		Altered porphyritic rhyolite dike on the left. The dike, which runs more or less parallel to the highway, is marked by a bleached-appearing, chalky band cutting through the darker, thin-bedded shale and limestone outcrops.
1.2	64.9	WP 64.85	Majors Place on the right. Named for a somewhat reclusive gentleman who operated a bar and associated tourist cabins here from the 1930s into the 1950s.
0.4	65.3		Intersection with U.S. 93, which follows the right hand fork. We follow U.S. 50, still combined with U.S. 6, to the left.

U.S. 93, one of the major north-south routes through Nevada, links Interstate 15 and Las Vegas with Interstate 80. The historical mining town of Pioche, county seat of Lincoln County, is 81 miles south. Cathedral Gorge State Park, another 12 miles south of Pioche, is open all year and has both campsites and picnic areas.

This is a good place to pull over and note some geographic landmarks. Spring Valley is ahead, extending north and south between the Schell Creek Range to our back, and the Snake Range ahead. Mount Moriah, in the northern Snake Range, is at 11:00. The road to the mining camp of Osceola is at 12:00. Windy Peak is at 12:30, and Bald Mountain is at 1:00. The smooth sided, treeless peak at 2:00 is Wheeler Peak, named for Captain George M. Wheeler who led Army Corps of Engineers surveys of lands west of the 100th meridian in 1875–89. Wheeler Peak, at an elevation of 13,063 feet, is the second highest peak in Nevada (the highest is Boundary Peak at 13,140 feet on the Nevada-California border about 150 miles southeast of Reno). The peak partially obscured by Wheeler Peak is Jeff Davis Peak, and the pointed peaks farther south are Baker Peak and Pyramid Peak. All of these peaks are composed mainly of Cambrian Prospect Mountain Quartzite. Farther to the south is Mount Washington, capped by Cambrian limestone.

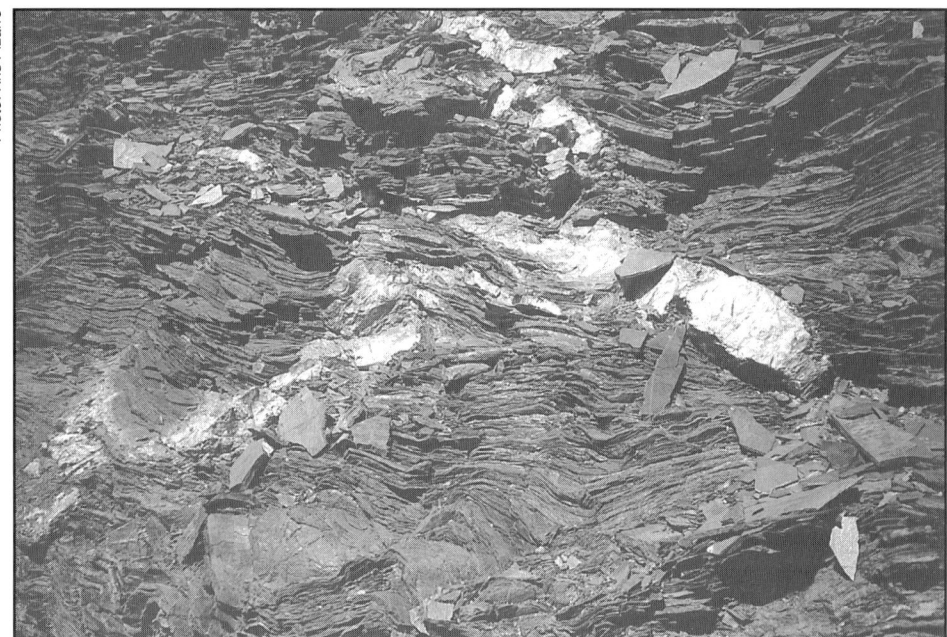

Photo: Kris Pizarro

White calcite veins formed along fractures in phyllite. Note that some of the calcite appears to have formed along bedding planes in the phyllite, while other calcite has formed along crosscutting fractures. This exposure is on the east side of Connors Pass.

Thick calcite vein in brecciated limestone, east side of Connors Pass.

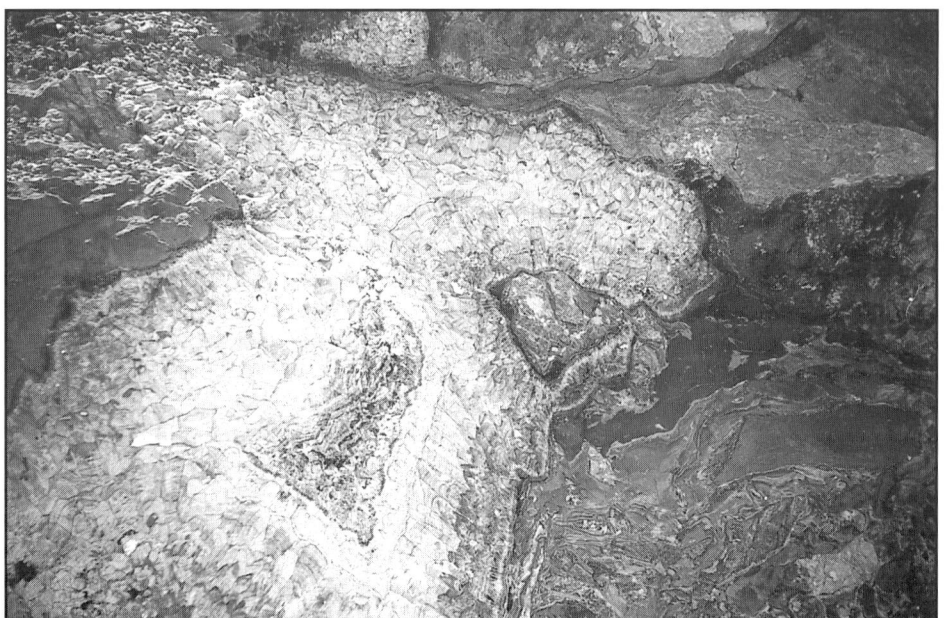

Photo: Kris Pizarro

interval	cumulative	milepost
1.5	66.8	

Intersection, State Route 893 on the left. This road travels north along the west side of Spring Valley for 53 miles, then connects with the road west over Schellbourne Pass. The first 40 or so miles are paved; the last stretch and the 15 miles over Schellbourne Pass to U.S. 93, north of Ely, are gravel. There are several ghost towns and abandoned mine camps in the Schell Creek Range with access from State Route 893. Roads to these areas are not maintained, however, and they should be attempted only with four-wheel-drive vehicles. There is a small U.S. Forest Service campground at Cleve Creek, about 15 miles north from our U.S. 50 intersection.

1.2	68.0	WP 69.0

For the next few miles we will be heading northeast across Spring Valley, with a great view of the Snake Range ahead.

Most of the rocks that make up the range from about 11:00 to 3:00 are Cambrian and Precambrian limestone, shale, and quartzite. To the south, from about 3:00 on, the high parts of the range are capped by gray Cambrian through Devonian limestones, which have a layered appearance. The older rocks to the north also have a layered appearance, but they are more massive and form steep, rugged cliffs.

0.6	68.6	

Rattlesnake Knoll, on right. The small hill about ¼ mile south of the highway is composed of rhyolite breccia. Faults cutting the breccia contain spotty, irregular veinlets of fluorite, and there are a few old pits and cuts left by prospectors.

0.4	69.0	

Wheeler Peak is the rugged, triangular-shaped mass at about 2:00. The highest peaks are within the Great Basin National Park, which includes about half of the western slope of the range and extends across the range almost to the eastern range front.

3.5	72.5	

Our route turns slightly to the left (north), and travels along the west front of the Snake Range. To the left, as the road bends, is an assortment of mining and milling equipment in varying stages of decay. This was a small milling operation constructed to treat ore from one or more of the mines at Osceola, or from the tungsten mines in the old Shoshone mining district about 25 miles to the south. To the right, a wide gravel road leads south along the range front. This connects with a paved road to several ranches and old mining camps.

The Osceola mining district is at 3:00. As we turn, several large dumps and some old equipment are visible partway up the large alluvial fan at the mountain front.

0.4	72.9	

Intersection, gravel road to Osceola.

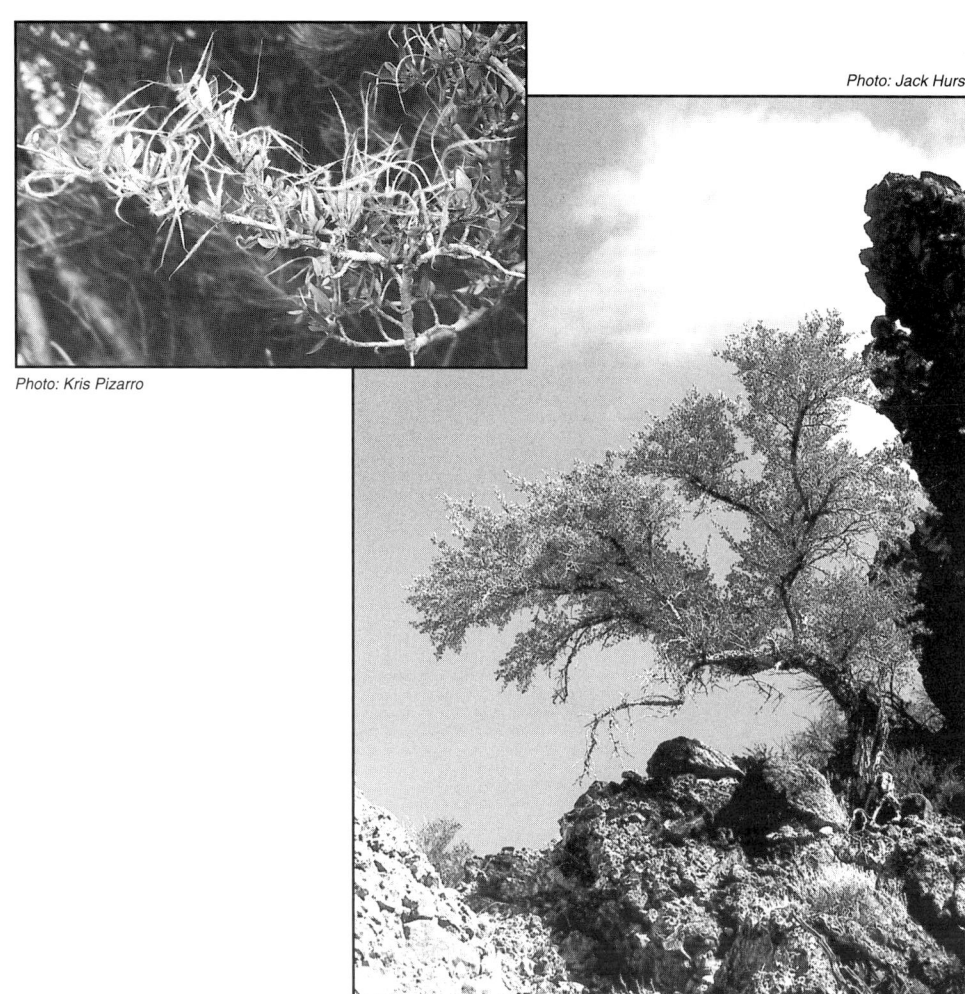

Photo: Jack Hursh

Photo: Kris Pizarro

Curlleaf mountain-mahogany (*Cercocarpus ledifolius*) evergreen shrub or small tree, 15–30' tall; twisted branches grow from short trunk; deeply furrowed, reddish-brown bark; leathery leaves ½–1" long, lanceolate, curled under at the edges.

Curlleaf mountain-mahogany grows on dry, rocky slopes, ridgetops, and rock outcrops. It often occupies a belt above piñon-juniper woodlands and commonly occurs in isolated pure stands. Mountain-mahogany provides food and cover for a variety of wildlife species.

Inconspicuous funnel-shaped flowers are produced in spring. Seeds mature in late summer and are about ¼" long with a 1½–3" long twisted tail covered with whitish hairs. These can be so densely packed that the tree appears to be frosted.

A couple of the best places on the U.S. 50 trip for seeing curlleaf mountain-mahogany are along the The Success Loop (side trip 7) and on the drive to the Wheeler Peak campground at Great Basin National Park (side trip 9).

PLACER GOLD AT OSCEOLA

Gold-bearing veins were discovered at Osceola in 1872, but the main production from the camp came from placer gold deposits found along the range front to the south (the area of dumps and old equipment visible from the road). The placers were discovered in 1877, and have been active on and off over the years even to the present time. Total gold production from the district has been about $3.3 million.

Placer mining at Osceola was hampered from the start by lack of water, and various ditches were constructed to bring water to the diggings from nearby canyons. The 18-mile-long Osceola Ditch, the longest of these, came from Lehman Creek on the east side of the range. Much of the eastern portion of this ditch is now within Great Basin National Park. If you visit the Park, there is a short trail, complete with signs and informational brochures, which can be walked to the trace of the old ditch.

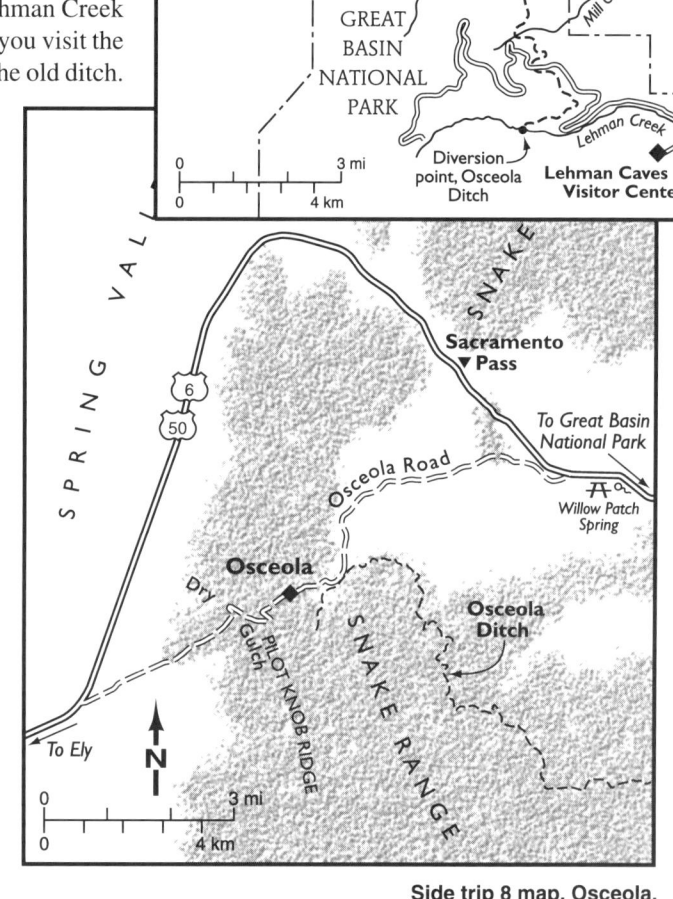

SIDE TRIP 8, OSCEOLA

The gravel road leading straight ahead at the turn goes to the ghost town of Osceola, about 3.5 miles ahead in the canyon, then crosses the Snake Range and reconnects with U.S. 50 about 2 miles east of Sacramento Pass. The road is gravel, but is usually in good shape and does not require four-wheel drive. There are old buildings (some still occupied—private property) and an interesting cemetery to explore. The slow mountain road provides the chance to view some mountain scenery at your own pace.

Side trip 8 map, Osceola.

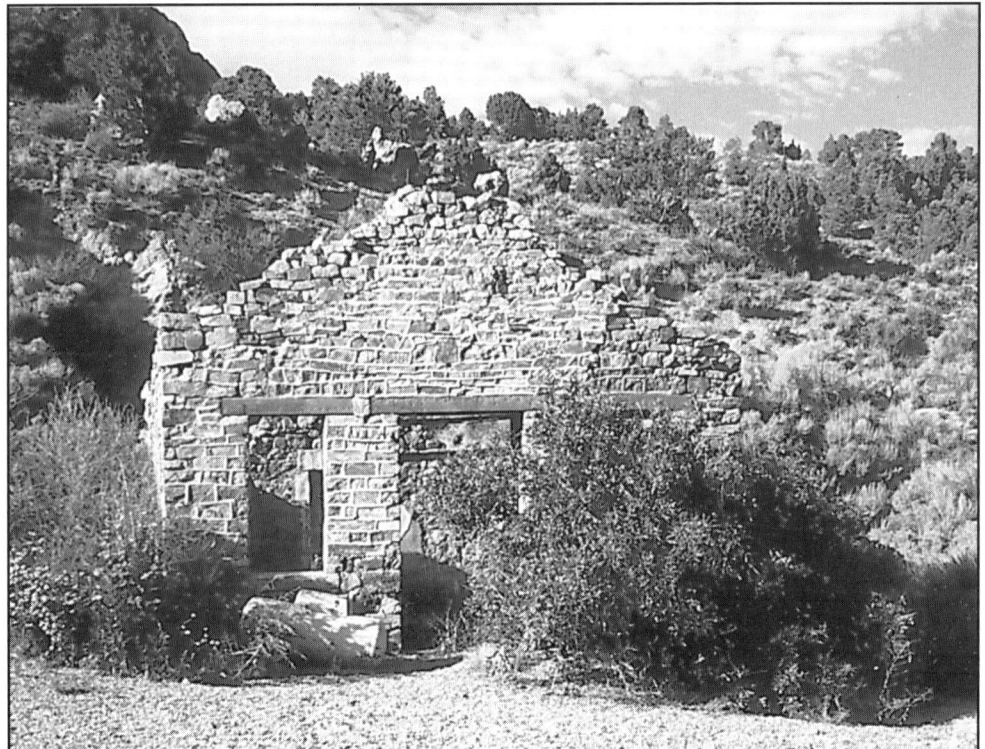

Remains of an old store building at the Osceola townsite.

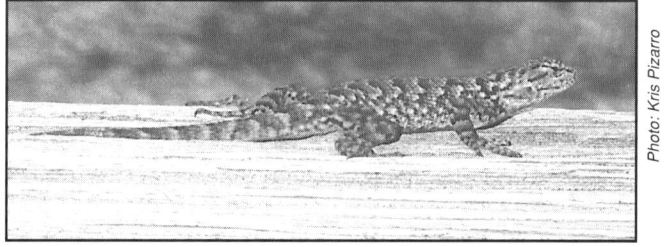

Western fence lizard (*Sceloporus occidentalis*), a commonly seen resident of Osceola.

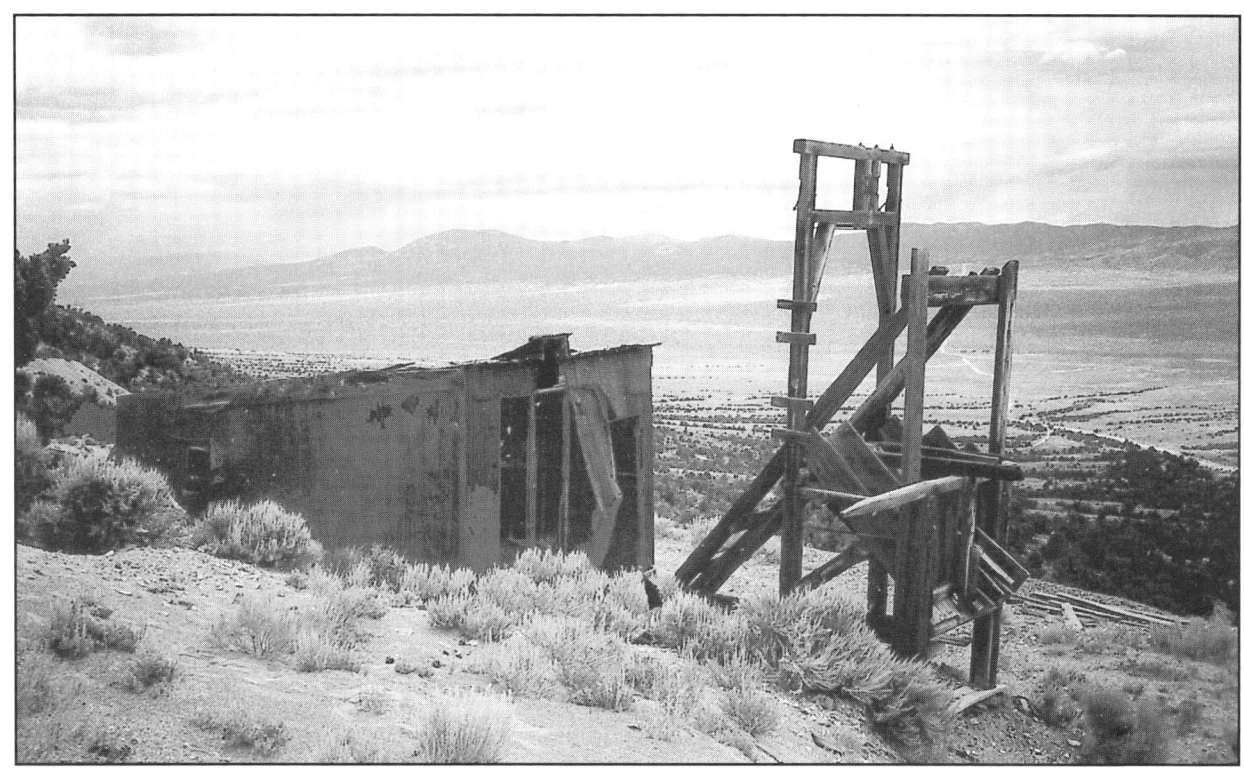

Headframe over a shaft at one of the Gold Exchange group of mines on Pilot Knob Ridge, Osceola district. View is to the west with Spring Creek Valley and the Schell Creek Range in the background.

Lower left photo: An open stope in the area of the Gold Exchange group of mines. As the ore was removed from this deposit, the timber posts (stulls) were installed to discourage the unstable roof (called the "back") of the opening from caving on the miners. Old mine openings such as this should NEVER be entered.

The pond at the rest stop/picnic area on the east end of Osceola Road near Sacramento Pass.

Photos: Kris Pizarro

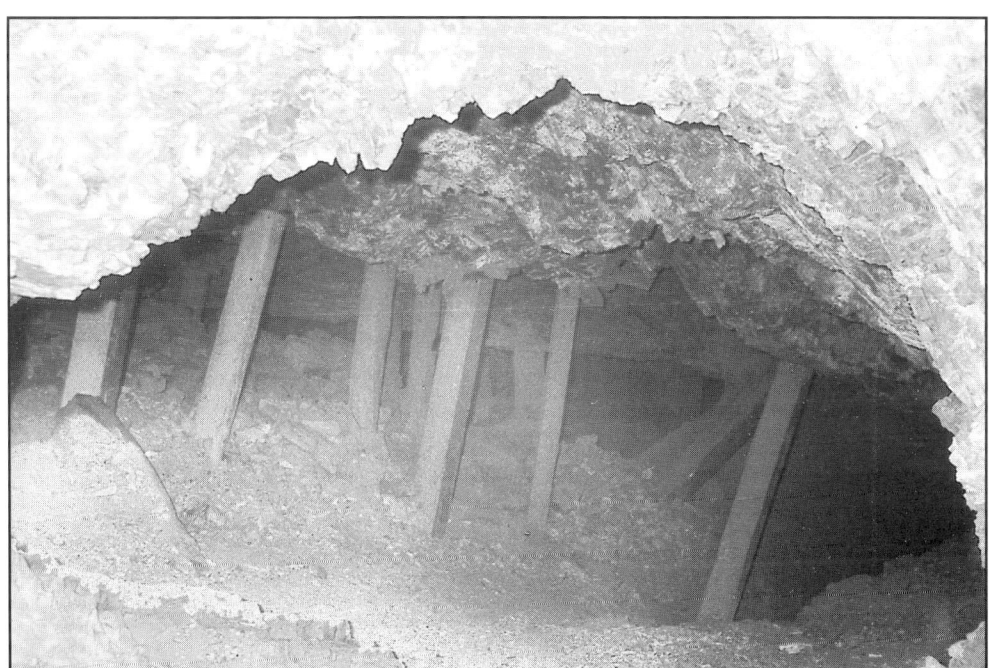

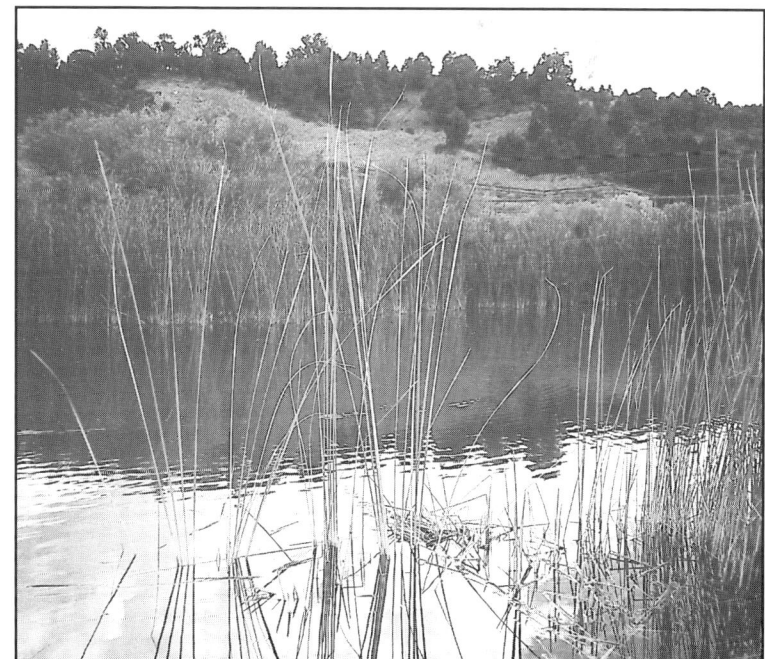

interval	cumulative	milepost	
3.6	76.5		Rose Cave is in the rugged part of the range at about 2:00. Osceola Arch is to the right.
4.6	81.1		Intersection with Spring Valley Road (east side of valley).
			The light-colored, barren, rounded hills on the west-trending spur to the north are Cambrian Prospect Mountain Quartzite. The wooded, cliffy part of the ridge farther to the east consists of Ordovician and Silurian limestone in the upper plate of the Snake Range décollement (a fancy name for a low-angle fault).
0.9	82.0	WP 82.00	Entering Turnley Canyon, entrance to Sacramento Pass.
			The small mine dumps in the low hills south of the highway are in the Sacramento mining district. Gold was discovered in this district in 1869, and small amounts of ore were produced until 1875. In 1915, tungsten ore was also found and mined for a year or so and again in 1941–1942.
0.7	82.7		At this point we are passing through some tilted Tertiary-age sedimentary rocks composed of cemented alluvial-fan gravels that are stained by iron oxide and contain boulders and pebbles of weathered Paleozoic rocks.
2.0	84.7		Sacramento Pass (elevation 7,154 feet). This pass marks the last mountain range crossed by U.S. 50 in Nevada. From here it's all downhill to the Utah border.
0.6	85.3		Mount Moriah is visible to the northeast. Wheeler Peak is to the south at 2:00.
1.7	87.0	WP 87.00	The gravel road to the right goes to Osceola (side trip through Osceola from the west side of the range rejoins U.S. 50 here).
0.6	87.6		Ghost town at 9:00 is Black Horse, relative latecomer in this area of generally late 1860s mining camps. The gold veins of Black Horse were not discovered until 1906. Gold on the underside of an overhanging ledge was found, so the story goes, by a rider on a black horse seeking refuge from a spring storm. Although very high grade, the veins played out by 1913 and the town was abandoned by 1914. Nothing remains at the site except mine dumps and a scattering of open shafts.
			Gold-bearing veins at the Black Horse mines occur in blocks of Cambrian limestone and shale that have been faulted over Precambrian quartzite and argillite. The fault is a low-angle thrust fault.
			To the north of the road here are west-dipping beds of reddish alluvial fan deposits, similar to those at Milepoint 82.7.

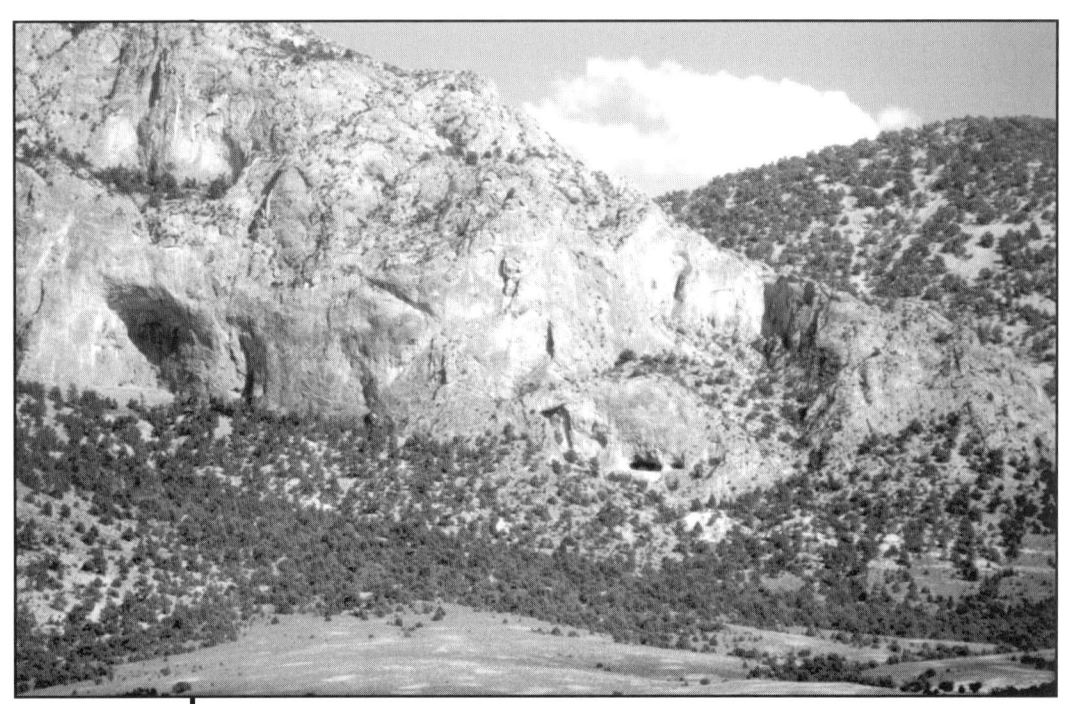

Photo: Kris Pizarro

Rose Cave near the base of a thick band of cliff-forming Ordovician limestone. The cave is the small, oval black opening just above tree line to the right of center of the photo. The white dump in the trees slightly to the right and below the cave is at the mouth of the now-sealed tunnel driven to access the guano resources of the cave.

BATS IN ROSE CAVE

Bats are beneficial members of the natural ecosystem. Bats are the major predator of night-flying insects, consuming as much as one-half their body weight in insects in a single night. Rose Cave has been home to several thousand bats for hundreds of years. During this time large deposits of bat guano accumulated in the cave.

Bat guano is a rich source of industrial and agricultural chemicals such as nitrate and phosphate, and it was mined from Rose Cave in the early 1920s. In about 1926, a tunnel was driven into the mountain front below the natural cave entrance and connected to the cave above to extract the guano more easily. Recently, researchers determined that this artificial entrance upset the natural environment of the cave, causing distress to the bat population. In October 1998, the tunnel was sealed to allow the cave environment to return to its natural state.

interval	cumulative	milepost	
3.0	90.6		Leaving the Snake Range, Snake Valley lies ahead, with the Confusion Range, in Utah, in the distance to the east.
4.1	94.7		Intersection, State Route 487. Baker, Lehman Caves, and Great Basin National Park are to the right. The small town of Baker, 5 miles south, serves as the gateway to the park and provides a quiet, friendly base of operations for exploring the area. There is a small motel, a couple of restaurants, a bar or two, limited groceries and supplies, and a couple of small antique and souvenir shops. Be sure and check out the metal sculptures in town and the "crows" and other creatures on fences along the road from Baker into Great Basin National Park.
7.1	101.8	WP 101.88	Nevada-Utah state line. U.S. 50 leaves Nevada at this point. Food, gas, motel, fossil trilobite specimens, and your last chance at slot machines are provided at the Border Inn. The slots are in Nevada, but the gas is sold in Utah where fuel taxes are lower. Delta, Utah, about 85 miles to the east, is the next town.

Photo: Kris Pizarro

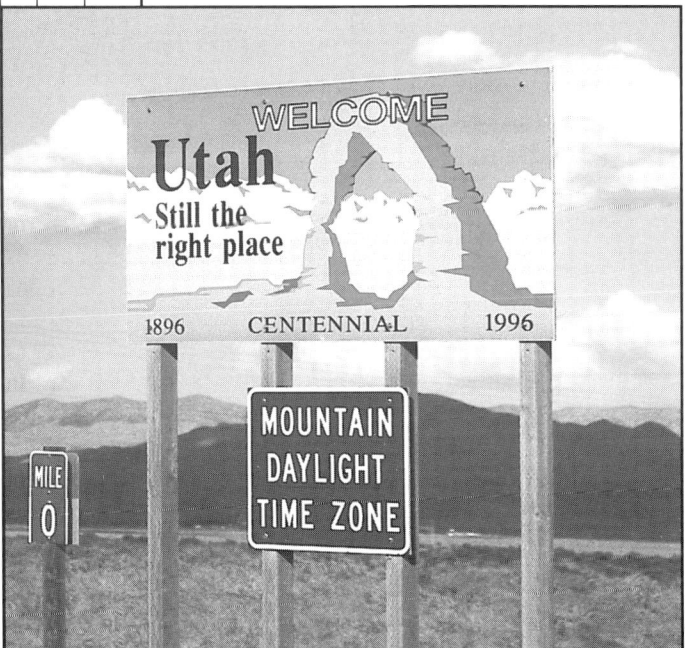

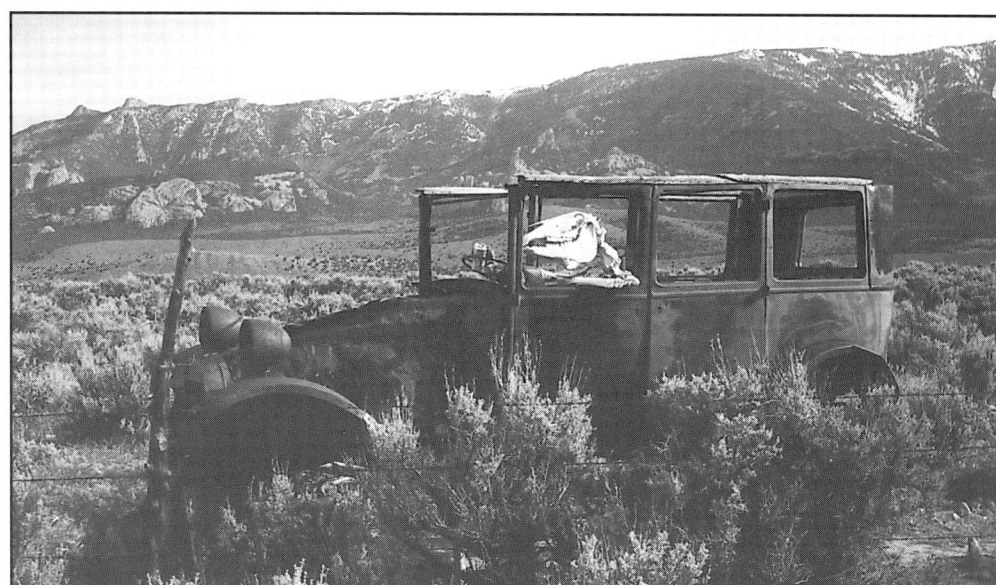

Photo: Kris Pizarro

Roadside art near the entrance to Great Basin National Park, "A Horse with No Mane."

SIDE TRIP 9, GREAT BASIN NATIONAL PARK

Great Basin National Park is Nevada's first national park, and is the only national park entirely within the state (Death Valley National Park is mostly in California). Lehman Caves National Monument was established in 1922. The caves and the surrounding area, including Wheeler Peak, became a national park in 1986. It was chosen to represent the Great Basin environment because examples of all of the life zones comprising northern and central Nevada are found within and around the park boundaries. There is forest of rare bristlecone pine in a high glacial valley on Wheeler Peak's north flank. Even higher in the valley, the only remaining glacier within the Great Basin occupies the north slope of a high ridge extending between Wheeler Peak and neighboring Jeff Davis Peak (plates 12a to 12g).

The entrance to Great Basin National Park is about 5 miles west of Baker. The visitor center is the former headquarters building at Lehman Caves, about ½ mile into the park. Brochures and other information are available there, and the center is well stocked with books on the area. There are five campgrounds within the park, two on Lehman Creek close to the park entrance, one high on the range near the base of Wheeler Peak, and others on Baker Creek and Snake Creek to the south. The Wheeler Peak campground is at an elevation of 10,000 feet, and there are trailheads leading to bristlecone pines and to the 13,063-foot summit of the peak. Campsites are first-come, first-served. Fishing, horseback riding, and mountain biking are permitted in designated areas. In the more remote southern half of the park, four-wheel-drive vehicles are recommended to access primitive campgrounds and hiking trails.

Great Basin National Park is one of the least visited of the national parks, and crowds usually aren't a problem except perhaps during the busiest of the summer weekends. If you have the time, a visit to the park is perhaps the best way to end your journey on U.S. 50, the "loneliest road" across Nevada.

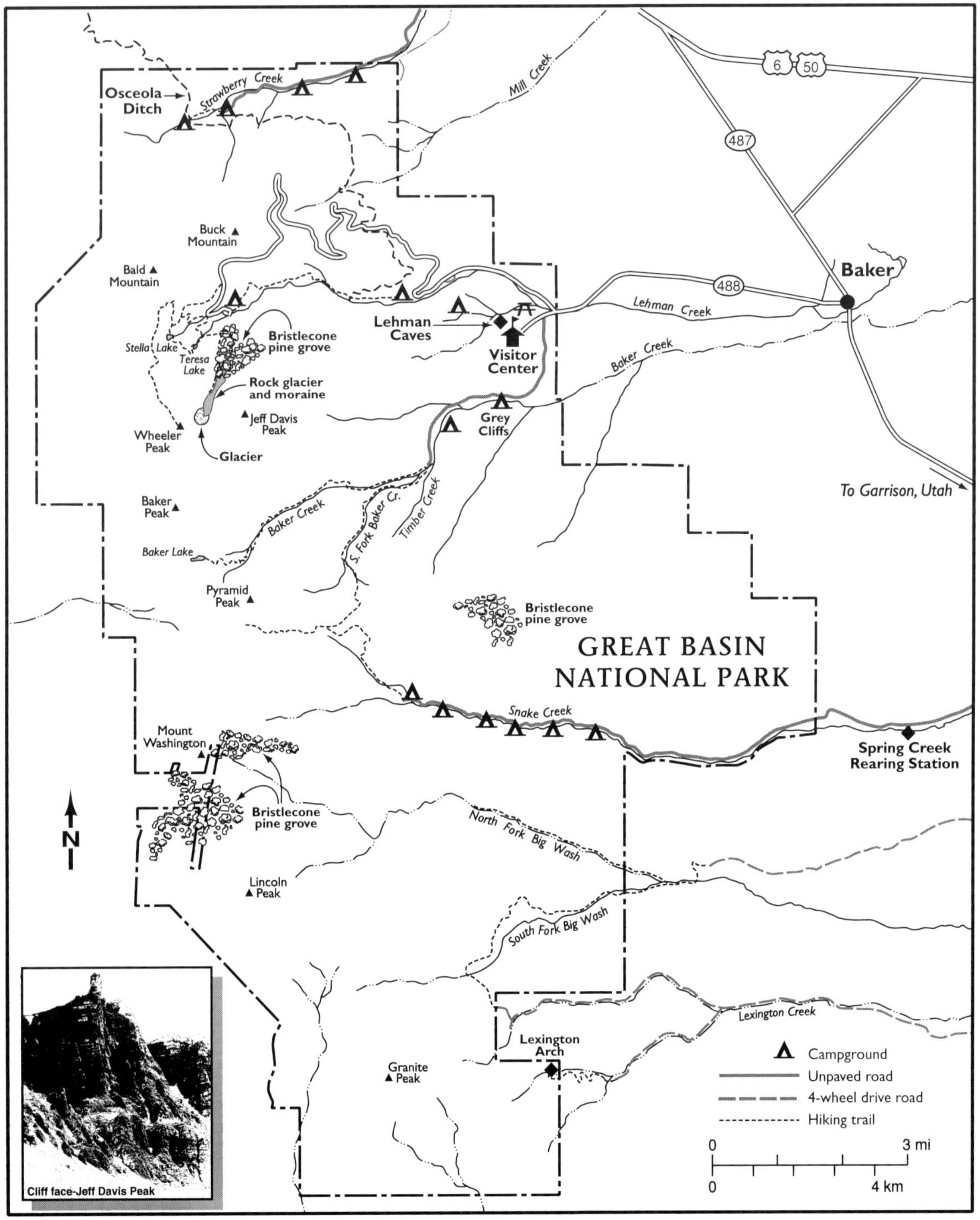

The map labels, read in their positions:

- Osceola Ditch
- Strawberry Creek
- Mill Creek
- 6 50
- 487
- Buck Mountain
- Bald Mountain
- Baker
- 488
- Lehman Creek
- Lehman Caves
- Visitor Center
- Stella Lake
- Teresa Lake
- Bristlecone pine grove
- Rock glacier and moraine
- Baker Creek
- Grey Cliffs
- Wheeler Peak
- Jeff Davis Peak
- Glacier
- Baker Peak
- Baker Lake
- S. Fork Baker Cr.
- Timber Creek
- To Garrison, Utah
- Pyramid Peak
- Bristlecone pine grove
- GREAT BASIN NATIONAL PARK
- Snake Creek
- Mount Washington
- Spring Creek Rearing Station
- Bristlecone pine grove
- Lincoln Peak
- North Fork Big Wash
- South Fork Big Wash
- Lexington Creek
- Lexington Arch
- Granite Peak

Legend:
- ⛺ Campground
- Unpaved road
- 4-wheel drive road
- Hiking trail

0 — 3 mi
0 — 4 km

Cliff face-Jeff Davis Peak

Side trip 9 map, Great Basin National Park.

N

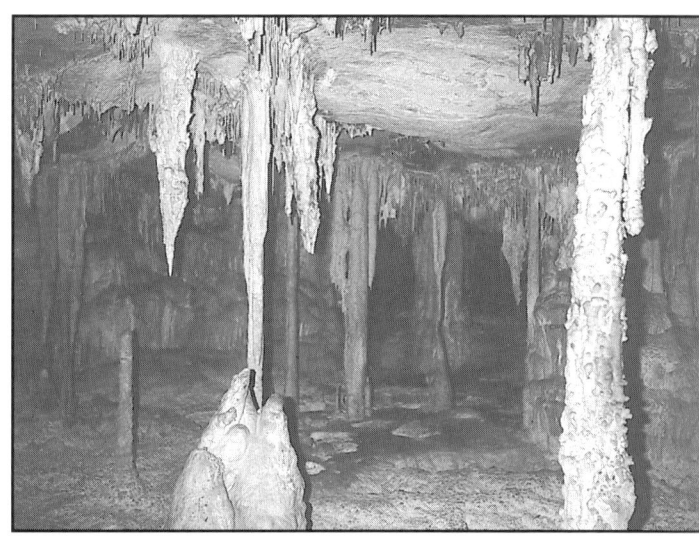

Photo: Kris Pizarro

Stalactites hanging from the limestone roof (above), and stalagmites extending from the floor (below) of a room in Lehman Caves, Great Basin National Park.

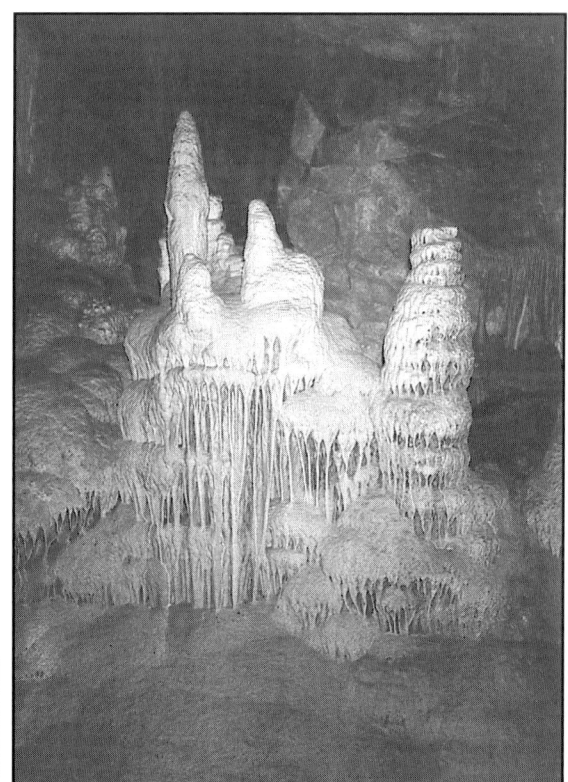

Photo: Kris Pizarro

Glacier and cirque beneath the north face of Wheeler Peak, Great Basin National Park. This is the only permanent ice found in the Great Basin. Rock rubble in the valley of the ice field is a possible rock glacier. Rocks that overlie permanent ice and move downhill as one mass with the ice. The rock face of the cirque is cut in Cambrian-age quartzite.

Trail to the glacier.

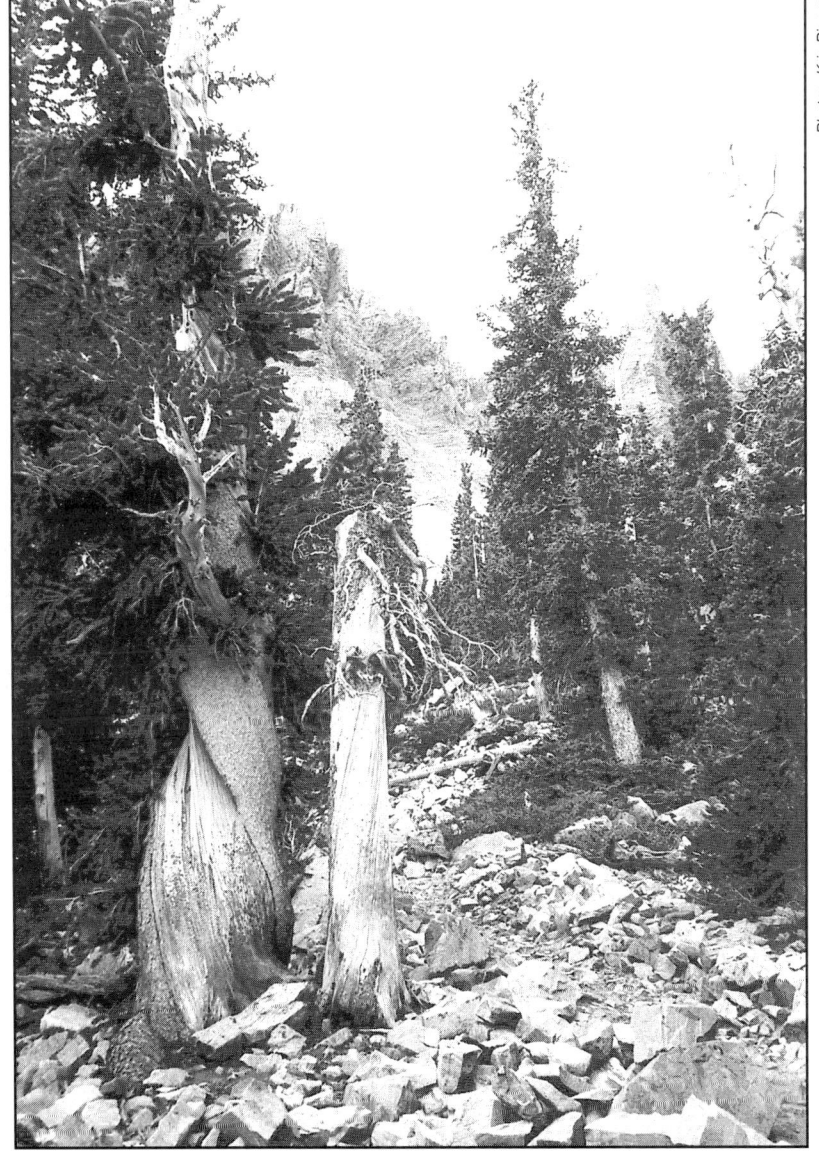

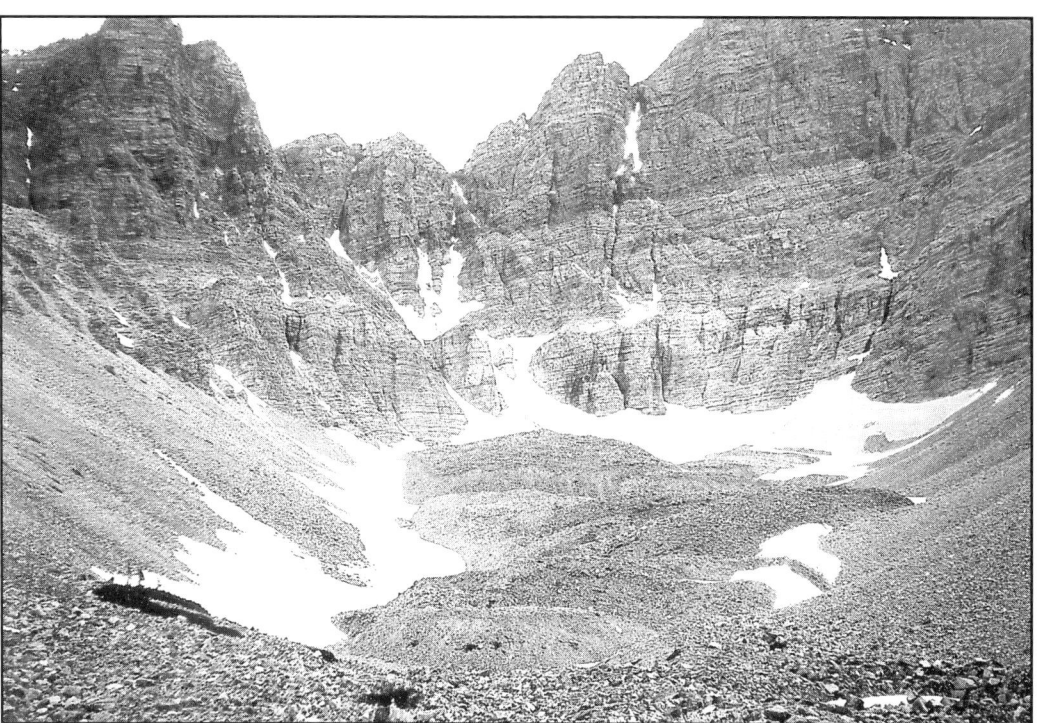

Photos: Kris Pizarro

Stella Lake with Wheeler Peak (right) and Jeff Davis Peak (left).

Photo: Jack Hursh

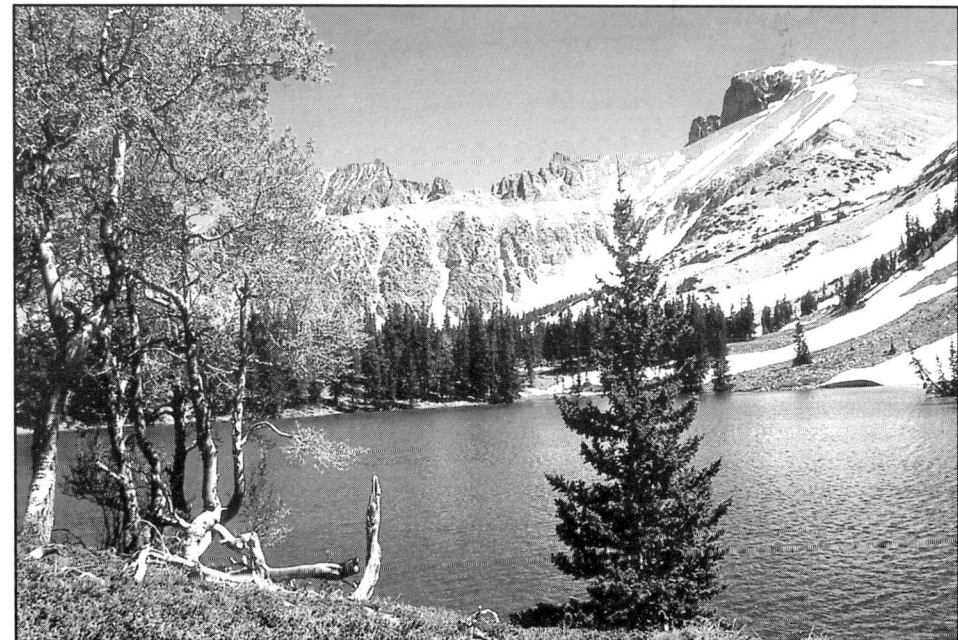

Bibliography

Abbe, D.R., 1985, Austin and the Reese River mining district, Nevada's forgotten frontier: University of Nevada Press, Reno, 117 p.

Austin, G.T., 1998, Definitive destination: Spring Mountains, Nevada: American Butterflies, v. 6, no. 4, Winter 1998, p. 4–11.

Basso, D., ed., 1981, Nevada historical marker guidebook: a guide to 234 historical markers, second edition: Falcon Hill Press, Sparks, Nevada, 88 p.

Bates, R.L., and Jackson, J.A., eds., 1995, Glossary of geology CD-ROM, third edition: American Geological Institute, Alexandria, Virginia.

Benedetto, F.M.F., Dean, D.A., and Durgin, D.C., 1991, Roadside geology and precious-metal mineralization along US 50, in Buffa, R.H., and Coyner, A.R., eds., Geology and ore deposits of the Great Basin, field trip guidebook compendium, volume 2: Geological Society of Nevada, Reno, p. 1124–1140.

Benson, J., 1995, The traveler's guide to the Pony Express Trail: Falcon Press, Helena, Montana, 140 p.

Bradley, P., 1987, Great Basin National Park, a traveler's guide to America's newest national park: Nevada Magazine, v. 47, no. 3, p. 38–44.

Brown, L., 1997, National Audubon Society nature guides: grasslands: Alfred A. Knopf, Inc., New York, 606 p.

Camp, C.L., 1981, Child of the rocks, The story of Berlin-Ichthyosaur State Park: Nevada Bureau of Mines and Geology Special Publication 5, 33 p.

Carlson, H.S., 1974, Nevada place names: University of Nevada Press, Reno, 250 p.

Castleman, D., 1991, Nevada handbook, fourth edition: Moon Publications, Inc., Chico, California, 473 p.

Chernicoff, S., and Venkatakrishnan, R., 1995, Geology, an introduction to physical geology: Worth Publishers, New York, 593 p.

Christopherson, R.W., 1998, Elemental geosystems, second edition: Prentice-Hall Inc., Upper Saddle River, New Jersey, 534 p.

Clark, J.L., 1993, Nevada wildlife viewing guide: Falcon Press Publishing Co., Inc., Helena and Billings, Montana, 87 p.

Cox, D., 1998, Tangled up in knots: Reno Gazette-Journal, Friday, September 4, 1998, p. 1C.

Crampton, B., 1974, Grasses in California: University of California Press, Berkeley and Los Angeles, 178 p.

Dreves, H., 1967, Geology of the Connors Pass quadrangle, Schell Creek Range, east-central Nevada: U.S. Geological Survey Professional Paper 557, 93 p.

Dunne, P., Sibley, D., and Sutton, C., 1988, Hawks in flight: Houghton Mifflin Co., Boston, 254 p.

Egan, F., 1985, Frémont, explorer for a restless nation: University of Nevada Press, Reno, 582 p.

Evans, R., 1997, Nevada's Highway 50, the loneliest road in America: VIA Magazine, v. 118, no. 5, p. 54–57, 75–76, 97.

Farquhar, F.P., 1965, History of the Sierra Nevada: University of California Press, Berkeley, Los Angeles, London, 262 p.

Fay, A.H., 1947, A glossary of the mining and mineral industry: U.S. Bureau of Mines Bulletin 95, 754 p.

Fiero, B., 1986, Geology of the Great Basin, a natural history: University of Nevada Press, Reno, 198 p.

Glass, M.E., and Glass, A., 1983, Touring Nevada, a historic and scenic guide: University of Nevada Press, Reno, 253 p.

Grater, R.K., 1978, Discovering Sierra mammals: Yosemite Natural History Association and Sequoia Natural History Association, 174 p.

Grayson, D.K., 1993, The desert's past, a natural prehistory of the Great Basin: Smithsonian Institution Press, Washington, D.C., 356 p.

Hall, E.R., 1946, Mammals of Nevada: University of California Press, Berkeley and Los Angeles, 710 p.

Hall, S.R., 1994, Romancing Nevada's past, ghost towns and historic sites of Eureka, Lander, and White Pine Counties: University of Nevada Press, Reno, 226 p.

Hamblin, W.K., and Howard, J.D., 1999, Exercises in physical geology: Prentice-Hall Inc., Upper Saddle River, New Jersey, 256 p.

Hardyman, R.F., Brooks, W.E., Blaskowski, M.J., Barton, H.N., Ponce, D.A., and Olson, J.E., 1988, Mineral resources of the Clan Alpine Mountains Wilderness Study Area, Churchill County, Nevada: U. S. Geological Survey Bulletin 1727-B, 16 p.

Harrison, C., and Greensmith, A., 1993, Birds of the world: Dorling Kindersly, Inc., New York, 416 p.

Hauntz, C., 1985, Road log/trip guide: Sediment-hosted gold deposits, Reno-Ely, in Slavic, G., ed., Geological Society of Nevada 1985 Meeting and Fall Field Trip Road Log: Geological Society of Nevada Special Publication 3, Reno, p. 33–39.

Henry, C.D., 1996, Geologic map of the Bell Canyon Quadrangle, western Nevada: Nevada Bureau of Mines and Geology Field Studies Map FS-11.

Henry, C.D., 1996, Geologic map of the Bell Mountain Quadrangle, western Nevada: Nevada Bureau of Mines and Geology Field Studies Map FS-12.

Herron, G.B., Mortimer, C.A., and Rawlings, M.S., 1985, Nevada raptors, their biology, and management: Nevada Department of Wildlife Biological Bulletin No. 8, 114 p.

Hinkle, G., and Hinkle, B., 1987, Sierra-Nevada lakes: University of Nevada Press, Reno, 383 p.

Hokanson, D., 1988, The Lincoln Highway, main street across America: University of Iowa Press, Iowa City, 159 p.

Hose, R.K., and Blake, M.C., 1976, Geology and mineral resources of White Pine County, Nevada: Nevada Bureau of Mines and Geology Bulletin 85, 105 p.

Jaeger, E.C., 1961, Desert wildlife: Stanford University Press, Stanford, California, 308 p.

John, D.A., 1994, Field guide to Oligocene-Miocene ash-flows and source calderas in the Great Basin of Nevada: U.S. Geological Survey Open-File Report 94-193, 44 p.

Knuepfer, P.L.K., 1997, Clustering, contagion, and other paleoseismic diseases: Geological Society of America Abstracts with Programs, v. 29, no. 6, 72 p.

Lambert, D., 1988, The field guide to geology: Cambridge University Press, Cambridge, 256 p.

Lanner, R.M., 1981, The piñon pine: a natural and cultural history: University of Nevada Press, Reno, 208 p.

Lewis, O., 1986, The town that died laughing, the story of Austin, Nevada, rambunctious early-day mining camp, and of its renowned newspaper, the Reese River Reveille: University of Nevada Press, Reno, 235 p.

Lewis, O., 1988, High Sierra country: University of Nevada Press, Reno, 291 p.

Longwell, C.R., Flint, R.F., and Sanders, J.E., 1969, Physical geology: John Wiley and Sons, Inc., New York, 685 p.

Lyons, W.A., 1997, The handy weather answer book: Visible Ink Press, Detroit, Michigan, 397 p.

Mack, R.N., 1986, Alien plant invasion into the Intermountain West: a case history, in Mooney, H.A., and Drake, J.A., eds., Ecology of biological invasions of North America and Hawaii: Springer-Verlag, New York, p. 191–213.

MacMahon, J.A., 1997, The National Audubon Society nature guides: deserts: Alfred A. Knopf, New York, 638 p.

Maturi, R.J., and Maturi, M.B., 1997, Nevada, off the beaten path: The Globe Pequot Press, Old Saybrook, Connecticut, 159 p.

McKee, E.H. Barton, H.N., Conrad, J.E., Ponce, D.A., and Benjamin, D.A., 1987, Mineral resources of the Desatoya Mountains wilderness study area, Churchill and Lander Counties, Nevada: U.S. Geological Survey Bulletin 1727-A, 15 p.

McPhee, J.A., 1980, Basin and Range: Farrar, Straus, Giroux, New York, 215 p.

Meeuwig, R.O., Budy, J.D., and Everett, R.L., 1992, Singleleaf pinyon (Pinus monophylla Torr. and Frém.) in Silvics of native and naturalized trees of the United States and Puerto Rico: U.S. Department of Agriculture Agricultural Handbook 654. v. 1, p. 380–384.

Molinelli, L. & Co., 1982, Eureka and its resources (reprint of 1879 publication): University of Nevada Press, Reno, 136 p.

Milne, L., and Milne, M., 1980, National Audubon Society field guide to North American insects and spiders: Alfred A. Knopf, New York, 989 p.

Moore, J.G., 1969, Geology and mineral deposits of Lyon, Douglas, and Ormsby Counties, Nevada: Nevada Bureau of Mines and Geology Bulletin 75, 45 p.

Moreno, R., 1987, The road less traveled by: Nevada Magazine, v. 47, no. 4, p. 48–51.

Moreno, R., 1991, The backyard traveler, 54 outings in northern Nevada: Carson City Children's Museum, Carson City, Nevada, 246 p.

Moreno, R., 1995, Ten great loneliest roads: Nevada Magazine, v. 55, no. 3, p. 90.

Morrison, R.B., 1964, Lake Lahontan: Geology of southern Carson Desert, Nevada: U. S. Geological Survey Professional Paper 401, 156 p.

Mozingo, H.N., 1987, Shrubs of the Great Basin: University of Nevada Press, Reno, 342 p.

Myrick, D.F., 1962, Railroads of Nevada and eastern California, Volume I: Howell-North, San Diego, California, 453 p.

Nicklas, M.L., 1996, Great Basin, the story behind the scenery: K.C. Publications, Inc., Las Vegas, Nevada, 48 p.

Nolan, T.B., Merriam, C.W., and Blake, M.C., Jr., 1974, Geologic map of the Pinto Summit Quadrangle, Eureka and White Pine Counties, Nevada: U.S. Geological Survey Map I-793.

Nolan, T.B., Merriam, C.W., and Brew, D.A., 1971, Geologic map of the Eureka Quadrangle, Eureka and White Pine Counties, Nevada: U.S. Geological Survey Map I-612.

Plummer, C.C., McGeary, D., and Carlson, D.H., 1999, Physical geology: The McGraw-Hill Companies, 577 p.

Press, F., and Siever, R., 1994, Understanding earth: W.H. Freeman and Company, New York, 593 p.

Purdue, M., 1999, Adventure guide to Nevada: Hunter Publishing, Inc., Edison, New Jersey, 217. P.

Purkey, B.W., and Garside, L.J., 1995, Geologic and natural history tours in the Reno area: Nevada Bureau of Mines and Geology Special Publication 19, 211 p.

Rigby, J.G., 1986, Road log and trip guide, in Tingley, J.V., and Bonham, H.F., Jr., eds., 1986, Precious-metal mineralization in hot springs systems, Nevada-California: Nevada Bureau of Mines and Geology Report 41, p. 3–83.

Roberts, R.J., Montgomery, K.M., and Lehner, R.E., Geology and mineral resources of Eureka County, Nevada: Nevada Bureau of Mines and Geology Bulletin 64, 152 p.

Rocha, G.L., 1999, Nevada's first permanent settlement: Nevada State Archives and Records, Historical Myth a Month, Myth #22, www.clan.lib.nv.us/docs/NSLA/ARCHIVES/myth/myth22.htm.

Ryser, F.A., Jr., 1985, Birds of the Great Basin: University of Nevada Press, Reno, 603 p.

Savage, C., 1995, Bird brains: the intelligence of crows, ravens, magpies, and jays: Greystone Books, Vancouver, British Columbia, 134 p.

Scott, S.L., ed., 1987, Field guide to the birds of North America, second edition: National Geographic Society, Washington, D.C., 464 p.

Seedorff, E., 1996, Road log from Utah border to Ely, Nevada in Green, S. M., and Struhsacker, E., eds., Geology and ore deposits of the American Cordillera: field trip guidebook compendium, Geological Society of Nevada, April 10–13, 1995, Reno/Sparks, Nevada, p. 79–85.

Schumacher, S., 1987, Loneliest road paved with gold: Reno Gazette-Journal, Sunday July 26, 1987, p. 1F.

Simpson, J.H., 1876, Report of explorations across the Great Basin of the Territory of Utah (1983 reprint): University of Nevada Press, Reno, 518 p.

Slemmons, D.B., 1966, Road log—Sparks to Fairview Peak and Dixie Valley earthquake areas, in Lintz, J., Jr., and Abdullah, S.K.M., eds., Guidebook for field trip excursions in northern Nevada, Cordilleran Section meeting, Geological Society of America, April 7–9, 1966, Reno, Nevada, p. A8–A43.

Smith, G.H., 1943, The history of the Comstock Lode, 1850–1920: Nevada Bureau of Mines and Geology Bulletin 37, 305 p.

Stewart, J.H., 1980, Geology of Nevada: Nevada Bureau of Mines and Geology Special Publication 4, 136 p.

Stewart, J.H., and Carlson, J.E., 1976, Cenozoic rocks of Nevada: Nevada Bureau of Mines and Geology Map 52.

Stewart, J.H., McKee, E.H., and Stager, H.K., 1977, Geology and mineral deposits of Lander County, Nevada: Nevada Bureau of Mines and Geology Bulletin 88, 106 p.

Storer, T.I., and Usinger, R.L., 1963, Sierra Nevada natural history: University of California Press, Berkeley, 374 p.

Tarbuck, E.J., and Lutgens, F.K., 1999, Earth: an introduction to physical geology: Prentice-Hall, Inc., Upper Saddle River, New Jersey, 638 p.

Tausch, R.J., West, N.E., and Nabi, A.A., 1981, Tree age and dominance patterns in Great Basin pinyon-juniper woodlands: Journal of Range Management, v. 34, no. 4, p. 259–264.

Taylor, R.J., 1992, Sagebrush country: a wildlife sanctuary: Mountain Press Publishing Co., 211 p.

Terres, J.K., 1980, The Audubon Society encyclopedia of North American birds: Alfred A. Knopf, Inc., New York, 1109 p.

Tingley, J.V., and Bonham, H.F., Jr., 1986, Road log and trip guide, in Tingley, J.V., and Bonham, H.F., Jr., eds., 1986, Sediment-hosted precious metal deposits of northern Nevada: Nevada Bureau of Mines and Geology Report 40, p. 3–51.

Tingley, J.V, Horton, R.C., and Lincoln, F.C., 1993, Outline of Nevada mining history: Nevada Bureau of Mines and Geology Special Publication 15, 48 p.

Titley, S.R., and Hicks, C.L., 1966, Geology of the porphyry copper deposits, southwestern North America: The University of Arizona Press, Tucson, 287 p.

Toll, D.W., 1996, The complete Nevada traveler: Gold Hill Publishing Co., Virginia City, Nevada, 256 p.

Townley, J.M., 1984, The Pony Express guidebook: across Nevada with the pony express and overland stage line: Jamison Station Press, Reno, Nevada, 57 p.

Trexler, D.T., and Melhorn, W.N., 1986, Singing and booming sand dunes of California and Nevada: California Geology, v. 39, no. 7, July 1986, p. 147–152.

Upadhaya, M.K., Turkington, R., and McIlvride, D., 1986, The biology of Canadian weeds: Canadian Journal of Plant Science, v. 66, p. 689–709.

Whitney, S., 1997, The National Audubon Society nature guides: western forests: Alfred A. Knopf Inc., New York, 670 p.

Whitson, T.D., Burrill, L.C., Dewey, S.A., Cudney, D.W., Nelson, B.E., Lee, R.D., and Parker, R., 1991, Weeds of the west: Western Society of Weed Science, Newark, California, 630 p.

Willden, R., and Speed, R.C., 1974, Geology and mineral deposits of Churchill County, Nevada: Nevada Bureau of Mines and Geology Bulletin 83, 95 p.

Williams, G.J., III, 1996, Hot springs of Nevada: Tree by the River Publishing, Carson City, Nevada, 72 p.

Wuerthner, G., 1992, Nevada mountain ranges: American and World Geographic Publishing, Helena, Montana, 96 p.

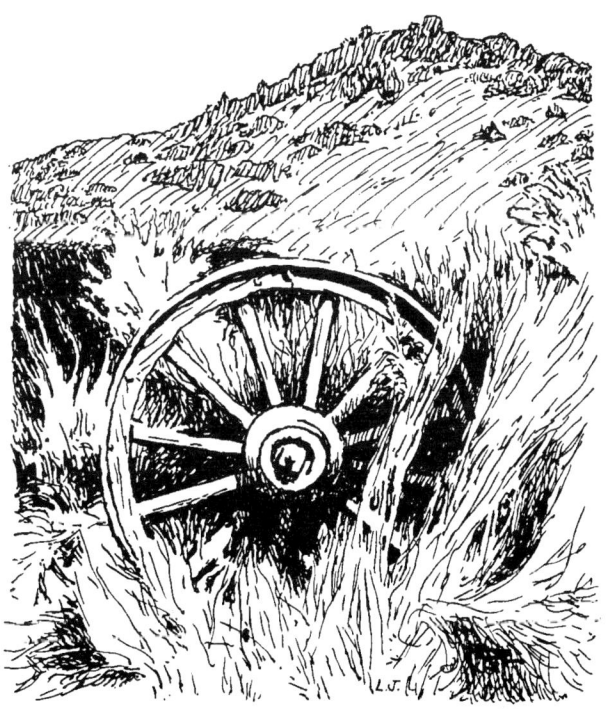

artwork: Larry Jacox

GLOSSARY

adit A nearly horizontal passage from the surface into a mine. An adit has only one opening, differing from a tunnel which is open at both ends.

air-fall tuff Deposit formed from shower-like falling of pyroclastic fragments from a volcanic eruption cloud.

alaskite a light-colored granitic rock composed mainly of quartz and feldspar with little or no associated dark minerals.

alkali flat A level area or plain in an arid or semiarid region, encrusted with alkali salts that became concentrated by evaporation and poor drainage; a salt flat. See also: playa.

alluvial Pertaining to alluvium.

alluvial fan A fan-shaped deposit of alluvium typically built where a stream leaves a steep mountain valley and runs out onto a lower-gradient surface.

alluvium A general term for clay, silt, sand, gravel, or similar unconsolidated material, deposited during comparatively recent geologic time by a stream or other body of running water.

alteration See altered, hydrothermal alteration.

altered Said of a rock that has undergone a change in its mineral composition, typically brought about by the action of hydrothermal solutions.

amalgam An alloy of mercury with another metal; in gold metallurgy, an alloy of gold and mercury, usually obtained by allowing gold-bearing minerals, after crushing, to come in contact with mercury.

andesite A fine-grained volcanic rock that solidifies from molten lava at the Earth's surface. It is intermediate in composition between basalt and rhyolite, and ranges in color from dark gray-green, to lighter gray, red, or brown.

anglesite A mineral, lead sulfate.

anticline A fold, generally convex upward, whose core contains the stratigraphically older rocks.

aplite A light-colored plutonic igneous rock characterized by a fine-grained granular (i.e., aplitic) texture, from a Greek word meaning "simple."

argentite A dark lead-gray cubic form of acanthite, silver sulfide (Ag_2S). Argentite is a valuable ore of silver, also known as silver glance; vitreous silver.

argillite A rock, derived either from mudstone (claystone or siltstone) or shale, that has been compacted and hardened more than mudstone or shale but is less clearly laminated than shale, and that lacks the cleavage distinctive of slate.

ash, volcanic Fine pyroclastic material produced by the explosive emission of hot, gas-charged lava from a volcanic crater or fissure that cools on its descent to the ground surface. It consists of fragments under 4 mm in diameter and is usually light gray.

ash-flow tuff Pyroclastic volcanic rock composed of formerly gas-charged volcanic ash that flowed as a "fiery cloud" down the side of a volcano or erupted from a caldera.

asthenosphere The weak, plastic, partly molten layer of the upper mantle directly below the lithosphere. It lies at a depth of 100 to 350 kilometers (60 to 220 miles) below the Earth's surface.

asymmetric Without symmetry.

Azoic (a) That earlier part of Precambrian time, represented by rocks in which there is no trace of life. (b) The entire Precambrian.

basalt A fine-grained igneous rock that solidified from molten lava at the Earth's surface. It is usually black or dark gray, due to the predominance of the minerals calcic plagioclase, olivine, pyroxene, and other dark-colored accessory minerals. It may have cavities (vesicles) that formed by trapped gas as the magma cooled.

basaltic andesite An igneous rock intermediate in composition between andesite and basalt. Its color is commonly darker than andesite and lighter than basalt.

basement The undifferentiated complex of rocks (generally igneous or metamorphic) that underlie surface rocks.

Basin and Range province A physiographic province in the western United States and northern Mexico that consists of fault-block mountains and intervening sediment-filled basins. The province generally lies between the Sierra Nevada on the west, the Columbia Plateau and Snake River Plain on the north, and the Colorado Plateau on the east. On the south, the province extends through southern Arizona and into northern Mexico. See page 18.

batholith A large body of intrusive igneous (plutonic) rock generally having an area of over 40 square miles. They are commonly produced by multiple intrusions.

beach terrace See strandline.

bedding Depositional layers or planes dividing sedimentary rocks of the same or different lithology.

bedrock Solid rock exposed at the Earth's surface or covered by unconsolidated material.

bleaching A lightening of the original color of rock; a surface effect caused by long exposure to weathering (including acid solutions) or a more penetrating effect caused when circulating solutions, generally hot, have altered the original chemical composition of the minerals forming the rocks.

blind lode A vein having no outcrop.

BLM An abbreviation for the U.S. Bureau of Land Management.

bonanza A miner's term for good luck, or a body of rich ore.

borates Oxide compounds of the element boron; includes the minerals borax, colemanite, and ulexite.

bornite A mineral, copper iron sulfide. Also known as peacock copper ore, it is characterized by an iridescent purple color on tarnished or weathered surfaces. A fresh fracture is red-brown.

breccia A coarse-grained rock composed of angular fragments of broken rock in a finer-grained matrix.

brecciated Rock that has been broken into angular fragments within a finer-grained matrix which may or may not cement the angular fragments.

calcite Calcium carbonate, the principal constituent of limestone.

caldera A large, bowl-shaped volcanic depression with a diameter many times greater than the included volcanic vent or vents. It may be formed by explosion or collapse.

Cambrian The earliest period of the Paleozoic era, thought to have covered the span of time between 544 and 495 million years ago; also, the corresponding system of rocks. It is named after Cambria, the Roman name for Wales, where rocks of this age were first studied.

Carlin-type gold Gold occurring as microscopic particles (up to 30 microns) that must be identified by chemical analysis as it is not recoverable by panning. The term is taken from its occurrence at Carlin, Nevada. Synonyms: disseminated gold, invisible gold, micron gold.

Cenozoic An era of geologic time, from the beginning of the Tertiary Period to the present. It is characterized by the evolution and abundance of mammals, advanced mollusks, birds, and angiosperms. The Cenozoic is considered to have begun about 65 million years ago.

cerussite A mineral (lead carbonate). An ore of lead.

chalcocite A black or dark lead-gray copper sulfide mineral (Cu_2S). It has a metallic luster, can be crystalline or massive, and is an important ore of copper; also known as copper glance or vitreous copper.

chalcopyrite A mineral (copper iron sulfide). An ore of copper.

chalk A soft, earthy, fine-textured, usually white to light gray or buff limestone of marine origin, consisting almost wholly (90–99 percent) of calcite, formed mainly by shallow-water accumulation of calcareous remains of floating microorganisms and of calcareous algae. The best known and most widespread chalks are of Cretaceous age, such as those exposed in cliffs on both sides of the English Channel.

chert A hard, extremely dense or compact, dull to semivitreous, sedimentary rock, consisting dominantly of microcrystalline quartz. It sometimes contains impurities such as calcite, iron oxide, and the remains of siliceous and other organisms. It has a tough, splintery to smoothly curved fracture, and may be white or variously colored gray, green, blue, pink, red, yellow, brown, and black. Chert occurs principally as nodules in limestones and dolomites, and less commonly as extensive layered deposits (bedded chert); it may be an original organic or inorganic precipitate or a replacement product.

cinder cone A conical hill formed by the accumulation of volcanic ash or cinders (generally of basalt composition) around a vent.

cinder, volcanic Uncemented, glassy, frothy-appearing, porous rock ejected from a volcanic vent.

clast A piece of broken rock or an individual constituent of sedimentary rock produced by the physical disintegration of a larger rock mass.

clastic rocks Rocks consisting of clasts or fragments of other rocks that have moved individually from their place of origin.

clay (a) An extremely fine-grained, natural sediment or soft rock composed of clay-sized (less than 0.002 mm) particles. (b) A group of silicate minerals that generally occur as platy particles in fine aggregates.

coal A readily combustible rock containing more than 50 percent by weight and more than 70 percent by volume of carbonaceous material formed from compaction and induration of variously altered plant remains.

columnar jointing Parallel, prismatic columns, polygonal in cross section, in basaltic flows and sometimes in other extrusive and intrusive rocks. It is formed as the result of contraction during cooling.

conglomerate A sedimentary rock consisting of rounded rock fragments, over 2 mm in diameter, set in a finer-grained matrix.

contact The place or surface where two different kinds of rocks come together.

contact metamorphic Said of a rock or mineral that has originated through the process of contact metamorphism.

continental crust The portion of the Earth's crust that underlies the continents and is relatively enriched in silica and alumina. See oceanic crust.

continental plate A plate or slab of the Earth's crust that consists partly or mainly of continental crust.

coquina A limestone composed wholly or chiefly of mechanically sorted fossil debris that is weakly to moderately cemented but not completely indurated; a porous light-colored limestone made up of loosely aggregated shells and shell fragments, such as the relatively recent deposits occurring in Florida and used for roadbeds and construction.

crop out See outcrop.

crust The outermost compositional shell of the Earth, 10 to 40 km (6 to 24 miles) thick, consisting mainly of relatively low density silicate rocks.

crystal A homogeneous, solid body of a chemical element, compound, or mixture, having a regularly repeating atomic arrangement that may be outwardly expressed by plane faces.

cyanidation A process involving the use of cyanide (potassium or sodium cyanide), especially in the recovery of gold and silver from ore.

décollement Detachment structure of rock strata owing to deformation, resulting in independent styles of deformation in the rocks above and below. It is associated with folding and with overthrusting.

decomposed granite The fragmental product of in-place granular disintegration of granite and granitic rocks; properly termed "grus"; informally called "DG."

deflation The sorting out, lifting, and removal of loose dry fine-grained particles by the turbulent eddy action of the wind, as along a sand-dune coast or in a desert; a form of wind erosion.

desert pavement A natural residual concentration of wind-polished, closely packed pebbles, boulders, and other rock fragments, mantling a desert surface where wind action and sheetwash have removed all smaller particles, usually protecting the underlying finer-grained material from further deflation.

desert varnish A thin dark shiny film or coating, composed of iron oxide accompanied by traces of manganese oxide and silica, formed on the surfaces of pebbles, boulders, and other rock fragments in desert regions after long exposure, as well as on ledges and other rock outcrops. It is believed to be caused by sweating of mineralized solutions from within and deposition by evaporation on the surface. Also called rock varnish.

detachment fault A low-angle thrust fault, see décollement.

Devonian A period of the Paleozoic Era thought to have covered the span of time between 418 and 362 million years ago; also, the corresponding system of rocks. It is named after Devonshire, England, where rocks of this age were first studied.

diamond A mineral, a crystalline form of carbon. The hardest natural substance known. Diamonds form under extreme temperatures and pressures and are found in breccias, pipes in igneous rocks, and alluvial deposits. Pure diamond is colorless or nearly so, color is imparted by impurities.

dike A tabular sheet of intrusive igneous rock that cuts across the structure of the intruded rock or cuts massive rock.

diopside A calcium-magnesium silicate mineral. It ranges in color from white to green; transparent varieties are used in jewelry. Diopside occurs in some metamorphic rocks, and is found as a contact-metamorphic mineral in limestones.

diorite A plutonic (intrusive) igneous rock composed essentially of sodium plagioclase and hornblende, biotite, or pyroxene. Small amounts of quartz and orthoclase may be present.

dip The angle in degrees between the horizontal and an inclined geologic plane, such as a bedding plane or a fault. Dip is measured in a plane that is perpendicular to the intersection of the plane with the horizontal.

discordant Said of a contact between an igneous intrusion and the surrounding rock that is not parallel to the foliation or bedding planes of the latter; or, in bedded rocks, strata of different ages that lack conformity or parallelism of bedding or structure.

disseminated gold See Carlin-type gold.

dolomite A carbonate sedimentary rock of which more than 50 percent by weight consists of the mineral dolomite (calcium-magnesium carbonate). Dolomite occurs in crystalline and noncrystalline forms, is clearly associated and often interbedded with limestone, and usually represents a postdepositional replacement of limestone.

doré Bullion consisting of a mixture of gold and silver, the metal obtained when gold-silver precipitate obtained by the cyanide process is melted and poured into bars.

drag mark A long, even groove or striation made by a solid body, such as a stone, dragged over a soft sedimentary surface, such as the wet surface of a desert playa.

dump A pile or heap of waste material removed from a mine.

earthquake A sudden motion or trembling in the Earth caused by the abrupt release of slowly accumulated strain.

eon The formal geochronologic unit of highest rank, next above era.

epidote A yellowish-green, pistachio-green, or blackish-green calcium-aluminum-iron silicate mineral. It commonly occurs as formless grains, masses, or as crystals in metamorphic rocks derived from limestones.

era The formal geochronologic unit next in order of magnitude below an eon.

erosion The group of related processes by which rock is broken down physically and chemically and the products removed from any part of the Earth's surface. It includes the processes of weathering, solution, and transportation.

escarpment A long, generally continuous cliff or steep face at the edge of a region of high local relief.

evaporation The process, also called vaporization, by which a substance passes from the liquid or solid state to the vapor state. The opposite of condensation.

extension In geology, horizontal expansion or pulling apart of the Earth's crust.

extrusive igneous rocks See volcanic rocks.

fan See alluvial fan.

fault A fracture or planar break in rock (which may be a few centimeters or many kilometers long) along which there has been movement of one side relative to the other. See page 8.

fault plane A fault surface that is more or less planar.

fault scarp The steep slope or cliff formed by a fault. Most fault scarps have been modified by erosion since the faulting.

fault surface The surface along which movement has taken place.

fault zone A fault, instead of being a single clean fracture, may be a zone as much as hundreds or even thousands of feet wide; the fault zone consists of numerous interlacing small faults or a complex zone of fault breccia.

fault-block mountains Mountains bounded on at least two opposite sides by faults. Common in the Basin and Range province.

feldspar A framework silicate mineral common in many rocks and making up 60 percent of the Earth's crust.

fissure An extensive crack, break, or fracture in rock.

flexure Hinge.

floodplain That portion of a river valley that is covered with water when the river overflows its banks.

flume An inclined channel, usually of wood and often supported on a trestle, for conveying water from a distance to be utilized for power, transportation, mining, logging.

fluorite Also know as fluorspar, calcium fluoride is a transparent to translucent mineral. Found in many different colors (often blue or purple), fluorite occurs in veins, usually as a gangue mineral associated with lead, tin, and zinc ores, and is commonly found in crystalline cubes with perfect cleavage. It is the principal ore of fluorine.

fold A curve or bend of a planar structure such as rock strata, bedding planes, foliation, or cleavage.

fool's gold A term used for pyrite, iron sulfide; often mistaken for gold because of its yellow color.

footwall The lower side of a fault plane, vein, lode, or bed of ore. So named because miners in underground developments along a vein stood on the "foot" wall. See page 8.

formation A body of rock distinctive enough on the basis of physical properties and of sufficient extent to constitute a basic unit for a geologic map.

fossil Any remains, trace, or imprint of a plant or animal that has been preserved in the Earth's crust since some past geologic or prehistoric time; loosely, any evidence of past life.

galena A bluish-gray to lead-gray mineral, lead sulfide. It frequently contains included silver minerals. Galena occurs in cubic or octahedral crystals, in masses, or in coarse or fine grains; it is often associated with sphalerite as disseminations in veins in limestone, dolomite, and sandstone. It has a shiny metallic luster, exhibits highly perfect cubic cleavage, and is relatively soft and very heavy. Galena is the most important ore of lead and one of the most important sources of silver.

garnet A group of minerals composed of calcium-magnesium-aluminum silicates containing variable amounts of iron, manganese, and chrome. Garnet is a brittle and transparent to subtransparent mineral, having a vitreous luster, no cleavage, and a variety of colors, dark red being the most common.

glacial Pertaining to, characteristic of, produced or deposited by, or derived from a glacier.

glacier A large mass of ice formed, at least in part, on land by the compaction and recrystallization of snow, moving slowly by creep downslope or outward in all directions due to the stress of its own weight, and surviving from year to year. Included are small mountain glaciers as well as ice sheets continental in size.

gneiss A foliated rock formed by regional metamorphism, in which bands of granular minerals alternate with bands in which minerals having flaky or elongate habits predominate.

gold A soft, heavy, yellow, mineral, the native metallic element Au. It is often naturally alloyed with silver or copper and occasionally with bismuth, mercury, or other metals, and is widely found in alluvial deposits (as nuggets and grains) or in veins associated with quartz and various sulfides. Gold is malleable and ductile, and is used chiefly for jewelry and as the international standard for world finance.

gold amalgam A variety of native gold containing mercury; an amalgam composed of gold, silver, and mercury.

gossan An iron-bearing weathered product overlying a sulfide deposit. It is formed by the oxidation of sulfides and the leaching of the sulfur and most metals, leaving behind iron oxides and rarely sulfates. Gossans are usually highly colored due to a high concentration of residual iron oxides; leached cappings generally contain dispersed specks of iron oxides, and therefore are less intensely colored.

granite A light-colored coarse-grained, igneous (plutonic) rock containing the minerals quartz and alkali feldspar, with lesser amounts of plagioclase and mica.

granitic rock Granite or a close relative, such as granodiorite or quartz monzonite.

granodiorite A coarse-grained, plutonic rock resembling granite and consisting of quartz, plagioclase, and potassium feldspar, with biotite mica, hornblende, or pyroxene as the mafic (dark-colored) minerals. The almost equal amounts of dark- and light-colored minerals give this rock a salt-and-pepper appearance.

gravel An unconsolidated, natural accumulation of rounded rock fragments resulting from erosion, consisting predominantly of particles larger than sand (diameter greater than 2 mm), such as boulders, cobbles, pebbles, granules, or any combination of these fragments.

grus See decomposed granite.

Great Basin Specifically refers to that part of the Basin and Range province that has no exterior drainage; the Great Basin is a hydrographic province, with no outlet to the sea. See page 18.

guano A phosphate or nitrate deposit formed by the leaching of bird (and bat) excrement accumulated in arid regions. It is processed for use as a fertilizer

gypsum A mineral consisting of hydrous calcium sulfate. It frequently forms thick, extensive beds interstratified with limestone, shale, and clay. Gypsum is soft and white or colorless when pure. It occurs massive (alabaster), fibrous (satin spar), or in crystals (selenite) Gypsum is used chiefly as a soil amendment, as a retarder in portland cement, and in making plaster of paris.

hanging wall The rock on the upper side of a fault, mineral vein, or mineral deposit. See page 8.

hanging wall split A vein that splits away from the main vein and passes into the hanging wall.

headframe The structure above a mine shaft supporting the sheave wheel (pulley) over which the hoisting cable passes.

hornblende A common rock-forming dark silicate mineral with a complex chemical formula.

horn silver Common name for the mineral cerargyrite, silver chloride. Horn silver was a common mineral found in the oxidized outcrops of many Nevada silver camps.

hot spring terrace Deposits of mineral that build up around the edges of hot springs. The minerals deposit from the water of the spring, and form benchs or terraces around the spring vent.

hydrographic province A province defined by its common drainage characteristics; a large area of connected basins. The Great Basin hydrographic province, for example, is a large area with no external drainage.

hydrothermal Literally, "hot water"; describes processes or ore deposits related to circulating subsurface water warmed by shallow magma or hot rock; hydrothermal fluids commonly contain dissolved minerals and gasses.

hydrothermal alteration Changes in rocks brought about by the addition or removal of materials through the medium of hydrothermal fluids; silicification, for example.

ichthyosaur Common name for Ichthyosauria, the sole order of the reptilian subclass Ichthyopterygia, of uncertain ancestry but of porpoiselike or sharklike body form as adaptation for life in the sea. Range, Middle Triassic to Late Cretaceous

igneous rock Rock formed by the cooling and consolidation of magma.

incandescent Emits visible light as a result of being heated, said of an ash flow or any pyroclastic matter that is glowing.

interbedded Interstratified. Occurring between or alternating with beds of a different material.

intermontane basin A basin situated between or surrounded by mountains, mountain ranges, or mountainous regions;

intrusion The process of emplacement of magma in preexisting rock; also, the igneous rock mass so formed within the surrounding rock.

intrusive Rock formed by the process of intrusion.

invisible gold See Carlin-type gold

iron pyrites See pyrite.

joint A fracture in rock along which no appreciable displacement has occurred.

Jurassic The second period of the Mesozoic Era (after the Triassic and before the Cretaceous), thought to have covered the span of time between 200 and 145 million years ago; also, the corresponding system of rocks. It is named after the Jura Mountains between France and Switzerland, in which rocks of this age were first studied.

lacustrine Pertaining to, produced by, or formed in a lake.

landform Any feature of the Earth's surface having a characteristic shape as the product of natural processes. Examples are continents, ocean basins, mountains, alluvial fans, sand dunes, and valleys.

landslide A general term covering a variety of rapid, mass movement processes downslope on the Earth's surface.

lateral a horizontal mine working; a mine working that branches right or left from a main working.

latite An extrusive (volcanic) igneous rock in which potassium feldspar and plagioclase are present in nearly equal amounts. Augite and hornblende are usually present and biotite may be present.

lava A general term for a molten extrusion; also, for the rock that is solidified from it.

lava flow   Magma that flows out of the Earth's surface.

leached capping   An iron-bearing weathered product overlying a sulfide deposit. It is formed by the oxidation of sulfides and the leaching of the sulfur and most metals, leaving hydrated iron oxides and rarely sulfates. Sometimes called a gossan but, since it most commonly forms over a deposit of disseminated sulfides, it is less intensely colored than gossan formed from massive sulfide bodies.

leeward   As used in this text, the side of a mountain range opposite from that which the storms arrive.

left-lateral fault   A strike-slip fault in which relative motion is such that to an observer looking directly at (perpendicular to) the fault, the motion of the block on the opposite side of the fault is to the left. See page 8.

life zone   A zone defined by a characteristic assemblage of plants.

limestone   A sedimentary rock consisting chiefly of calcium carbonate, mainly in the form of the mineral calcite.

lithosphere   The rigid, outermost layer of the Earth, 50 to 200 km (30 to 120 miles) thick, encompassing the crust and upper mantle.

lode   A tabular deposit of valuable mineral between definite boundaries; a fissure in country rock filled with mineral; a deposit consisting of a zone of veins or veinlets; a mineral deposit in consolidated rock as opposed to placer deposits.

lower-plate rocks   In Nevada, generally refers to rocks of early Paleozoic age that are structurally below the Roberts Mountains thrust fault.

Ma   Million years old or million years ago (mega-annum).

maar   A low-relief, broad volcanic crater formed by shallow explosive eruptions. It is surrounded by a crater ring and may be filled by water.

magma   Molten rock (together with any suspended crystals and dissolved gases) in the mantle or crust. Igneous rocks are formed when magma cools and consolidates.

magma chamber   A reservoir in the Earth's crust occupied by magma.

magmatic   Of, pertaining to, or derived from magma.

magnetite   Magnetic iron ore. A black mineral (iron oxide). A frequent minor mineral in igneous rocks.

mantle   The zone of the Earth below the crust and above the core, which is divided into the upper mantle and the lower mantle, with a transition zone between.

manto   A flat-lying, bedded deposit; either a sedimentary bed or a replacement strata-bound orebody.

matrix   The finer-grained material enclosing, or filling the interstices between, the larger grains or particles of a sediment or sedimentary rock; in igneous rocks, refers to the groundmass.

Mesozoic   One of the large division or eras of geologic time, following the Paleozoic Era and succeeded by the Cenozoic Era, comprising the Triassic, Jurassic and Cretaceous Periods.

metallic ore   Valuable mineral deposit from which a metal such as gold, silver, copper, lead, iron, or tungsten is recovered.

metamorphic rock   Rock whose original textures or mineral components, or both, have been transformed to new textures and components as a result of one or several of the following: high temperatures, high pressure, and chemically active fluids.

metasedimentary rock   A sedimentary rock that shows evidence of having been subjected to metamorphism.

micron gold   See Carlin-type gold.

mineral   A naturally occurring, inorganic, solid element or compound with a definite composition or compositional range and a regular internal structure.

Miocene   The fourth of the five epochs into which the Tertiary Period is divided.

Mississippian   A period of the Paleozoic Era (after the Devonian and before the Pennsylvanian), thought to have covered the span of time between 362 and 323 million years ago; also, the corresponding system of rocks. It is named after the Mississippi River valley, in which there are good exposures of rocks of this age.

moraine   A mound, ridge, or other accumulation of unsorted and unstratified sediment and larger clasts of rock deposited by a glacier. For example, an end moraine is one produced at the front of a glacier.

mud crack   An irregular fracture in a crudely polygonal pattern, formed by the shrinkage of clay, silt, or mud, generally in the course of drying under the influence of atmospheric surface conditions.

mudstone   A rock that includes clay, silt, siltstone, claystone, shale, and argillite. The term is used when precise rock identification is in doubt or when a deposit consists of an indefinite or variable mixture of clay, silt, and sand particles.

normal fault   An inclined fault along which the upper side has moved downward relative to the lower side. See page 8.

obsidian   A black or dark-colored volcanic glass. Usage of the term goes back as far as Pliny, who described the rock from Ethiopia. Obsidian has been used for making arrowheads, other sharp implements, jewelry, and art objects.

oceanic crust   That type of Earth's crust that underlies the ocean basins. It is relatively enriched in silica and magnesia. See continental crust.

oceanic plate   A tectonic plate of the Earth's crust that underlies an ocean.

Oligocene   An epoch of the early Tertiary Period, after the Eocene and before the Miocene; also, the corresponding worldwide series of rocks.

Ordovician   The second earliest period of the Paleozoic Era (after the Cambrian and before the Silurian), thought to have covered the span of time between 495 and 442 million years ago; also, the corresponding system of rocks. It is named after a Celtic tribe called the Ordovices.

ore   The naturally occurring material from which a mineral or minerals of economic value can be extracted at a reasonable profit.

orebody   Generally a solid and fairly continuous mass of ore which can be distinguished by form or character from adjoining rock.

outcrop   That part of a geologic formation that appears at the surface of the Earth.

Paleozoic   An era of geologic time, from the end of the Proterozoic Eon to the beginning of the Mesozoic Era, or from about 544 to about 251 million years ago.

perlite   A volcanic glass, with a large percentage of water, that has cracked due to contraction during cooling resulting in a concentric, shelly texture.

petroglyph   Writings or drawings chipped or scraped into rock surfaces or faces. Usually the markings are made in the veneer of dark rock varnish, exposing underlying lighter-colored rock for good visual contrast.

Phanerozoic   That part of geologic time represented by rocks in which the evidence of life is abundant, i.e., Cambrian Period and later.

phyllite   A metamorphosed rock, intermediate in grade between slate and mica schist. Minute crystals of sericite and chlorite impart a silky sheen to the surfaces of cleavage (or schistosity). Phyllites commonly exhibit corrugated cleavage surfaces.

placer   A surficial mineral deposit formed by mechanical concentration of mineral particles from weathered debris. The common types are beach placers and alluvial placers. The mineral concentrated is usually a heavy mineral such as gold.

plate tectonics   The processes or mechanisms by which the Earth's lithosphere (upper crust) is broken up into a series of rigid plates that move over the asthenosphere (lower crust, upper mantle).

playa   The flat, vegetation-free, lowermost area of a desert basin, where water gathers after a rain and evaporates.

Pleistocene   The earlier of the two epochs in the Quaternary Period; the other epoch is the Holocene.

Pliocene   The last epoch in the Tertiary Period.

pluton   A body of igneous rock that formed at considerable depth beneath the Earth's surface by consolidation of magma.

plutonic rock   Magma that penetrated into or between other rocks and solidified at relatively great depth below the Earth's surface. Intrusive rock emplaced at depths below volcanic processes.

porphyritic   Term applied to igneous rock that has coarse crystals (phenocrysts) in a finely crystalline or glassy groundmass (the material between the coarse crystals).

porphyry   Igneous rock containing conspicuous phenocrysts in a fine-grained groundmass. The resulting texture is described as porphyritic.

porphyry copper deposit   A large body of rock, typically porphyry, that contains disseminated chalcopyrite and other sulfide minerals. Such deposits are mined in bulk on a large scale, generally in open pits, for copper and by-product molybdenum. Most deposits are 3 to 8 km across, and of low grade (less than 1 percent copper). Distribution of sulfide minerals changes outward from dissemination to veinlets and veins.

Supergene enrichment has been very important at most deposits, as without it the grade would be too low to permit mining.

Precambrian   All geologic time from the formation of the Earth to the start of the Cambrian Period of the Paleozoic Era. More than 90 percent of the Earth's estimated 4.5 billion years is Precambrian.

precious metal   A general term for gold, silver, or any of the minerals of the platinum group

Proterozoic   The eon from 2500 to 544 million years ago.

pulp   A term for pulverized ore mixed with water; also applied to dry, crushed ore.

pyrargyrite   A dark-red, gray, or black silver-antimony sulfide mineral; also known as ruby silver. It is an important ore of silver.

pyrite   A common, pale-bronze or brass-yellow, iron sulfide mineral. Pyrite has a brilliant metallic luster and an absence of cleavage, and has been mistaken for gold (hence the name, "fool's gold"). Pyrite is the most widespread and abundant of the sulfide minerals and occurs in all kinds of rocks, or as a common vein material associated with many different minerals.

pyroclastic   A general term applied to volcanic materials that have been explosively or aerially ejected from a volcanic vent. Also, a general term for the class of volcanic rocks made up of these materials, including volcanic ash. A volcanic rock texture; composed of fragments.

quartz   Crystalline silica, an important rock-forming mineral: SiO_2. Quartz is the commonest gangue mineral of ore deposits, forms the major proportion of most sands, and has a widespread distribution in igneous (esp. granitic), metamorphic, and sedimentary rocks. It has a vitreous to greasy luster, a smoothly curved fracture, an absence of cleavage. Quartz will scratch glass easily, and cannot be scratched by a knife.

quartz monzonite   A granular plutonic rock that resembles and is related to granite. Its major constituents are potassium feldspar, plagioclase and quartz, with minor quantities of biotite, hornblende, apatite, and zircon.

quartzite   A metamorphic rock consisting mainly of quartz and formed by recrystallization of sandstone,

or a very hard but unmetamorphosed sandstone, consisting chiefly of quartz grains that have been so completely and solidly cemented with secondary silica that the rock breaks across or through the grains rather than around them.

Quaternary   The second period of the Cenozoic Era, following the Tertiary; also, the corresponding system of rocks. It began about 1.6 million years ago and extends to the present.

rain shadow   The region of diminished precipitation on the lee side of a mountain or mountain range, where the rain and snow are noticeably less than on the windward side.

reef limestone   A limestone consisting of the remains of active reef-building organisms, such as corals and sponges, and of sediment-binding organic constituents, such as calcareous algae.

relict   a feature of an earlier rock that has persisted in a later rock that has been otherwise changed by metamorphic processes.

replacement   The process by which one mineral or substance takes the place of some earlier different substance, often preserving the earlier structure or form; specifically replacement ore where metal sulfides, such as galena, take the place of earlier rock-forming carbonate minerals, such as calcite.

resurgent caldera   A caldera in which the caldera block, following subsidence, has been uplifted, usually in the form of a structural dome. See page 51.

retort   A vessel with a long neck used to distill mercury from amalgam.

reverse fault   A fault on which the hanging wall appears to have moved upward relative to the footwall. See page 8.

rhyolite   A fine-grained igneous (volcanic) rock with the same chemical composition as granite.

rhyolitic   Having the characteristics of a rhyolite.

right-lateral fault   A strike-slip fault in which relative motion is such that to an observer looking directly at (perpendicular to) the fault, the motion of the block on the opposite side of the fault is to the right. See page 8.

riparian Of, relating to, or growing on the bank of a river, lake, or other water body.

ripple mark An undulatory surface or surface sculpture consisting of alternating subparallel small-scale ridges and hollows. It is produced on land by wind action and under water by currents or by the agitation of water in wave action, and generally trends at right angles or obliquely to the direction of flow of wind or water.

rock Any naturally formed, solid aggregate of one or more minerals.

rock cycle A sequence of events involving the formation, alteration, destruction, and reformation of rocks as a result of such processes as magmatism, erosion, transportation, deposition, lithification, and metamorphism. A possible sequence involves the crystallization of magma to form igneous rocks that are then broken down to sediment as a result of weathering, the sediments later being lithified to form sedimentary rocks, which in turn are altered to metamorphic rocks.

rock varnish A thin, dark, shiny coating consisting mainly of manganese and iron oxides, formed on the surfaces of stones and rock outcrops in various climatic regions (from hot deserts to glacial regions) after varying lengths or exposure. Also called desert varnish.

ruby silver See pyrargyrite.

San Andreas Fault A major strike-slip fault zone extending from the north central coast of California through the Gulf of California.

sand A rock fragment or particle of a size between that at the lower limit of visibility of an individual particle with the unaided eye and that of the head of a small wooden match, being somewhat rounded by abrasion in the course of transport. The material is most commonly composed of quartz resulting from rock disintegration, and when the term "sand" is used without qualification, a siliceous composition is implied; but the particles may be of any mineral composition or mixture of rock or mineral fragments, such as "coral sand" consisting of limestone fragments.

sandstone A sedimentary rock made up of sand-sized particles.

scarp See fault scarp.

scheelite A white to brownish-white calcium tungstate mineral ($CaWO_4$). It is found in veins associated with quartz and in contact metamorphic deposits; it fluoresces yellow to blue-white. Scheelite is an ore of tungsten.

schist A crystalline rock formed by metamorphism, that can be readily split into thin flakes or slabs.

sediment Solid fragmental material that originates from weathering of rocks and is transported or deposited by air, water, or ice, or that accumulates by other natural agents, such as chemical precipitation from solution or secretion by organisms, and that forms in layers on the Earth's surface at ordinary temperatures in a loose, unconsolidated form; e.g. sand, gravel, silt, mud, till, loess, alluvium.

sedimentary rock A consolidated accumulation of rock and mineral grains and organic matter or a rock formed by chemical or organic precipitation.

shaft A vertical mine working, extending from surface down; can be vertical or inclined. Used to move men and equipment in and out of the mine, and to move ore out of the mine.

shale A very fine-grained, laminated, sedimentary rock made up of clay- and silt-sized particles. Shale tends to break along parallel planes.

shear zone A zone in which shearing has occurred on a large scale so that the rock is crushed and brecciated.

sheared Highly deformed (crushed or stretched out) due to differential movement of rock bodies that have been forced to slide past each other along a plane or zone.

silica Silicon dioxide (SiO_2).

silicate A chemical compound or mineral made up of silica tetrahedra, which are arranged with metal elements to form chains, sheets, or frameworks. Many common rock-forming minerals are silicates.

siliceous Containing abundant silica.

sill A tabular igneous intrusion that is parallel to the structure in the surrounding rock.

silt A sediment in which most of the particles are smaller than fine sand and larger than clay, generally between 0.002 and 0.05 mm in diameter.

siltstone A clastic sedimentary rock composed predominantly of silt-sized particles.

sink A depression containing a central playa or saline lake with no outlet, as where a desert stream comes to an end or disappears by evaporation.

sinter A chemical sediment deposited by a hot spring. Usually the term refers to siliceous sinter, a deposit of amorphous silica. Calcareous spring deposits, of either hot or cold springs, are commonly called travertine. See also tufa.

slate A compact, fine-grained metamorphic rock that can be split into slabs and thin plates (called slaty cleavage). Most slate was formed from shale.

solution cave A cave formed in a soluble rock.

sphalerite A mineral (zinc iron sulfide). The principal ore of zinc.

spheroidal weathering The successive loosening of concentric shells of decayed rock from a solid rock mass as a result of weathering.

stamp mill An apparatus in which rock is crushed by descending iron pestles (stamps) generally grouped in units (batteries) of five per mortar. Amalgamation (collection with mercury) is usually combined with crushing to recover gold and silver from the crushed rock.

stock A body of intrusive (plutonic) rock that is similar to but smaller than a batholith, having a surface exposure of less than 40 square miles.

stockwork An ore deposit of such a form that it is worked in floors or stories. It may be a solid mass of ore, or a rock mass so interpenetrated by small veins of ore that the whole must be mined together.

stope An underground excavation from which ore has been extracted.

strandline Former shore or beach line, as around a lake.

striated Solid rock that has had parallel grooves cut into it during movement along fault planes or of glacial ice.

strike The direction of a horizontal line in the plane of an inclined rock unit, joint, fault or other structural surface. It is perpendicular to the dip.

strike-slip fault A high-angle fault along which displacement has been horizontal.

subduction The process whereby a slab of oceanic lithosphere is descending beneath another plate (continental or oceanic) and forced down into the mantle of the Earth.

sulfide A mineral compound by the combination of sulfur with a metal, such as galena (lead and sulfur), or pyrite, (iron and sulfur).

sump The low point of a basin in which the drainage water of an area collects.

supergene orebody A mineral deposit or enrichment formed near the surface, commonly by descending solutions; during the supergene processes of mineral deposition, near-surface oxidation produces acidic solutions that leach metals, carry them downward, and reprecipitate them, thus enriching sulfide minerals already present and forming a supergene orebody.

syncline A fold in which the core contains the stratigraphically younger rocks; it is generally concave upward.

tabular Slab-like; having a table-like surface.

talus An accumulation of coarsely broken rock debris from rockfalls or slides that forms an apron sloping outward from the cliff supplying the material.

tectonic Of, or pertaining to, the rock structures and external forms resulting from the deformation of the Earth's crust.

tailings The residuum after most of the valuable mineral has been extracted from an ore; the part rejected in milling an ore.

Tertiary The first period of the Cenozoic era (after the Cretaceous of the Mesozoic era and before the Quaternary), thought to have covered the span of time between 65 and 1.6 million years ago. It is divided into five epochs: the Paleocene, Eocene, Oligocene, Miocene, and Pliocene.

tetrahedrite A steel-gray to iron-black copper-iron-antimony sulfide mineral. It often contains zinc, lead, mercury, cobalt, nickel, or silver replacing part of the copper. Tetrahedrite is an important ore of copper and sometimes a valuable ore of silver.

thrust fault A low-angle reverse fault with the fault plane dipping less than 45 degrees.

transform fault A special variety of strike-slip fault along a plate boundary; a plate boundary that ideally shows pure strike-slip displacement.

travertine A dense, finely crystalline massive or concretionary limestone, of white, tan, or cream color, often having a fibrous or concentric structure, formed by rapid chemical precipitation of calcium carbonate from solution in surface and ground waters, as evaporation around the mouth or in the conduit of a spring, esp. a hot spring.

tufa A chemical sedimentary rock composed of calcium carbonate, deposited from solution in the water of a spring or lake or from percolating groundwater.

tuff A rock formed of compacted volcanic fragments. A general term for all consolidated pyroclastic rocks.

tunnel Strictly speaking, a passage in a mine that is open to the surface at both ends. It is often used incorrectly as a synonym for adit, which has only one opening to surface; or drift, which is driven underground within a mine and has no openings to surface.

turquoise A copper phosphate mineral. Turquoise is blue, blue-green, or yellowish green; when sky blue it is valued as the most important of the nontransparent gem materials. It usually occurs as masses and irregular veins in the zone of surface oxidation and leaching, and is commonly found in the leached capping zones of porphyry copper deposits.

ulexite A white borate mineral. It forms rounded masses of extremely fine needle-like crystals and is usually associated with borax in saline crusts on alkali flats in arid regions.

upper-plate rocks In Nevada, generally refers to rocks of early Paleozoic age that are structurally above the Roberts Mountains thrust fault.

upthrown block The block or mass of rock on that side of a fault which has been displaced relatively upward.

vein A mineral-filled fault or fracture.

veinlet A narrow vein or occurrence of ore in rock or gangue. A thin filling or intrusion in rock.

vent The conduit or orifice through which volcanic materials (lava, gas, and water vapor) reach the Earth's surface.

volcanic ash See ash, volcanic.

volcanic neck The solidified material filling a vent or pipe of a volcano. The hard igneous rock may resist erosion better than the mountain mass originally encompassing it and eventually stand alone as a column, tower, or crag.

volcanic rock Igneous rocks derived from magma or magmatic materials that are poured out or ejected (extruded) at or very near the Earth's surface.

volcano A vent from which magma, gas, and ash are erupted; also, the usually conical structure built by such eruptions.

volcano-tectonic trough A large-scale trough, usually linear, that is controlled by both tectonic and volcanic processes.

wall rock The rock forming the walls of a vein or lode; rock into which a pluton is intruded.

wash A shallow streambed with steep sides cut into unconsolidated sediments. This kind of streambed usually carries water only after brief, local precipitation.

Washoe pan process The process of treating silver ores by grinding in pans or tubs with the addition of mercury, and sometimes of chemicals such as blue vitriol and salt. Named for the Washoe district (an early name for the Comstock) where it was first used.

wave-cut terrace See strandline.

western facies rocks Generally refers to siliceous rocks (quartzite, shale, and chert) deposited during early Paleozoic time in deep waters along what was then the coast of North America. Deposited in the western part of the ocean basin flanking the continent, they are called western assemblage or western facies rocks.

window An eroded area of a thrust sheet that displays the rocks beneath the thrust sheet.

zeolite A generic term for a large group of white or colorless (sometimes red or yellow) hydrous aluminum silicate minerals. Zeolites occur as well-formed crystals in cavities in basalt and as minerals in the sediments of saline lakes and the deep sea, and especially in beds of tuff.

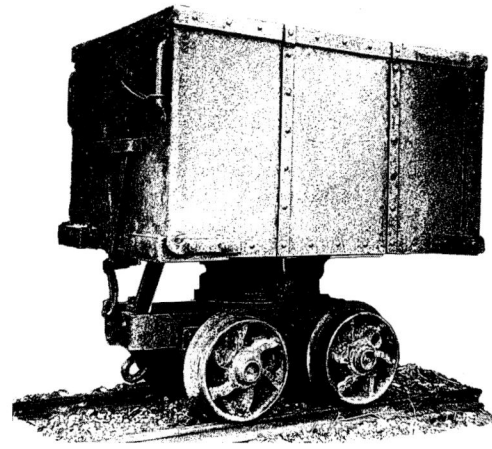

INDEX

131

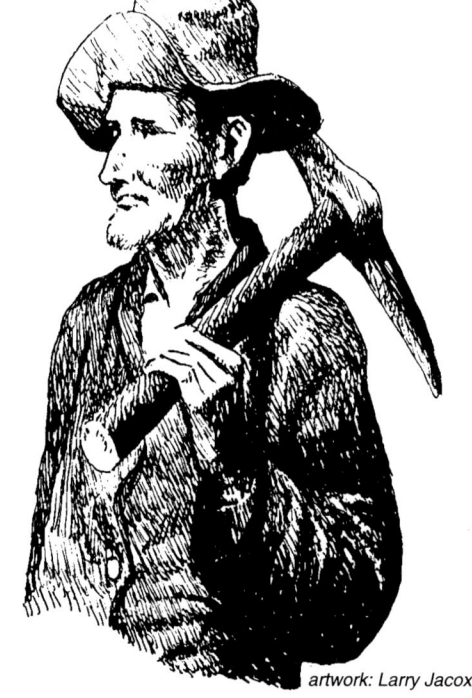

artwork: Larry Jacox

Back cover: Looking southwest from Cold Springs with Fairview Peak in the left distance and the Clan Alpine Mountains on the right. *Photo: Jack Hursh*